Kevin Redford

Historical Ecology in the Pacific Islands

Historical Ecology in the Pacific Islands

Prehistoric Environmental and

Landscape Change

Edited by Patrick V. Kirch and Terry L. Hunt

Yale University Press

New Haven and London

Copyright © 1997 by Yale University. All rights reserved. This book may not be reproduced, in whole or in part, including illustrations, in any form (beyond that copying permitted by Sections 107 and 108 of the U.S. Copyright Law and except by reviewers for the public press), without written permission from the publishers.

Set in New Caledonia type by Northeastern Graphic Services, Inc., Hackensack, New Jersey.

Printed in the United States of America

Library of Congress
Cataloging-in-Publication Data

Historical ecology in the Pacific Islands : prehistoric environmental and landscape change / edited by Patrick V. Kirch, Terry L. Hunt.
    p.   cm.
  Based on papers presented at the XVIIth Pacific Science Congress held in Honolulu in 1991.
  Includes bibliographical references and index.
  ISBN 0-300-06603-1 (cloth : alk. paper)

1. Man, Prehistoric—Oceania—Congresses. 2. Land settlement patterns, Prehistoric—Oceania—Congresses. 3. Landscape changes—Oceania—History—Congresses. 4. Human ecology—Oceania—History—Congresses. 5. Biotic communities—Oceania—History—Congresses.
I. Kirch, Patrick Vinton.
II. Hunt, Terry L.   III. Pacific Science Congress (17th : 1991 : Honolulu, Hawaii)
GN871.H57   1997
304.2′0995—dc20   96-15733
                                          CIP

A catalogue record for this book is available from the British Library.

The paper in this book meets the guidelines for permanence and durability of the Committee on Production Guidelines for Book Longevity of the Council on Library Resources.

10 9 8 7 6 5 4 3 2 1

In memory of F. Raymond Fosberg,
foremost scholar of Pacific Island botany and ecology

Contents

	Preface	xi
	List of Contributors	xiii
1	Introduction: The Environmental History of Oceanic Islands *Patrick V. Kirch*	1
2	The Impact of Pleistocene Hunters and Gatherers on the Ecosystems of Australia and Melanesia: In Tune with Nature? *Jim Allen*	22
3	From Horticulture to Agriculture in the New Guinea Highlands: A Case Study of People and Their Environments *Jack Golson*	39

- **4** Extinctions of Polynesian Birds: Reciprocal Impacts of Birds and People — 51
  David W. Steadman

- **5** Landscape Catastrophe and Landscape Enhancement: Are Either or Both True in the Pacific? — 80
  Matthew Spriggs

- **6** The Historical Ecology of Ofu Island, American Samoa, 3000 B.P. to the Present — 105
  Terry L. Hunt and Patrick V. Kirch

- **7** Coastal Morphogenesis, Climatic Trends, and Cook Islands Prehistory — 124
  Melinda S. Allen

- **8** Changing Landscapes and Sociopolitical Evolution in Mangaia, Central Polynesia — 147
  Patrick V. Kirch

- **9** Environmental Change and the Impact of Polynesian Colonization: Sedimentary Records from Central Polynesia — 166
  Annette Parkes

- **10** Human Occupation and Environmental Modifications in the Papeno'o Valley, Tahiti — 200
  M. Orliac

- **11** Pre-Contact Landscape Transformation and Cultural Change in Windward O'ahu — 230
  Jane Allen

- **12** Hawaiian Native Lowland Vegetation in Prehistory — 248
  J. Stephen Athens

- **13** Prehistoric Polynesian Impact on the New Zealand Environment: Te Whenua Hou — 271
  Atholl Anderson

Epilogue: Islands as Microcosms of Global Change?   284
Patrick V. Kirch

References   287

Index   321

Preface

In 1961, at the Tenth Pacific Science Congress in Honolulu, Raymond Fosberg chaired a symposium entitled "Man's Place in the Island Ecosystem" (Fosberg, ed., 1963). The symposium was a milestone in anthropology and human biogeography, for it took advantage of the analytical power of the ecosystem concept, applying this to the role of humans in island environments. In 1961, however, there was as yet little understanding of the *long-term history* of human impacts on island ecosystems, and thus the emphasis tended to be on the recent past (that is, since European arrival in the Pacific), and on synchronic relationships between humans and island ecosystems. Anticipating the return of the Seventeenth Pacific Science Congress to a Honolulu venue some 30 years after the Fosberg symposium, Terry Hunt and I spawned the idea of a session devoted to revisiting the concept of "man's place in the island ecosystem." In contrast to the 1961 symposium, however, which was dominated by cultural anthropology, the emphasis would be on the advances in knowledge spearheaded by archaeology and its allied paleobiological sciences. Our invitations to field-workers who were at the forefront of research in island paleoecology were enthusiastically accepted, and in June 1991 we convened a highly stimulating day-long symposium that truly confirmed how far our understanding of the role of humans in shaping island ecosystems had come in a mere 30 years.

This volume presents 12 papers delivered in Honolulu and serves as a representative overview of the current state of research and knowledge on the subject. With one exception, the original participants were able to revise their contributions as chapters for this book. We thank all who participated in the 1991 session, especially those who have collaborated in preparing this volume. In addition, I thank Jean Thomson Black of Yale University Press for her enthusiasm and assistance in bringing the volume to fruition. Thanks are also due to Matt McGlone and two anonymous readers for Yale University Press, whose careful reviews of the draft chapters were much appreciated.

Patrick V. Kirch

Contributors

Jane Allen, Ogden Environmental and Energy Services Co., Honolulu, Hawaii. Dr. Allen specializes in the prehistory of Oceania and Southeast Asia and has conducted fieldwork both in Hawaii and in Malaysia. For the past several years she has directed several major field projects in the Hawaiian Islands, yielding important new evidence for early intensive agriculture and landscape change.

Jim Allen, Department of Archaeology, La Trobe University, Bundoora, Victoria, Australia. Professor Allen is a noted authority on the prehistory of Australia and the New Guinea region. He organized and directed the international Lapita Homeland Project in the Bismarck Archipelago and more recently has led a major field effort in Tasmania.

Melinda S. Allen, Department of Anthropology, University of Auckland, Auckland, New Zealand. Dr. Allen is a specialist in the prehistory and prehistoric human ecology of Polynesia. She has done fieldwork in Hawaii and the Cook Islands and has published extensively on her research.

Contributors

Atholl Anderson, Division of Archaeology and Natural History, Research School of Pacific and Asian Studies, Australian National University, Canberra. Professor Anderson is widely recognized as one of the leading authorities on New Zealand prehistory and on the extinction of the moa.

J. Stephen Athens, International Archaeological Research Institute, Inc., Honolulu, Hawaii. Dr. Athens is a professional archaeologist with considerable field experience in the Hawaiian Islands and in Micronesia, where he has carried out studies of the famed Nan Madol site on Pohnpei Island. Much of his recent work in the Pacific has been concerned with reconstructing the history of environmental changes during the period of human occupancy.

Jack Golson, Division of Archaeology and Natural History, Research School of Pacific and Asian Studies, Australian National University, Canberra. Professor Golson virtually pioneered modern archaeology and interdisciplinary prehistorical research in Melanesia, and he is renowned throughout the world for his seminal work on agricultural origins and human landscape change in New Guinea.

Terry L. Hunt, Department of Anthropology, University of Hawaii, Honolulu. Professor Hunt has carried out fieldwork in Papua New Guinea, Fiji, Samoa, and Hawaii. Among his research interests are long-distance exchange between island communities and the reconstruction of prehistoric environments.

Patrick V. Kirch, Department of Anthropology, University of California, Berkeley. A member of the National Academy of Sciences and the American Academy of Arts and Sciences, Professor Kirch holds the Class of 1954 Chair for Distinguished Teaching at Berkeley and is internationally recognized for his archaeological and anthropological research throughout Oceania. He has carried out field projects in Papua New Guinea, the Solomon Islands, Samoa, Tonga, the Cook Islands, and Hawaii.

Michel Orliac, Laboratoire d'Ethnologie Préhistorique, Centre National de la Recherche Scientifique, URA-275, Paris. Mr. Orliac is one of the leading French archaeologists who in recent years have reinvigorated the field of archaeology in French Polynesia. His research is characterized by a strong emphasis on prehistoric human ecology.

Annette Parkes, School of Geography and Earth Resources, University of Hull. Dr. Parkes is a geographer who specializes in palynological methods to reconstruct prehistoric vegetation and ecological changes. She has carried out field research in the Cook Islands and in French Polynesia.

Matthew Spriggs, Division of Archaeology and Natural History, Research School of Pacific and Asian Studies, Australian National University, Canberra. Dr. Spriggs is noted for his pioneering research on prehistoric agriculture and human-induced landscape change in Melanesia. He has carried out fieldwork in eastern Indonesia, Vanuatu, Papua New Guinea, and the Hawaiian Islands.

David W. Steadman, Florida State Museum, Gainesville. Dr. Steadman is the world's leading authority on the prehistoric bird life of the Polynesian Islands, having carried out extensive research throughout this region. His seminal research has revolutionized our understanding of avifaunal diversity in the Pacific and of the role of humans in bird extinctions.

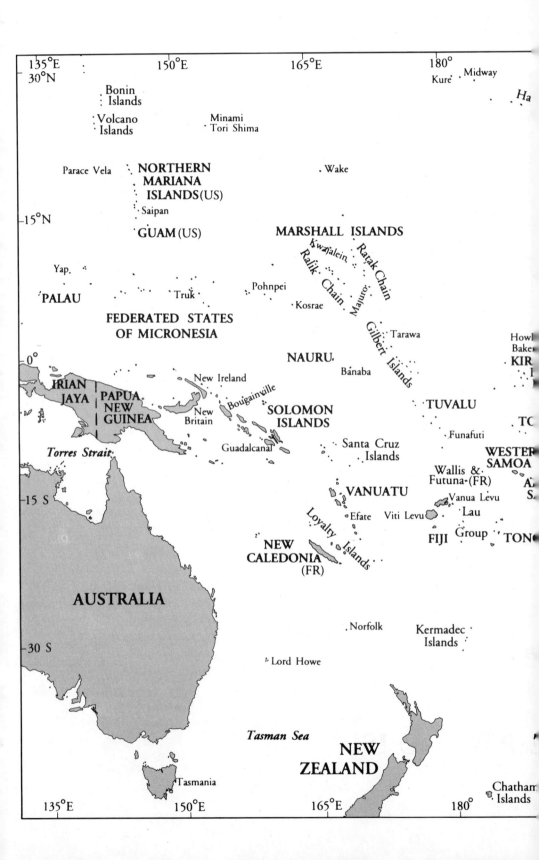

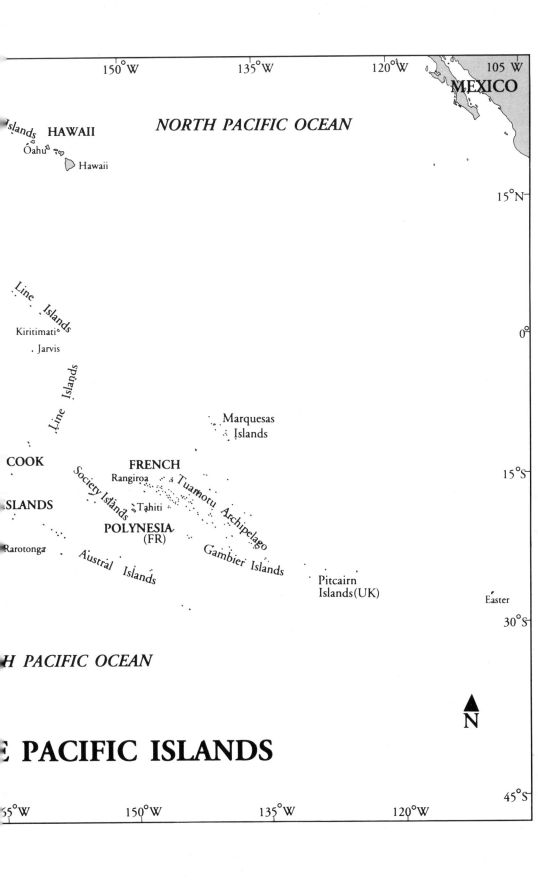

1
Introduction
The Environmental History of Oceanic Islands

Patrick V. Kirch

In the history of the natural sciences, islands occupy a privileged niche. Island faunas inspired Charles Darwin and Alfred Russell Wallace in their independent formulations of evolution and natural selection. Robert MacArthur and Edward O. Wilson's pioneering work on biogeographical theory derived from island models. Islands have also been key settings for the development of anthropology, a discipline that sits (sometimes uncomfortably) at the interface between the natural and the social sciences. The inventor of modern field ethnography, Bronislaw Malinowski, worked out his methods in the Trobriand Islands, and other prominent anthropologists (Margaret Mead, Raymond Firth, and Marshall Sahlins, to name a few) have found island societies ideal testing grounds for exploring human social variability.

It is the contention of the contributors to this volume that islands are likewise exemplary localities in which to advance the young but rapidly growing field of *historical ecology,* an interdisciplinary approach combining the perspectives and methods of both the natural sciences and anthropology. Carole Crumley succinctly defines historical ecology as "the study of past ecosystems by charting the change in landscapes over time" (1994:6). A critical tenet of historical ecology, moreover, is the recognition that there are virtually no landscapes anywhere on the globe today that have not been modified and affected, at least in part, by human actions. Historical ecology is thus the study of the complex, historical interactions between human populations and the ecosystems they have inhabited. As Bruce Winterhalder maintains, this approach requires an "epistemological commitment to the temporal dimension in ecological analysis. . . . The structural and functional properties of organisms, communities, and ecosystems must be sought in their history because they are only partly revealed in their extant form" (1994:40).

Island ecosystems—especially those of the Pacific region—offer several advantages for historical ecological studies. The many thousands of islands scattered throughout the Pacific provide a highly varied set of circumscribed ecosystems: in size they range from the near-continental islands of New Guinea and New Zealand down to such diminutive islets as Nihoa or Necker in the leeward Hawaiian archipelago; in geology, from complex island arc (the Bismarck Archipelago), to mid-ocean basalt (Hawaii), to upraised limestone and coral atolls (Mangaia); in climate, from temperate to tropical; in biota, from highly diverse (New Guinea) to impoverished (Easter Island). Moreover, human populations have colonized and inhabited these islands over highly varied time spans and in widely differing modes of adaptation and economic use. The Near Oceanic Islands were settled by hunting-and-gathering peoples more than 35,000 years ago (see Chapter 2), yet Polynesian agriculturalists did not reach New Zealand or Hawaii until the past 1,000 or 2,000 years (see Chapters 11, 12, 13).

Because islands are defined in part by their *isolation*—and this is particularly true of the central Pacific—they tended to have relatively stable natural ecosystems before the advent of humans. Human arrival on central Pacific islands was often therefore an event with dramatic consequences for ecosystems characterized by a biota lacking defenses against large, omnivorous vertebrates (and the portmanteau biota these new two-legged vertebrates carried with them; see Crosby 1986). As Raymond Fosberg (1963a) stressed some years ago, Pacific islands are by and large fragile environments, highly susceptible to disturbance. Because of this susceptibility, interactions between island environments and humans are all the more readily detectable in the historical record, whether that record be revealed by proxy measures of pollen spectra in sediment cores or by the skeletal remains of endemic fauna from archaeological sites.

In contrast to earlier and long-held notions that indigenous, preindustrialized populations in the Pacific had little impact on islands, from their very first incursions these populations began to change island ecosystems in diverse ways. Indeed, it is becoming clear that hardly any islands in the Pacific—even those that had no human populations at the time of European contact—were without human impact at some point in prehistory. Thus, for example, remote Henderson Island, thought to be a rare example of a "pristine" insular ecosystem, has recently revealed an archaeological record of Polynesian occupation spanning some five to six centuries, a period that resulted in the extinction of several endemic birds and in the introduction of rats and other exotic biota (Weisler 1994).

This volume represents the interdisciplinary research between archaeologists and natural scientists in the Pacific Islands over the past decade or so. To appreciate the significance of this new research, however, it is essential to provide some historical background. The modern conception that the preindustrial, indigenous peoples of the Pacific played a major role in shaping island landscapes and environments—which is the thrust of this collection of essays by key scholars involved in generating that knowledge—has not come easily. For years, various implicit and fundamental assumptions concerning the place of *Homo sapiens* (or at least of certain human cultures) in the natural world helped to mask the evidence that from the very first arrival of humans on their shores, islands have been immensely and irreversibly changed. Of course, it has long been recognized that the Pacific Islands have undergone dramatic impacts on their natural biota, in part as a consequence of animal and plant introductions since the arrival of Europeans. What until quite recently has not been generally admitted, however, is that even the preindustrial peoples of the Pacific region—which on some Near Oceanic islands have an antiquity of 35,000 or more years—have also been significant engines for ecological change. The dominant perception of most twentieth-century anthropologists toward Pacific cultures and their environments has been to see the environment largely as a changeless backdrop, a stage canvas if you will, against which the more interesting social dramas are played out.

The implicit assumption that indigenous Pacific populations had little significant impact on island ecosystems—and as a consequence that one need not pay heed to the history of ecological change on islands before the nineteenth century—has also underlain much biological research in the region. A classic example can be found in the highly influential monograph of Robert MacArthur and E. O. Wilson, *The Theory of Island Biogeography* (1967), which inspired a generation of biogeographers and evolutionary biologists. Drawing heavily on the work of Pacific systematists and field naturalists, MacArthur and Wilson hypothesized patterned regularities between island size (area) and taxonomic diversity, a hypothesis they tested quantitatively. Unfortunately, whether the numbers of bird or mollusk species known to museum curators were in fact accurate indications of the "natural"—prehuman occupation)—situation on Oceanic islands was an assumption that went unquestioned. With the hindsight of recent work on avian extinctions on Pacific islands (see Chapter 4; also Steadman 1995), this assumption now appears incredibly naive. In the Hawaiian Islands, for example, the endemic avifauna was reduced by more than half through human-induced extinctions during the period of Polynesian occupation (Olson and James 1984, 1991). The marvelously speciated Hawaiian drepaniid birds known to classic ornithology (before the recent archaeological and paleontological discoveries) are thus a mere remnant of the diverse fauna that had evolved on the islands before human arrival.

How and why did such implicit assumptions that the pre-European inhabitants of Pacific islands were simply actors on a changeless stage arise? It is instructive to ask this question briefly as background to the research initiatives of the past few decades, some results of which are reported in this volume. As with any historical perspective on Pacific natural history, we must begin with the great exploratory voyages of the late eighteenth century, which first introduced the Western intellect to the Oceanic world.

"OTAHEITE": MYTHS OF HUMANS AND NATURE

The French navigator Louis de Bougainville arrived off Tahiti in April 1768, unaware that the island had been "discovered" only the previous June by Captain Wallis in H.M.S. *Dolphin*. Predisposed by the writings of Enlightenment philosophers like Jean-Jacques Rousseau to find in the Great South Sea the true *homme naturel*, Bougainville compared the Tahitians to ancient Greeks: "I never saw men better made" (quoted in Smith 1985:42). Not only were the people godlike; the environment that sustained them was a perfect Eden, overflowing with "natural" abundance. "I thought I was transported into the garden of Eden," wrote Bougainville. As historian Bernard Smith observes, Bougainville's depiction of Tahiti "stamped itself permanently upon the imagination of Europe. The country was so rich, the air so salubrious that people attained to old age without its inconveniences" (1985:42).

A year after Bougainville's visit, the British explorer James Cook was dispatched by the Royal Society to this Nouvelle Cythère, as the French had dubbed it, to observe the transit of Venus. Accompanying Cook was Joseph Banks, later to become president of the Royal Society, and a major influence on British science well into the next century. Cook and Banks immediately adopted Bougainville's perspective. On the relationship between the Tahitians and their island environment, Cook wrote: "In the article of food these people may almost be said to be exempt from the curse of our forefathers; scarcely can it be said that they earn their bread with the sweat of their brow, benevolent nature hath not only supply'd them with necessarys but with abundance of superfluities" (Beaglehole, ed., 1968:121). This vision of Paradise on earth held no room for humanity as manipulator or modifier of the landscape. Rather, the landscape was God-given in its abundance, and the Tahitians were its carefree recipients.

These first perceptions of "Otaheite" and the islands of the South Sea changed only subtly during the course of Cook's second voyage (1772–75). Banks's classical allusions and "noble savage" sensibilities were replaced now by the somewhat more modern perspective of the Forsters (Johann Reinhold and Georg), who as Glacken observes first noticed "changes which man makes in the natural environment" (1967:702). Knowledgeable botanists building on the work of Solander on the first voyage, the Forsters recognized that the breadfruit (*Artocarpus altilis*), Tahitian Chestnut (*Inocarpus fagiferus*), and Vi Apple (*Spondias dulcis*) trees of the Tahitian landscape were not part of an original Garden of Eden but had been brought to the islands by humans who planted them. J. R. Forster commented on other aspects of native manipulation of the environment, as in his description of valley irrigation: "The natives had made several wears [sic] of stone across the river in order to raise & to stem the water & by that means introduce it into their plantations of *Taro* or *Arum esculentum* Linn" (Hoare, ed., 1982:341). The general impression left by Forster, nonetheless, remains one of man in harmony with nature, "increasing the beauty of the earth and its usefulness to him" (Glacken 1967:702). The Tahitians "joined art with nature" to meld their salubrious environment. Visually, the Forsters' perspective is well conveyed in the landscape paintings and engravings of the expedition artist William Hodges (see Smith 1985:64–65; 1992:111 and passim).

Other late eighteenth-century voyagers, however, continued to perpetuate the Garden of Eden myth of Polynesian landscapes. Captain Edward Edwards, sent to apprehend the *Bounty* mutineers, wrote of Tahiti in 1791: "And what poetic fiction has painted of Eden, or Arcadia, is here realized, where the earth without tillage produces both food and clothing, the trees loaded with the richest of fruit, the carpet of nature with the most odiferous flowers" (Edwards and Hamilton 1915:108–9). The mutineer Morrison, who had spent two years living among the Tahitians, also advanced this image of the noble savage who was spared the sweat of labor on the land: "As every part of the island produces food without the help of man, it may of this country be said that the curse of Eden has not reached it, no man having his bread to get by the sweat of his brow nor has he thorns in his path" (Morrison 1935:152).

If Cook's voyages and those of his contemporaries introduced the Western world to the Pacific Islands and their natural history, it was in the nineteenth century that the systematic investigation of islands and their human inhabitants burgeoned. During the nineteenth century a science of ethnology (and the foundations of modern anthropology) emerged out of dilettante observations of "native customs." Yet the perceived role of humans within the natural history of islands long remained much as the Forsters perceived it.

It comes as a surprise to find how little changed from the Forsters' portrait of Tahiti was Charles Darwin's perception formed more than a half-century later, during the *Beagle*'s famous voyage round the world. Like his naturalist predecessors, Darwin also remarked on aspects of native cultivation, yet he envisioned these human modifications of nature as confined to "a fringe of low alluvial soil, accumulated round the base of the mountains" (1957:367). On inland forays, Darwin carefully described such anthropogenic vegetation as "consisting almost exclusively of small dwarf ferns, mingled, higher up, with coarse grass" (1957:369). That such a "singular" vegetation resembling to his eye the Welsh Hills should occur "so close above the orchard of tropical plants on the coast" surprised Darwin, but it did not seem to occur to him to that these fernlands may have been an artifact of human agency. Likewise, on his traverse into the central mountain peaks of Tahiti he found himself in a veritable garden of "wild" bananas (*Musa* spp.), ava (*Piper methysticum*), taro (*Colocasia esculenta*), ti (*Cordyline terminalis*), and yams (*Dioscorea* spp.), yet he compared the scene "with an uncultivated one in the temperate zones." Darwin gives away his perceptions of Tahitian people and nature most clearly at the end of his description of this mountain scene: "I felt the force of the remark, that man, at least savage man, with his reasoning powers only partly developed, is the child of the tropics" (1957:374).

W. A. Bryan's *Natural History of Hawaii* (1915) epitomizes the continuity of nineteenth- to early twentieth-century perceptions of indigenous peoples as part of a harmonious and artful blend with nature. This view is neatly summed up by the title "Tranquil Environment of Hawaii and Its Effect on the People" given to the chapter in which Bryan promulgated the view that the "gentle demands of their environment" molded the Hawaiian people, "producing a patient, tranquil, self-reliant mind" (1915:37). Although he fully credited and described the agricultural pursuits of these still "noble savages," he did not consider whether such industry might have significantly influenced the islands' environment. Rather, to Bryan it was only with the "remarkable agricultural transformation of the Hawaiian Islands" through the advent of capitalized, plantation-mode production of sugar and other cash crops that real effects on plant and animal life appeared (1915:269 and passim).

"MAN'S PLACE IN THE ISLAND ECOSYSTEM"

Scholarship on Pacific natural history and anthropology flourished in the early twentieth century, including contributions by the precursors of modern interdisciplinary field research. Under the guiding hand of geologist Herbert E. Gregory, the Bishop Museum launched its Bayard Dominick Expeditions in 1920 and 1921, covering Tonga, Tahiti, the Austral Islands, and the Marquesas. Recognizing that the economic life of the indigenous people was closely linked to the plant world, each Bayard Dominick field party included a botanist as well as an ethnologist and an archaeologist. The fruits of this collaboration are most clearly evident in the publications emanating from the Marquesan team, yet critical distinctions are evident in the strikingly different ways that the botanist Forest Brown and the ethnologist Edward S. C. Handy treated the relationship between the Marquesan people and their environment. In his description of "native agriculture," Brown stressed the extensive agricultural modifications necessitated by "growing needs of the population," a significant recognition of demographic processes in island ethnography. This expansion of indigenous cultivation resulted, in Brown's opinion, in destruction of the "original vegetation" (1931:24–25). In short, the botanist's field laboratory had been significantly changed by prehistoric Marquesan actions. In contrast, Handy's brief treatment of the Marquesan environment from the ethnologist's perspective inverts the whole relationship between people and nature. "The marked influence of environment on the people and on their culture" became for Handy the issue, and he interpreted it primarily as a one-way molding of Marquesan culture to fit the "massive, strong, and vigorous" features of the land and climate (1923:8–9). This divergent set of perspectives from botanist and ethnologist previews the prevailing attitudes of natural scientists and anthropologists toward this issue during much of the twentieth century.

Brown's botanical insights about the Marquesan landscape were further developed by the entomologist A. M. Adamson (1939:26–27), who explicitly recognized that the prehistoric human inhabitants of the Marquesas had introduced "a large number" of plants as well as animals (some inadvertently), and had destroyed "a considerable amount of forest." Indeed, the very staghorn (*Gleichenia linearis*) fernlands that puzzled Darwin on Tahiti were finally recognized by Adamson as artifacts of human agency.

By the mid-twentieth century, field naturalists exploring the central Pacific islands had come to a working position that prehistoric humans had contributed significantly to environmental change. Evidence for such change, however, had largely been adduced indirectly from contemporary vegetation patterns or comparative studies of animal and plant distributions (such as the inferences of C. M. Cooke [1926] regarding the spread of certain terrestrial gastropods by Polynesians). There was practically no concern with reconstructing the *historical processes* of environmental change, or interest in seeking direct paleontological or fossil evidence for prehistoric biotic communities. Moreover, the impacts of Europeans were judged to be of an order of magnitude greater than those occurring during the prehistoric era. In reviewing the history of biotic extinctions in the Hawaiian Islands, Zimmerman (1948:172–177) dismisses the cumulative impact of Polynesian occupation in a short paragraph, thereafter dealing at some length with the historically recorded extinctions that occurred after European arrival in 1778.

While Adamson, Zimmerman, Merrill, and other naturalists did much to advance the biological understanding of Polynesians as a force in the transformation of island vegetation communities, the ethnographic perspective typified by Handy (1923) remained unchanged among anthropologists well into the 1950s. For example, archaeologist Robert C. Suggs, who pioneered stratigraphic excavations on Nukuhiva Island (Marquesas) in 1956 and 1957, and whose explanatory models were more sophisticated than those of his contemporaries, nonetheless did little more than reiterate Handy's views on the influence of the rugged Marquesan environment on the indigenous culture (Suggs 1961:13–15). Suggs's lack of interest in human interactions with the environment is reflected in his near total neglect of biological remains recovered from his excavations (which are relegated to a one-page appendix listing the presence or absence of pig, dog, rat, and cat at his sites). Even as archaeologists elsewhere in Polynesia during the 1950s were beginning to pay attention to faunal remains recovered in their sifting screens (as with the program of excavations in Hawaii directed by Kenneth P. Emory of the Bernice P. Bishop Museum), the implications for human paleoecology were generally bypassed as being of lesser interest than issues of cultural origins and settlement chronologies.

There were, however, some exceptions to the prevailing anthropological perspective relegating environment to an uninteresting and changeless backdrop. Harold J. Wiens's classic work *Atoll Environment and Ecology* (1962:454–62) includes an explicit consideration of "man's effects" on reef and land morphology, on vegetation patterns and soil, on the composition of land and bird fauna, and on marine fauna, and Wiens situates these effects in the context of population pressure on limited resources. It is telling, however, that Wiens was a geographer (rather than an anthropologist) who drew extensively on the interdisciplinary research of many natural scientists funded by the post–World War II Pacific Science Board's program in Micronesia.

A dramatic upsurge of interest among anthropologists and archaeologists in an "ecological approach" marked the 1960s. Inspired by the theoretical perspectives of such pioneers as Julian Steward and Leslie White, a new generation of anthropological field-workers began to look explicitly at the complex web of interactions between indigenous peoples and their island ecosystems. Typical of this new perspective was the work of Andrew Vayda and his students in New Guinea (e.g., Vayda 1961; Bowers 1968; Rappaport 1968). The emphasis, however, was largely on *synecology,* within a functionalist paradigm, rather than on a longer-term, dynamic perspective that incorporated paleoecological data.

The post–World War II shift to an explicit interest in the ecology of human cultures in the Pacific reached a culmination of sorts in the symposium organized by botanist Raymond Fosberg at the Tenth Pacific Science Congress held in Honolulu in 1961, published as a landmark volume under the title *Man's Place in the Island Ecosystem* (Fosberg, ed., 1963). Although Tansley (1935) had defined the concept of *ecosystem* nearly a quarter-century earlier, it was only beginning to be applied in such fields as anthropology and geography in the early 1960s. In his introductory essay, "The Island Ecosystem," Fosberg (1963a:5) pointed to "isolation and limited size" as essential features of insularity, as opposed to continentinality. Fosberg went on in two brief paragraphs to define certain significant characteristics of island ecosystems, and the implications of these for human advent:

> Some of the more significant characteristics of the island ecosystem are relative isolation; limitation in size (space resource); limitation in, or even absence of certain other resources; limitation in organic diversity; reduced inter-species competition; protection from outside competition and consequent preservation of archaic, bizarre, or possibly ill-adapted forms; tendency toward climatic equability; extreme vulnerability, or tendency toward great instability when isolation is broken down; and tendency toward rapid increase in entropy when change has set in.
>
> It is probable that no island ecosystem was ever completely stable. The limited size makes even relatively small changes capable of rather profound general effects; in other words, the buffering effects of great size and diversity are lacking. However, it is likely that, before the advent of man, many or most of the older island ecosystems had reached such relative stability that changes were mostly very slow. In most respects organisms present had evolved into an effective equilibrium with their environments. Closed biotic communities had developed that made difficult the unaided invasion of new organisms. [Fosberg 1963a:5]

The symposium itself brought together a remarkable mixed group of scholars and scientists, ranging from biologists (E. C. Zimmerman, T.

Dobzhansky) to anthropologists (A. P. Vayda, R. A. Rappaport, G. P. Murdock), geographers (C. J. Glacken, M. Bates, K. B. Cumberland), and historians (O. H. K. Spate). Throughout their papers and discussions they grappled with understanding the complex interactions and intricate "feedback loops" between human populations (both ancient and modern) and their island environments. The field data upon which these scholars were able to draw, however, was by and large confined to a very short segment of Pacific Islands history—the period of European contact from which both anthropological and biological observations were available. Archaeological research in the Pacific was still in its infancy at the time of the Tenth Pacific Science Congress, and the symposium contributors made little reference to the potential of archaeological or paleoecological information for reconstructing the longer-term history of human interactions with the environment.

Hampered by this restriction of their data sets to the recent past, the Fosberg symposium contributors almost uniformly stressed the tremendous impact wrought in island ecosystems by the advent of Europeans, and they down-played the extent of earlier, indigenous impact. Anthropologist G. P. Murdock, for example, suggested (without any archaeological data for support) that the advent of "pre-agricultural" human populations in western Oceania had no more impact in magnitude than the "natural introduction of any one of a considerable number of other new species" (Murdock 1963:151; cf. Chapter 2). With regard to agriculturally based populations, such as the Polynesians, Murdock admitted that forest clearance would have led to some reduction in natural organic diversity, but he felt that this biotic impoverishment was offset by biotic enrichment through the introduction of cultivated plants. Thus, "there was little net change in organic diversity, but there were substantial changes in the organization and distribution of biotic communities" (1963:151). Murdock's propositions indeed seemed reasonable in the context of the available data but, as we shall see, have turned out in the light of a vastly improved archaeological and paleoecological record to be far off the mark.

HISTORICAL ECOLOGY: ARCHAEOLOGY AND INTERDISCIPLINARY RESEARCH

By the close of the 1960s, an "ecological perspective" had become pervasive and influential in Western academic archaeology (see Trigger 1989), contributing to changed research orientations in Pacific Islands archaeology as well. A brief review of the changing practice of archaeology in the Hawaiian Islands from 1950 to the late 1970s illustrates these trends. Pioneering stratigraphic excavations of the 1950s and early 1960s incorporated a concern for dietary reconstruction, and thus the systematic recovery and collection of dietary remains (see, e.g., Emory and Sinoto 1961), but this research was driven first and foremost by issues of Hawaiian cultural origins, chronology, and material culture change. Only after Roger Green introduced the "settlement pattern" approach into Hawaiian archaeology in the late 1960s did questions concerning prehistoric human adaptation to the island environment emerge as part of the central research agenda. In several projects implemented in the late 1960s to early 1970s (e.g., Kirch and Kelly 1975; Yen et al. 1972) the ecological focus of the "New Archaeology" began to manifest itself. Much initial work centered on defining the variability of ancient Hawaiian agricultural production, and this led inevitably to broader concerns with human impacts on the environment. In Halawa Valley, for example, geomorphological and faunal studies of colluvial fan deposits yielded evidence for forest destruction and increased erosion rates resulting from early Polynesian shifting cultivation (Kirch 1975b).

Perhaps the most dramatic shift toward the recognition of prehistoric human effects on the Hawaiian environment came with the archaeological excavation of extinct bird bones and endemic land snails from limestone sinkholes at Barber's Point, O'ahu. Interdisciplinary collaboration with avian paleontologists Storrs Olson and Helen James (1982a, 1984; see also Christensen and Kirch 1986) resulted in a spate of new avifaunal discoveries, and in the realization that the number of avifaunal extinctions during the period of Polynesian occupation of the islands (pre-1778) actually exceeded the historically documented extinctions occurring after European intrusion. This changing archaeological perspective on prehistoric Hawaiian ecology was summarized by Kirch (1982).

In New Zealand, the occurrence of avifaunal extinctions (particularly of the flightless moa) before European contact was documented by the late nineteenth century, but the role of Polynesians in these extinctions—and their general impact on the New Zealand landscape—has had a long history of debate. Cumberland (1962), a geographer, maintained that Maori agriculture and burning had dramatically altered the New Zealand vegetation, opening vast tracts of forest to grass and fernlands. Yet others have steadfastly held to climate change as the main cause of these vegetation changes (e.g., Grant 1994). Not until extensive palynological records began to be compiled in the 1960s and 1970s did the weight of evidence swing in favor of the anthropogenic impact hypothesis (McGlone 1983; Anderson and McGlone 1992). And, as in Hawaii, not until the theoretical and methodological paradigms of the New Archaeology began to dominate archaeological practice in New Zealand universities did the direct prehistoric evidence of massive faunal exploitation and extinction get full play (see Anderson 1989b).

Beyond the main academic centers of Hawaii and New Zealand, archaeological investigation of other Pacific islands throughout the 1970s and 1980s contributed to the accumulation of evidence that prehistoric humans had to be counted as a significant force in altering island ecosystems. Kirch on Futuna (1975a) and Spriggs on Aneityum (1981) both used geomorphological evidence to reconstruct sequences of deforestation, erosion, and consequent alluvial deposition on the coastal plains of tropical high islands. In their joint study of Tikopia, Kirch and Yen (1982) documented an extensive array of human effects on and inputs into that small island's ecology, including forest clearance, erosion and deposition, soil modifications, and faunal extinctions. Also noteworthy was the increased application of palynological methods, with Hope and Spriggs (1982) finding proxy signals of anthropogenic deforestation in Aneityum. Hughes et al. (1979) revealed similar sequences in Fiji, and Flenley and King (1984; Flenley et al. 1991) documented dramatic deforestation within the period of Polynesian occupation on Easter Island. The Easter Island scenario was perhaps the most striking paleoecological discovery, as the island's vegetation had once included large *Jubea* palms and other trees contrasting so strikingly with Skottsberg's earlier views on the Rapanui landscape as a natural "oceanic steppe." During the 1980s, Oceanic archaeologists also became increasingly sophisticated in their treatment of zooarchaeological evidence, and collaboration with avian paleontologists, such as Olson and James (1982a), Steadman (1989a; 1995), and Balouet and Olson (1987), led to major revisions in our knowledge of prehistoric bird distributions on central Pacific islands.

On the large, near-continental island of New Guinea both archaeological and palynological research in the 1970s and 1980s led to a new understanding of the degree of prehistoric human impacts, especially in the Highlands. Palynological and geomorphological studies by Powell (1980, 1982), J. and G. S. Hope (G. S. Hope 1976, 1982; G. S. and J. Hope 1976; Hope, Golson, and Allen 1983), Oldfield (1977; Oldfield, Appleby, and Thompson 1980) and others provided extensive evidence for the extension of grasslands during the Holocene, as a result of human activities. Combined with faunal evidence for the extinction of endemic marsupial "megafauna" (e.g., Gillieson and Mountain 1983), and with Golson's remarkable evidence for very early Holocene agronomic modifications of Highland swamps (Golson 1977), these new interdisciplinary findings demonstrated that the New Guinea Highlands as known to twentieth-century ethnography and biogeography were the product of a lengthy historical sequence of environmental change, in which humans had played a fundamental role (see Chapter 3).

In sum, the anthropological view of "man's place in the *prehistoric* island ecosystem" has undergone a dramatic shift, largely as a consequence of archaeological or interdisciplinary studies. Spurred by the theoretical perspectives of the New Archaeology and of an ecological anthropology, and by significant methodological and technical advancements (in recovery and identification of plant and animal remains, in palynology, and in geomorphologically informed archaeological excavations), Pacific archaeologists and their natural science collaborators have accumulated an impressive array of evidence that the indigenous islanders wrought significant and lasting changes to their environments. It has become evident that explaining the cultural evolution of island societies requires an ecological component or context (e.g., Kirch 1984:123–51), while on the biological side, biogeographical studies cannot rely solely on contemporary or historically documented distributional data (e.g., Steadman 1989a).

The issue of prehistoric anthropogenic impacts on insular ecosystems has now become incorporated into the research agendas of most Pacific archaeologists and collaborating natural scientists. Indeed, several projects launched during the late 1980s had the primary and explicit aim of reconstructing sequences of late Holocene environmental change. These projects include Kirch and Steadman's study (Kirch et al. 1991, 1992) on Mangaia and S. Allen's (1992a) on Aitutaki in the Cook Islands (see Chapters 7 and 8); Flenley and Parkes's palynological research in the Cook and Society Islands (see Chapter 9); and Kirch and Hunt's geomorphologically oriented investigation of the To'aga site in American Samoa (Kirch and Hunt, eds., 1993; see Chapter 6). Paleoenvironmental investigations have likewise become incorporated into the standard research designs adopted for much "contract archaeology" (or culture resource management) in Hawaii (see Chapter 11) and New Zealand.

CURRENT THEMES IN PACIFIC ISLANDS HISTORICAL ECOLOGY

By way of specific introduction to the chapters that follow, I conclude with comments on several key themes that can be traced throughout the volume and that seem to me to be critical to current research in Pacific Islands historical ecology.

Natural versus Anthropogenic Change

As the history of natural science in the Pacific shows, recognition of the role of indigenous peoples in modifying island ecosystems has been slow in coming. Yet the geographer Patrick D. Nunn (1991:50) makes the curious, and to our mind unfounded, assertion that most writers have attributed environmental change on Pacific islands to human impact. Nunn (1991) rightly points to a variety of natural phenomena, such as climate changes, sea-level changes, and tectonic changes, that have also altered Pacific environments in the Holocene, but we reject his claim that "the importance of direct human impact [has been] overstressed" (1991:1).

Perhaps one of the most challenging issues facing historical ecologists in the Pacific region (and, indeed, in other parts of the globe) is the problem of disentangling the effects of natural processes from those of anthropogenic origin. Often this becomes a matter of distinguishing proximate and ultimate causation. For example, alluvial filling in of island valley mouths may be interpreted as the result of natural processes of sediment transport and deposition. When changing fluxes in sediment budget are interpreted diachronically, however, it may be found that increased sediment loads were the result of forest clearance for shifting cultivation on inland valley slopes. There is also the matter of direct and indirect human impacts on the environment. The endemic land-bird fauna that we now know to have been so much richer on central Pacific islands before human advent only partly succumbed to the direct effects of predation for food or feathers. Probably of greater import was the disturbance and reduction of habitats, as described so clearly by Anderson for New Zealand (see Chapter 13).

A number of the essays in this volume consider these issues of disentangling the relative contributions of natural and anthropogenic processes in changing island environments. Hunt and Kirch (Chapter 6) look at a particularly small island—the geologically youthful high island of Ofu in the Manu'a group of American Samoa. They show that during the approximately 3,000 years that Ofu was inhabited by Polynesians, the island's coastal terraces underwent substantial landscape transformations, and that these can only be accounted for with a complex model incorporating both natural and cultural inputs. In a parallel study, Melinda Allen (Chapter 7) examines the "almost-atoll" of Aitutaki in the southern Cook Islands, for which there is similar evidence for significant mid-to-late Holocene changes in sea level. While the sea-level changes in both Ofu and Aitutaki are largely due to natural causes, they have important implications for human settlement patterns and resource exploitation, as Allen demonstrates. Jane Allen (Chapter 11) also explicitly addresses the matter of causality in the transformation of windward valley landscapes on O'ahu. Although such natural phenomena as the Little Ice Age and the Little Climatic Optimum may have contributed to landscape changes (as argued by Nunn 1991), Allen finds that the lack of synchronization in the valley sequences argues against such natural processes being dominant. Rather, it was the actions of humans engaged in highly intensive forms of agriculture and land use that seem to have been most significant in shaping the windward Hawaiian landscape during the past 1,000 years.

Anthropogenic Impacts on Island Ecosystems

At the Fosberg symposium in 1961, anthropologist G. P. Murdock opined that the impact of pre-agricultural human populations on Oceanic islands had been no more than that of any other colonizing species. All of the contributors to this volume show how far we have come from that naive perspective (see also Dodson, ed., 1992). This issue is addressed by Jim Allen in Chapter 2, on the basis of extensive new archaeological data from New Ireland in the Bismarck Archipelago and from Tasmania. These widely separated islands—one tropical, the other temperate—were both settled more than 35,000 years ago as early modern humans moved out from Asia, across Wallacea, and into "Sahul" or "Greater Australia." Although human numbers were doubtless small, with a subsistence economy based on technologically simple hunting and gathering, these early people nonetheless affected their environments in ways that are archaeologically detectable. In the Bismarck Archipelago, for example, the "translocation" of animal species like *Phalanger* began at least 20,000 years ago. By the terminal Pleistocene these populations were experimenting widely with various tuberous and nut-bearing plants, which by the mid-Holocene would comprise a distinctly Melanesian complex of domesticates, critical to the ability of later Lapita colonists to disperse to the far-flung islands of Remote Oceania.

Golson (Chapter 3) continues the historical narrative of Near Oceania with his essay on the New Guinea Highlands during the Holocene, a period that witnessed an apparently independent development of root-crop horticulture as early as any agricultural "origins" known elsewhere in the world. These cultural innovations were intimately linked with a number of irreversible transformations of the Highlands landscape, in particular the expansion of anthropogenic grasslands. In more recent periods, the intensification of pig husbandry, followed by the "Ipomoean Revolution" precipitated by the introduction of the sweet potato (*Ipomoea batatas*), led to further reorganizations of humans on the New Guinea landscape with, as Golson notes, implications that are still being worked out by contemporary populations.

Without doubt one of the most remarkable discoveries of the past two decades of interdisciplinary collaboration between archaeologists and paleontologists has been the demonstration that the diversity of bird life on Remote Oceanic islands before initial human colonization was vastly greater than hitherto imagined. Only in New Zealand, where the large bones of extinct moa had been impossible to ignore in pioneering archaeological excavations, had it been established that the prehistoric avifauna had been far more diverse than that known to historical ornithology (see Cumberland 1963). As I have already noted, archaeological and paleontological work in Hawaii during the 1970s and 1980s produced a stunning array of previously unknown taxa, ranging from numerous extinct species of the Drepaniidae to the large, flightless *Thambetochen* geese that only a millennium ago had roamed the dry, leeward forests of Moloka'i and O'ahu (Olson and James 1982b). As archaeofaunal assemblages from various central Polynesian islands, such as the Marquesas, Society, and Cook Islands, began to be studied by avian paleontologists, it became clear that Hawaii and New Zealand were not exceptional cases but part of a general pattern of avifaunal extinctions during the period of Polynesian occupation and settlement (Steadman 1989a, 1995). In Chapter 4, David Steadman—who has been primarily responsible for developing this new database on central Polynesian bird life—explores in depth the reciprocal relationships that human and avian populations have had on Oceanic islands. His analysis shows that the relationship was not one-way, although the massive extinction records for numerous islands make it clear that the historically recorded diversity of Pacific bird life is a mere remnant of what existed a scant 3,000 years ago.

Being almost without exception agricultural peoples, Pacific islanders initiated the clearing of island forests and the conversion of natural vegetation associations to culturally managed landscapes. These activities perhaps more than any other had far-reaching impacts on island environments. Spriggs (Chapter 5) documents some of the effects evident in depositional sequences on several islands. In Chapter 8, Kirch reports on some of the key results of a major interdisciplinary investigation of paleoenvironmental change on the island of Mangaia, also in the southern Cook Islands. This research project, directed jointly with David Steadman, has provided a 7,000-year-long sequence of vegetation change revealed by pollen and geochemical data from deep cores in the island's valley-bottom sediments. Evidence for dramatic impacts on the island's vegetation following upon Polynesian colonization at about 2400 B.P. is matched in Mangaia with a faunal record of massive avian extinctions during the past 1,000 years.

In Chapter 9, Annette Parkes compares the vegetation histories revealed through pollen analysis at two lake sites, one on Atiu in the southern Cooks, and the other on Mo'orea in the Society group. The deep Atiu core in many respects parallels the Mangaian case, with dramatic human-induced changes in the local flora after Polynesian settlement at about 1300 B.P. For Mo'orea the situation is less clear, in part because the Temae core did not produce such a long sequence, and it is uncertain whether a truly prehuman component is present. Pollen records obtained from sediment cores have also been used by J. Stephen Athens (Chapter 12) to reconstruct, for the first time, the sequence of massive vegetation change that began with Polynesian colonization of the Hawaiian Islands. Athens's cores from O'ahu reveal a striking transformation of the lowland zone, with the removal of a former dryland forest dominated by *Pritchardia* palms and a now nearly extinct endemic shrub, *Kanaloa kahoolawensis*. Zimmerman's speculative remarks, made nearly a half-century ago, that the Polynesians had caused a "rapid retreat" of the Hawaiian lowland forests (1948:172) have finally been accorded an empirical basis by Athens's research.

Working within the Society Islands on the large high island of Tahiti, Michel Orliac described a diversity of paleoecological evidence from the Putoa Rockshelter. Orliac's study is a sort of "microcosmic" examination of the impact of humans on a highly localized landscape, in this case a limited sector of the Papeno'o Valley. Orliac's chapter demonstrates how far archaeologists have come in the past few years in fully exploiting the range of excavated materials for paleoenvironmental reconstruction: sediment samples, faunal remains, wood charcoal, pollen, and terrestrial gastropods when analyzed in concert allow the highly detailed reconstruction of local environmental change.

Environmental Evidence for Human Colonization

An issue that continues to drive much research and debate in Pacific archaeology is the chronology of human colonization and settlement of the many islands and archipelagoes. This is a matter of concern not only to the prehistorian but to historical ecologists as well, for accurate data on the time of arrival of humans in specific islands is obviously essential for determining whether particular environmental transformations detected in sediment cores or pollen spectra could have an anthropogenic cause. In the broadest terms, the chronology of human settlement in the Pacific commences with late Pleistocene movement of people into Near Oceania (New Guinea, the Bismarck Archipelago, and the Solomon Islands), with radiocarbon dates establishing a human presence on at least several islands by about 35,000 B.P. Beyond the eastern terminus of the Solomon Islands, however, we as yet have no firm evidence of human settlement before the rapid expansion of the Lapita cultural complex, beginning about 3600 B.P. (Kirch and Hunt, eds., 1988). (A debate continues, however, with regard to New Caledonia, specifically as to whether earthen tumuli there reflect a pre-Lapita population or, alternatively, are the incubation mounds of an extinct species of megapode; see Green 1988.) The Western Polynesian region including Fiji, Tonga, and Samoa was settled by at least 3000 B.P. Exactly when the Polynesians ventured beyond Samoa or Tonga to discover the Cook, Society, Austral, Tuamotu, and Marquesas archipelagoes that make up the core of central Eastern Polynesia, however, remains a matter of some controversy. Moreover, the use of paleoenvironmental evidence has become a key part of this debate.

The problem stems in part from the difficulties of finding *direct* archaeological evidence—in situ dwelling sites—for the earliest phases of colonization and settlement of a particular island. Given that founding human populations were often small in number, early sites will be correspondingly rare. When this is compounded by the high probability that early sites have been deeply buried (as at To'aga, see Chapter 6), eroded by streams or tsunami, or destroyed by modern agricultural and urban developments (as on windward O'ahu), discovery of direct archaeological evidence for the early phases of settlement (when population densities were low) will be difficult, especially on the larger high islands. Instead, it may be more realistic to look for *proxy measures* of human presence on islands, such as an order-of-magnitude increase in microscopic charcoal influxes in sediments. Such indicators should be present and detectable as background signals in a catchment when humans began to create significant disturbance, even if their actual dwelling sites are no longer archaeologically detectable (see Kirch and Ellison 1994).

The Mangaia case (Chapter 8) has recently become a flash point in the debate over the timing of Polynesian movement into the central Eastern archipelagoes. Based on extensive coring, combined with geochemical and pollen analysis, Ellison (1994) and Kirch (Kirch et al. 1991, 1992; Kirch and Ellison 1994) maintain that Polynesians had reached Mangaia by about 2400 B.P. and that their activities on the island's fragile, central volcanic hill set up a chain of environmental responses that are readily detectable in the island's sedimentary sequences. Spriggs and Anderson (1993; see also Anderson 1991) have questioned this interpretation, arguing that only direct evidence of human habitation (on Mangaia, at about 1000 B.P.) is admissible, and that the colonization of Eastern Polynesia did not occur until about 1300 to 1000 B.P. This debate will not likely be resolved soon, for the issues are complex, and much new field data are called for. Nonetheless, it is certain that paleoenvironmental evidence will play a critical role.

Environmental Change and Human Society

Historical ecology consists of more than simply cataloging the varied impacts and effects of humans in a landscape over time; its aim is ecological understanding, including the complex and reciprocal connections linking human populations with the myriad other life forms that share their world. This is where anthropology and paleoecology must come together, for in transforming their island environments, indigenous peoples also were compelled to change their technologies, economies, societies, and even ideologies. Mangaia exemplifies the case of certain mid-sized Polynesian islands (including also Easter Island and Mangareva) in which human-induced ecosystem transformations had major implications for the indigenous cultures themselves. As deforestation and landscape alteration channeled Mangaian agriculture increasingly toward intensive irrigation of the narrow valley bottoms for the cultivation of taro, both their sociopolitical structure and their religious ideology were reshaped (see Chapter 8). The Eastern Polynesian god Rongo, elsewhere typically a deity of agriculture and fertility (typically of *dryland* agriculture, for example, in Hawaii, where Rongo is manifest as Lono), became in Mangaia a dual god of war and irrigation to whom human sacrifice was made after each war for control of the precious irrigated lands.

In his succinct analysis of the Holocene period in the New Guinea Highlands, Golson (in Chapter 3) likewise demonstrates how far-reaching the consequences of shifting cultivation were for the cultures of these highland valleys. The systems of pig husbandry and elaborate exchange networks that social anthropologists have taken as so characteristic of New Guinea Highlands societies only came together in recognizable form within the past two millennia, as an adaptation to the environmental changes that were initiated with horticulture in the early Holocene. Further transformations in patterns of settlement, population density, and linguistic diversity were to stem from the introduction of the sweet potato (*Ipomoea batatas*) within the past three to four centuries.

A critical and politically sensitive issue with regard to impacts on island environments wrought by indigenous peoples is whether such changes should be labeled as "degradation" or "catastrophic" or whether, from the economic viewpoint, they should be assessed as forms of "landscape enhancement." In some modern Pacific island societies, these are not mere academic matters but hotly debated issues related to contemporary claims for land repatriation and sovereignty (e.g., Trask 1993). Spriggs boldly confronts these issues in Chapter 5, rightly pointing out that the research of archaeologists and natural scientists is not without relevance to contemporary political agendas.

Island Ecosystems: Fragile or Resilient?

A final question concerns the fundamental nature of island ecosystems themselves: Are they inherently fragile and hence vulnerable to human actions, or do they have a measure of resilience? And, are all insular environments the same in this regard, or do they vary in their susceptibility to anthropogenic impact? If they do vary, can we identify those factors that favor resilience? These questions are not without relevance to contemporary efforts to save what remains of the endemic floras and faunas of such places as Hawaii and Tahiti.

Fosberg's proposition that "the thing that most distinguishes islands, at least oceanic islands . . . is their extreme vulnerability, or susceptibility, to disturbance" (1963b:559) would seem on the whole to be amply borne out by the various studies reported in this volume. But the degree of impact that humans have had does appear to vary widely. Island size, which at first glance might appear to be a critical factor, seems not to be so critical. Although some smaller islands have been thoroughly disturbed, especially through deforestation (Easter Island, Mangaia), others have retained significant portions of their original vegetation and fauna (Ofu Island, see Chapter 6). Isolation is perhaps a more important variable, for those islands of the central and eastern Pacific that were the most isolated seem to have often suffered most strongly the consequences of human invasion, even though the time span of human intervention there is only on the order of one to two millennia. Here the contrast with the islands of Near Oceania (especially the Bismarck Archipelago, see Chapter 2) is remarkable, for these large high islands have not had nearly the same impact from human occupation, even though humans have been in place there for a order of magnitude longer.

Certainly, understanding relative resilience or susceptibility also requires that we pay close attention to different practices of land use. Here the two cases of Hawaii and New Zealand make for an instructive comparison, as is evident in the contributions of Allen (Chapter 11) and Anderson (Chapter 13). On windward O'ahu, intensive agricultural activities, particularly shifting cultivation, led to significantly increased rates of erosion and sedimentation in lowland valleys and embayments. As population increased and the sociopolitical system became more adept at organizing and mobilizing large-scale labor, however, these negative effects were offset by massive terrace construction schemes. Hillslopes became stabilized, and the landscape was converted to highly productive land use. At the opposite corner of the Polynesian triangle, Atholl Anderson discusses the Polynesian colonization of New Zealand—Te Whenua Hou, "The New Land"—and the impacts it had on these large, temperate islands. Anderson points out that these impacts were not everywhere identical throughout this vast and ecological varied landscape and that a close consideration both of chronology and of "regional patterns of economic opportunity" are necessary to make sense of the paleoecological data. Nonetheless, throughout the leeward forests the destruction of habitat and natural faunal resources was tremendous, leading in turn to wholly new modes of economic organization.

These, in brief, are some of the exciting new approaches, perspectives, and problems that currently engage interdisciplinary researchers in the field of historical ecology as it is developing in the Pacific Islands. I return briefly, in the Epilogue, to the relevance of historical ecology for contemporary problems of ecological change.

2

The Impact of Pleistocene Hunters and Gatherers on the Ecosystems of Australia and Melanesia: In Tune with Nature?

Jim Allen

One consequence of the growth of Green politics and worldwide awareness of environmental degradation is a developing notion that humans took a wrong turn when they intensified food production by developing advanced systems of agriculture and animal husbandry. Those of us who mark undergraduate essays increasingly encounter the view that, in contrast, hunter-gatherers were in tune with nature and conserved the landscapes they occupied in a more-or-less pristine state. This rise of neo–noble savage philosophy embraces qualities of romanticism and timelessness that bring new meaning to the half-century-old adage in Australian anthropology that Australian Aborigines were an unchanging people in an unchanging landscape. The popular pragmatic response is that hunter-gatherers had neither the population densities nor the necessary technology to have made a significant impact on the environment.

Each view is partly true, partly false. Although the scale of impact obviously does not approach that which industrial nations have imposed on the Pacific in the past two centuries, indications of hunter-gatherer alterations to Australian and Melanesian ecosystems are available in a variety of evidence reviewed below. Equally, there is no doubt that recent hunter-gatherers in Australia and hunter-fisher-horticulturalists in Melanesia had an intimate understanding of their environments and possessed many strategies that prevented the over-exploitation and depletion of their natural resources. But as White and Flannery (1991:8) have recently pointed out, these were strategies honed and refined over some 40,000 years. Given what we know of environmental changes during this period, continual modifications and adaptations of strategies have also been an integral part of hunter-gatherer life. In order to explore these issues, I examine two case studies, one from the Pleistocene of Tasmania, the other from the Pleistocene of island Melanesia.

SETTING THE BACKGROUND

A date of about 40,000 years B.P. for the initial colonization of Australia and New Guinea is generally accepted. The oldest radiocarbon dates available still fall short of this date, but three sites dated by the thermoluminescence (TL) method equal or exceed this date (Jim Allen 1989; Roberts et al. 1990a). From any assumed time of this arrival until about 8,000 years ago, Papua New Guinea and Australia formed a single landmass, commonly referred to as Sahul or Greater Australia. The Melanesian Islands in the Bismarck Archipelago (and those farther east) were always separate from the northeastern limits of this continental land, but the additional water barriers that separated the nearer large islands of New Britain and New Ireland provided little obstacle or delay in their settlement. At its southern end Tasmania was an island from about 50,000 B.P., and it was then joined to mainland Australia intermittently from about 37,000 B.P. and continuously from about 25,000 B.P. until 10,000 B.P. (Blom 1988; Sim 1989:4-5). Since that time its island status has been maintained, and its inhabitants have been isolated from mainland Australia.

Previous models of the initial colonization of Greater Australia (e.g., Birdsell 1957, 1977; Bowdler 1977; Horton 1981; White and O'Connell 1982) have tended to emphasize the minimum requirements for that colonization, such as the shortest water crossings, the smallest biologically viable group sizes, and the fewest ecological adaptations (Jim Allen 1989). Newer research, however, allows consideration of alternative models. Irwin (1991) has reviewed the question of Pleistocene sea travel from island Southeast Asia to Greater Australia in terms of distance and intervisibility between islands, the angle of the target island, weather conditions, and marine technology, and he has tested these predictions against the available data. Irwin determined that intervisibility could be maintained at all stages, concluding that the available evidence suggested that the colonization of Greater Australia and the nearer

Melanesian Islands was "undoubtedly systematic," in contrast to an earlier emphasis on this being a chance event. "On balance, it seems likely there were many interisland crossings and, with growing experience, the number of intentional crossings increased and covered an expanding field, but how and when this happened . . . is not known" (Irwin 1991:18-9). Irwin also acknowledged the possibility of return voyages from east to west. Even though a computer simulation by Wild (1985, cited in Irwin 1991) suggested that it was difficult to sail from Australia to Timor at any time of the year, more favorable conditions might occur farther north. It is also unclear what effect, if any, the absence of the Torres Strait during the Pleistocene may have had on weather and water currents.

The chronology of settlement is also revealing, as there is little suggestion of a clear time gradient for the earliest settlement in various parts of Australia. A clear primacy for the settlement of the site of Malakunanja II in the Northern Territory is claimed by Roberts et al. (1990a) on the basis of a TL date of 50,000 + B.P., but comparative TL dates are largely lacking elsewhere in Australia and Melanesia. Restricting consideration to radiocarbon dates only, there are at present approximate dates of 32,000 B.P. on the northwest coast of Western Australia (Morse 1988 and pers. comm.; O'Connor 1989), 38,000 in southwest Western Australia (Pearce and Barbetti 1981), 36,000 in western New South Wales (Balme and Hope 1990), 35,000 in Tasmania, 37,000 in northern Queensland (I. Anderson 1991), 35,000 on the northern coast of New Guinea (Gorecki 1991 and pers. comm.), and 32,000 in southern New Ireland (Allen et al. 1988).

The general similarity of these dates has been suggested to be a convergence caused by the problems of contamination of carbon samples older than about 25,000 years (see, e.g., Roberts et al. 1990b:95). But if they are in fact close to accurate then we need either to allow an unknown but reasonably long period of human occupation that remains archaeologically invisible, in order to build up sufficient population levels to occupy all these regions, or to allow that larger numbers of colonists than previously supposed entered Australia about 40,000 years B.P. On balance I currently prefer the latter reconstruction, partly because it is consistent with the implications of Irwin's conclusions but also because if there is a long period of archaeological invisibility in the initial occupation of Greater Australia, we need to find an explanation for humans becoming archaeologically visible virtually simultaneously across all of Greater Australia.

Accruing more data to help choose between the "long" and "short" chronologies is important on a number of levels apart from the simple fact of "when." For example, the question pertains directly to debates about the biological diversity of Pleistocene humans in the region, the variability among early artifact assemblages across Greater Australia, or, more pertinent to our topic here, the role of humans in late Pleistocene faunal extinctions. If the short chronology holds, we must contemplate Greater Australia—a single landmass in the order of 10 million km^2—having all of its major ecological zones (with the possible exception of the central desert) occupied by humans in as little as 3,000 years. When we consider further that these zones stretch from less than 1° south of the equator to almost 44° S, under the short chronology it is difficult to avoid the conclusion that this land-based advance was as systematic and deliberate as Irwin has claimed for the maritime advance. Equally it implies dynamic levels of adaptability not previously associated with late Pleistocene hunter-gatherers in Greater Australia.

IMPACTS ON FAUNA: THE TASMANIAN EVIDENCE

When humans reached Tasmania more than some 35,000 years ago, they were in the latitudes of Patagonia and thus farther south than humans had ever been before. Whereas the initial human landfall in Greater Australia occurred on tropical beaches where many of the plant and marine foods were familiar, humans now found themselves in cold and wet southwestern Tasmanian mountains, hunting wallaby and wombat and platypus. In its essentials, this hunting pattern would continue in this region for 25,000 years, first as the climate descended toward the maximum cold of the last ice age, about 18,000 years ago, and subsequently as it warmed. As rising world temperatures released the water trapped in the polar ice caps, seas rose, and Tasmania once again became an island about 10,000 years ago. At the same time, rainforests reclaimed the mountains and valleys of Southwest Tasmania, apparently driving out the game animals and thus their human predators (Kiernan et al. 1983).

In understanding both the adaptability of Pleistocene hunter-gatherers and their impact on the ecological systems of Greater Australia, Tasmania provides an important case study for several reasons. First, because we have a better understanding of when Tasmania was settled, we have little or no invisible record to contend with. On all available evidence it is improbable that Tasmania was settled while the 50,000 B.P. land bridge provided access. Watercraft access while Tasmania was subsequently an island seems only a remote possibility, given no obvious availability of buoyant raw materials to construct watercraft and the fact that the boisterous west-to-east weather patterns predominant in the Bass Strait were quite different from the predictable wind and current patterns in the "voyaging corridor" of the archipelagoes west of

New Guinea (Irwin 1991:14). Also, there is no suggestion that Holocene Aborigines ever made this crossing, from either direction. Thus we can place initial settlement somewhere after 37,000 B.P., as lowering sea levels exposed the higher parts of the Bassian plain. Currently Warreen Cave in the Maxwell River valley has an oldest (but not initial) occupation date of 34,790 ± 510 years B.P. (Beta-42122, Eth-7665), and four further sites with equally good sequences date to about 30,000 B.P. (Cosgrove et al. 1990).

Second, Tasmania now possesses 23 dated Pleistocene human occupation sites, of which 21 are in Southwest Tasmania, and perhaps 40 more known sites in this region are inferred to be of similar age. Most of the excavated sites have yielded long and artifactually rich sequences. Since most sites are limestone caves or rockshelters, the preservation of animal bone in particular is excellent. Both in terms of the density of Pleistocene sites in one region and the richness of the artifact assemblages, this region currently offers the best samples in Australia.

Third, humans have not occupied these cave sites or the general region for the past 10,000 years. This unique event in world prehistory, coupled with the fact that most of these sites have formed a hard calcium-carbonate crust on their surfaces, has resulted in sites undisturbed by any major interference beyond water erosion.

Finally, their extreme climatic setting, already alluded to, provides a good basis for measuring the adaptability of these Pleistocene hunters. About 18,000 B.P. some of these sites experienced a mean annual average temperature as low as 2°C, which suggests that low temperatures were frequently well below zero. Some sites were as little as 200-300 m below glacial limits, so that it is assumed that people were hunting up to the snow line. Occupation in some of the altitudinally highest sites was seriously reduced around the last glacial maximum, but site sequences at altitudes only 100-200 m lower show a continuity of occupation through this period. At present it is not certain that altitude differences are the key to this pattern, but if they are, we are less likely to be looking at the limits of human tolerance to cold than at delicate alterations to the wider ecosystems surrounding these sites, which were nonetheless sufficient to alter the human use of particular valleys or parts of valleys.

A striking feature of the excavated faunal sequences, and one common to all sites, is the predominance of a single species, Bennett's Wallaby (*Macropus rufogriseus*), which consistently represents more than 85 percent of the hundreds of thousands of animal bones recovered from these sites. Studies of the ecology of these medium-sized wallabies (Hocking and Guiler 1982; Southwell 1987; Johnson 1987; Driessen 1988) indicate that while they possess a wide alitudinal tolerance, they maintain small home ranges of 15 to 20 ha and may not shift centers of activity more than 30 m over several years. They are commonly found along the borders of forests and grasslands or alpine herbfields, where they have easy access to both protective cover and food. In a detailed synthesis of the extensive available evidence on soils, geology, microclimatic changes, and palaeobotany, Cosgrove (1991; see also Cosgrove et al. 1990) has recently questioned the earlier idea that Pleistocene grasslands and herbfields were extensive in Southwest Tasmania, and he has suggested instead that they were restricted to areas where fertile and well-drained soils occur on limestone geology. Combining this model with the ecology of the Bennett's Wallaby has led Cosgrove to suggest that the larger communities of this animal would be located in environmentally predictable and reliable patches in the southwest landscape; in contrast, in the southeast of Tasmania such patches were more likely dispersed and less predictable. It is entirely plausible that the reliability of Bennett's Wallaby as a resource outweighed the climatic rigors of periglacial Tasmania in any Pleistocene hunting group's cost-benefit analysis of the situation.

Two points are pertinent for the present discussion. The first is that the overall species patterning with its consistent emphasis on Bennett's Wallaby, across sites separated by at least 160 km and through sequences spanning 20,000 years, clearly reflects a deliberate hunting strategy. This is not the archaeological patterning of opportunistic hunting, but whether we are permitted to view these data as evidence for the targeting of this particular species or whether this patterning is the result of hunters concentrating on resource-rich patches where Bennett's Wallaby was dominant is somewhat moot. It is difficult to believe that hunters would not have taken either the Gray Kangaroo (*M. giganteus*), now locally absent but extant elsewhere in Tasmania, or the extinct *M. titan,* should they have encountered them. Although both were physically larger than Bennett's Wallaby, they were neither ferocious nor probably more difficult to hunt. Yet with the single exception of a cuboid bone thought to be either *M. giganteus* or *M. titan* among the 983 bones recovered from Beginners Luck Cave (Murray et al. 1980:147), there are no positive identifications of these species in the examination of about 500,000 bones from these sites. (The question of whether *M. titan* is the large Pleistocene form of *M. giganteus* [Flannery 1990b:46] is not relevant here and is passed over.) Given that among the remaining species present in these sites both wombat and platypus are certainly human prey, it is most parsimonious to conclude that larger macropods were absent.

The second point is that if the ecology of Bennett's Wallaby has not changed since the Pleistocene, and if Cosgrove's environmental reconstruction is accurate, then these animals would tend to be "tethered" to their discrete grassland patches. On first principles this would seem to make them vulnerable to over-predation by humans and to possible extinction. It is obvious from the archaeological record that despite this potential vulnerability, this species survived some 25,000 years of human predation at a time when extreme climatic oscillations were placing different stresses on survival. Although various conclusions can be (and have been) drawn from these sorts of data, there is no reason necessarily to attribute a conservation ethic to these hunters. Equally possible, and more pragmatic, are explanations invoking any or all of a range of variables, such as high numbers of animals, their robust breeding biology, low numbers of hunters, small hunting groups, or high mobility of hunting groups predicated on the availability of other necessary but scarce resources, such as carbohydrates. Perhaps movement to the next patch when returns began to diminish was recognized as the easiest strategy. The point here is that such survival of species may also imply the presence of deliberate hunting strategies (of whatever kind) on the part of their human predators.

IMPACTS ON FAUNA: PLEISTOCENE FAUNAL EXTINCTIONS

The Tasmanian evidence reviewed here provides few suggestions that Pleistocene hunters had a lasting impact on the Tasmanian fauna. But in Tasmania, as in other parts of Greater Australia and other regions that modern humans colonized, the extinction of a range of mammals, often referred to as *megafauna* in the late Pleistocene context, coincides with the arrival of humans. Whether there is a causal connection between the two events has been a major debating point in Australia no less than in the United States and is the subject of a voluminous literature (as a starting point see Flannery 1990b and Martin and Klein 1984). As Grayson (1984; see also comment in Flannery 1990b:61) has observed, the arguments have become so entrenched that either side can now employ the same evidence in its own support. Only the broad outline of the arguments is supplied here.

Flannery (1990b:46-47, Table 1, Fig. 1) lists 57 mammal species over 10 kg in weight that were extant at the time of the human colonization of Greater Australia. Of these, all 19 species between 100 and 2,000 kg are now extinct, as are 23 of the 38 mammal species between 10 and 100 kg. One of these, the Tasmanian Thylacine (*Thylacinus cynocephalus*) is the only mammal on this list to have become extinct since the arrival of Europeans (although other vertebrate species have become extinct during the same period).

Those who implicate humans in the extinction of these animals argue that they had survived since the Tertiary in a stable fashion, spread out over diverse environments from Papua New Guinea to Tasmania. They had survived marked climatic oscillations equal to or greater than those that occurred at the end of the Pleistocene, and their extinctions were, in geological time, virtually synchronous. The only new factor in the equation was the arrival of humans (Wright 1986a). Added to this, some of the extinct animals left unfilled environmental niches, so that competitive exclusion (Witter 1978) cannot be argued here as it can be later in the Holocene in the case of the dingo. This canine is thought to have been introduced to Australia in the past 4,000 years. The arrival of the dingo coincided with the disappearance on the Australian mainland of the thylacine and the Tasmanian Devil (*Sarcophilus harrisii*). Both survived in Tasmania, then an island not reached by the dingo, until European arrival in the early nineteenth century. The thylacine, a predator of sheep, subsequently succumbed to guns and dogs; the devil, a scavenger of carrion, remains extant.

Those who do not implicate humans appeal instead to climatic fluctuations and suggest that late Pleistocene aridity was the major cause of the loss of these species (Horton 1980). Proponents of this view emphasize that so far as one can accept the chronology of Australian extinctions at all, at least some species coexisted with humans for more than 10,000 years. This is important when seen in conjunction with the second major argument, which points to the absence of kill sites involving extinct animals and the distinct scarcity of their bones in any human sites in Australia or Papua New Guinea. In this context the long coexistence argument is seen to counter the view that human impact was so complete that the general absence of megafaunal remains in the oldest sites is the result of a "blitzkrieg" destruction so swift as to be undetectable in the archaeological record.

The Tasmanian evidence, while inconclusive, may yet be instructive. In none of the Pleistocene sequences from human sites in Tasmania has a single bone of any extinct mammal yet been positively identified. As I stated earlier, the evidence has involved the close examination of about half a million bones, many of them largely complete, from fauna-rich and extremely well dated sequences, which in four cases extend back to 30,000 B.P. and in one case to 35,000 B.P. However, even allowing the provisos put forward earlier for accepting a date of less than 37,000 B.P. for human arrival in Tasmania, the proponents of blitzkrieg could still argue that the Tasmanian data do not refute it, with perhaps a minimum of 2,000 years of hunting to destroy the larger mammal species not accounted for in this record.

As in the remainder of Australia, chronologies for the extinction of the Pleistocene fauna in Tasmania remain confused. Earlier claims for the possible coexistence of some megafaunal species and humans until the end of the

Pleistocene had been made on data from Beginners Luck Cave and the nearby Titans Shelter in the Florentine Valley (Goede and Murray 1977:9; Murray et al. 1980:151-52). These claims have subsequently been questioned by the same authors (Goede and Murray 1979:52; Goede and Bada 1985:161). The application of three different dating techniques (^{14}C, electron spin resonance, and aspartic acid racimization) to samples from these and other sites (Goede and Bada 1985) has thrown up sufficient dating discrepancies to cast doubt on the associations of humans and megafauna both in the Florentine Valley and in northwest Tasmanian caves. Beyond this the best dating evidence is a single ^{14}C date of greater than 47,000 B.P. at three standard deviations on a piece of wood found in stratigraphic association with a *Diprotodon* mandible at Pulbeena Swamp in northwest Tasmania (Banks et al. 1976:166).

Even so, it is equally difficult to claim with any certainty that humans and megafauna did not coexist in Tasmania. As Cosgrove (pers. comm.) himself points out, the small areas of grassland suggested by his patch model may not have been the areas in which hunters would have encountered these megafaunal grazers and browsers in large (or indeed any) numbers. Many animals still extant in other parts of Tasmania, including the Gray Kangaroo, Brushtail Possum, thylogale, and potoroo, remain entirely or largely absent in the archaeological faunal sequences reported here, and if we conclude that they were not encountered by these hunters, then the same argument must also be allowed to extend to the megafauna. We know too that both humans and megafaunal species existed in some of the same valleys, such as the Florentine, even if our chronologies do not yet allow claims for contemporaneity.

Given the uncertainties of these data, no clear conclusions emerge. At the same time, however, it is fair to ask in this case whether the absence of evidence may indeed be evidence of absence. At present there is no evidence at all to implicate humans in the extinction of large Pleistocene mammals in Tasmania, which, unless humans were there prior to 37,000 B.P., possibly disappeared before humans arrived. The argument might also be advanced that no single explanation for extinctions need necessarily hold for all of Greater Australia. Finally, if the Bennett's Wallaby was as vulnerable as I have suggested here, its nonextinction at the hands of humans might suggest that the strategies of these early hunters in Tasmania did not include blitzkrieg.

IMPACTS ON VEGETATION

Modification of habitat by fire has been suggested as an intermediate role for humans in megafaunal extinctions. Curr (1883; quoted by Singh and Geissler 1985:438) observed of the Aborigine that "living principally on wild roots and animals, he tilled his land and cultivated his pasture with fire." Influenced by such extensive ethnographic evidence as this for the widespread use of fire by Aborigines to "clean" the country and encourage new plant growth that would, in turn, attract game animals, Jones (1969) coined the evocative phrase "firestick farming." Despite criticisms of this term by Anderson (see his comment in Flannery 1990b:64), it has been widely accepted and widely used. It is difficult, however, to demonstrate the widespread influence of human fires on vegetation in the Pleistocene. Horton (1982:249, see also his comment in Flannery 1990b:60) suggested that Aboriginal burning was merely a substitute for natural fires and that had Aborigines never reached Australia, the distributions of plants and animals would have been similar to those existing at the time of European arrival.

Broad-scale habitat maps of temperate Australia produced by Dodson et al. (1991:121-23) for various periods between 28,000 B.P. and the present indicate climate as the major element of change. While human interference by fire may have affected local floristic compositions and the mosaics of communities in some regions (Dodson et al. 1991:129), the evidence for this is most likely to appear in pollen cores. There are several interesting examples of this, but equally some cores indicate little or no influence by fire. The two most frequently cited cases of evidence of burning seen in pollen cores are those from Lake George, near Canberra, and Lynch's Crater in northern Queensland.

The Lake George core provides a pollen sequence that exhibits a major change in zone F. For the first time in the sequence the dominant but fire-sensitive *Casuarina* was replaced by the fire-tolerant *Eucalyptus*. At the same time, the amounts of charcoal in the sequence increased dramatically, and these fluctuating but mainly high levels continued to the present. Singh and Geissler (1985:424-25) dated zone F by reconstructing temperature curves from the pollen diagrams and correlating these with glacials and interglacials. By comparing these curves with the stages of the deep-sea ^{18}O paleotemperature record of Shackleton and Opdyke, they concluded that zone F dates from the Last Interglacial, somewhere between 128,000 and 75,000 B.P. They concluded further that the dramatic changes seen to occur in zone F are related to increased numbers of fires, best attributed to the arrival of humans. However, they also noted (1985:439) the discrepancy between the dates they proposed and the accepted arrival dates for humans in Greater Australia. Subsequently Wright (1986b) constructed a depth-age regression on the available ^{14}C dates and proposed that zone F may be younger, at around 60,000 B.P. Flannery (1990:52) argued that an increase in the fire regime could have led to greater progressive sedimentation rates, which would make zone

F younger still. There are hints of this in the radiocarbon-dated part of the curve, down to the top of zone D; Singh and Geissler (1985:437) offer sedimentation estimates of 5 cm^3 per 1,000 years for zone C and 7.2 cm^3 per 1,000 years for zones A and B, although we need also to consider the difficulties of precision in such calculations. Beyond this independently dated part of the sequence, Singh and Geissler's estimated sedimentation rates depend circularly on their age estimates for the stratigraphic zones. In the case of Wright's calculations the same problem pertains. Five (or indeed 50) radiocarbon dates going back to 26,000 B.P. from the top 185 cm of lake deposits are an insufficient basis on which to prophesy whether dates a further 230 cm into the deposit will be younger or older than a straight-line regression estimate. The assumption of constant sedimentation rates is not warranted (see also Grayson's similar comment in Flannery 1990b:61).

The evidence of fire increase in the pollen core from Lynch's Crater is similar to that from Lake George, and the chronology is less contentious. An almost continuous sequence spanning 140,000 years is divided into eight phases. During phase 4, dated to between 38,000 B.P. and 26,000 B.P., there is a progressive shift from moist araucarian forests to sclerophyll vegetation dominated by *Eucalyptus* and *Casuarina*. Again this shift is accompanied by a marked increase in the frequency of charcoal particles. Although one climatic explanation of this variation in the fire regime might be that this change occurred during an arid period, Singh et al. (1981:47) argue that "contrary to expectations from earlier arid periods, forests of *Araucaria* all but disappeared and the previously common *Dacrydium* became extinct." They contend that this change in vegetational composition would not be expected under the impact of natural fires and climatic change only, and they squarely invoke human burning practices as the explanation.

THE VIEW FROM MELANESIA

The equivocal nature of the Australian data in terms of the possible impacts that human arrivals may have had on otherwise naive ecosystems is in contrast, in many respects, to the clearer evidence from Melanesia, and at present it is difficult to pinpoint why this is so. Certainly the quality of the Melanesian data is better in terms of their relevance to the questions of human impact, but it seems also likely to be related to two other factors.

First, researchers in Australia have found it difficult to come to terms simultaneously with both the great time depth and the huge area of Pleistocene Australia. In contrast, the smaller size and sharper geographical definition of Melanesia has allowed researchers to conceptualize it in different terms. This is true even of Papua New Guinea, part of the whole of Greater Australia but still able to be compartmentalized by latitudinal boundaries. Our perceptions of what Pleistocene Australia might have looked like move directly from the grossest units (deserts, coastlines, humid tropics, periglacial uplands) to the finest (pollen core, archaeological site) and rarely manage to depict middle geographical units. These were the regions where people operated, which we can divide off from adjacent regions where sometimes the same groups and sometimes different groups operated, and so on. Only at this level, the level at which hunter-gatherer groups operated, do we begin to comprehend the place of hunter-gatherers in Pleistocene ecosystems.

The second factor stems directly from the first. Unable to isolate this appropriate scale, either environmentally or archaeologically (e.g., typologically), research has tended to homogenize Australian Pleistocene prehistory. A prime example is the Australian Core Tool and Scraper Tradition. A facetious but nonetheless accurate definition of this tradition might be that it is what is left over after one removes the Australian Small Tool Tradition (a geographically widespread, typologically diverse, but chronologically tight tradition dating to the second half of the Holocene) from the entirety of Australian lithic assemblages, both Pleistocene and Holocene. The Pleistocene in Greater Australia is an archaeological record where the evidence for economy and subsistence is often scarce. If lithic technology is seen to be both simple and similar in deserts, along coastlines, in humid tropics, and in periglacial uplands, an inevitable consequence will be to view the human role in Pleistocene ecosystems as passive rather than dynamic and its impact as being minor, fire and megafaunal extinctions notwithstanding.

In Melanesia, the mindset of Pleistocene researchers is different. For some years there has been clear evidence that in the Highlands of Papua New Guinea the end of the Pleistocene also saw plant manipulation reach sufficiently complex levels to be labeled horticulture. This was accompanied by significant forest clearance by humans as well as hydraulic management (drainage and irrigation) of valley bottoms (Golson 1988; see also Chapter 3). In the absence of any evidence for either a migration of agriculturalists or a diffusion of agricultural practices into New Guinea at this time, it has been— and still is— reasonable to assume a long history of incipient environmental management leading to the local evolution of food-production systems there. Equally, it has long been argued that those human adaptations to the restricted resources of the island world of the Pacific that enabled the eventual colonization of Remote Oceania should be sought in the larger Melanesian islands of the Bismarck Archipelago. Ultimately such views have implied an inherent dynamism in the Pleistocene colonization of Melanesia, so that evidence of the human manipulation of Pleistocene ecosystems there has not been seen as surprising.

IMPACTS ON FAUNA AND VEGETATION: THE NEW GUINEA HIGHLANDS EVIDENCE

The site of Nombe at 1,720 m above sea level, occupied at about 25,000 B.P. or earlier (Mountain 1983), adds a postscript to the megafaunal extinctions debate. Its earliest levels contain five extinct species of megafauna in association with stone artifacts, but whether the animals were human prey or whether humans shared the site with other predators is uncertain. This evidence demonstrates, however, that in this part of Greater Australia early humans and animals that later became extinct were certainly contemporaries.

Equally interesting, Nombe is one of the Highlands sites that contain the distinctive tool known as the waisted blade but more accurately described by Groube (1986:172) as a hafted ax. This tool type is widely distributed in both the Highlands and the lowlands of Papua New Guinea and in the Melanesian Islands, and it appears to have a temporal distribution spanning the period from colonization into the early Holocene. Groube (1988:298-302) argues that this tool was used in forest manipulation virtually from the time of human arrival. Such manipulations as trimming and canopy thinning, ring-barking, and, with the aid of fire, some tree felling were used to create small disturbed patches for promoting the growth of useful food plants. Although these specific uses remain hypothetical, Groube (1988:296-97) maintains that the forms, wear-marks around the waisting, edge damage, and breakage patterns on the tools are consistent with such uses. Supporting evidence comes from a *Pandanus* swamp adjacent to the Pleistocene site of Kosipe where Hope (1982) recovered palynological evidence for forest clearance at 30,000 B.P. This clearance is assumed to be the work of humans.

IMPACTS ON FAUNA AND FLORA: THE BISMARCK ARCHIPELAGO EVIDENCE

Golson's (1971:124) observation that the move from Southeast Asia to Greater Australia took humans from one of the richest land faunas in the world to one of the poorest is accentuated when we consider the move from Greater Australia into island Melanesia. The physical achievement of reaching the Bismarck Archipelago reflects a challenge no different from that accomplished by the settlement of Greater Australia. It was, however, a qualitative shift of some magnitude because humans for the first time encountered the depauperate land faunas that characterize the Pacific Islands. The numbers of both mammal and bird species decrease dramatically across the narrow water barrier that separates New Guinea and New Britain (Allen et al. 1989:555), and it was the hunter-gatherers of more than 30,000 years ago moving into these nearer Oceanic islands who began the long process of adaptation that facilitated the eventual colonization of the Pacific.

In spite of the immense time gap between the beginning of this colonization and the one that would finally populate the distant corners of Polynesia, there are common elements in the strategies of both these Oceanic explorations. The ability to make water crossings in order to expand local resource bases is a common element of the later period, when evidence for the export and import of raw materials and manufactured items is repeatedly documented. This expansive and broad spectrum strategy is equally reflected in the repeated movement of obsidian from Mopir on New Britain to Matenbek on New Ireland about 20,000 years ago (Summerhayes and Allen 1993; Allen et al. 1989:554-55). The transport of this raw material over a distance of about 350 km in a straight line also involved a water crossing of about 30 km between New Britain and New Ireland. This is the oldest evidence we have of deliberate journeys to and fro across such a water barrier, and it suggests that Melanesian coastal canoe travel was common at least by this date, if not from the period of earliest arrivals.

A common characteristic of the later colonization, and one mentioned by other authors in this volume, is the manner in which the initial colonization of islands in Remote Oceania often saw the heavy impact of humans on naturally available foods in the initial stages of settlement, the period during which it is presumed that food-production systems involving horticulture, arboriculture, and animal husbandry were being established on newly settled islands.

During the Pleistocene occupation of New Ireland there seem to have been comparable impacts. The three oldest sites in New Ireland, Matenkupkum (ca. 32,000 B.P.; see Gosden and Robertson 1991), Buang Merabak (ca. 32,000 B.P.; see Balean 1989), and Matenbek (ca. 20,000 B.P.), all contain large *Turbo* shells in their earliest deposits. At the first and last of these at least (the data from Buang Merabak are not sufficiently detailed to tell) not only does this single species predominate, but the early examples of it are mainly large individuals. The shells suggest that the reefs in front of these sites were discriminately collected by early colonists in New Ireland. During the terminal part of the Pleistocene and into the Holocene the food-shell remains from these and other New Ireland sites show neither the selectivity for large species nor the presence of large individuals of those species. This I interpret as reflecting the effects of continuing human predation on natural populations.

The principal difference between this impact and the later impacts of colonizers of islands in Remote Oceania is the much slower speed of this early one. We are unable to differentiate either the size or frequency of this species between any periods during the first 10,000 to 12,000 years of deposition in Matenkupkum or between these levels and the earliest levels in Matenbek. We thus assume very low levels of human predation throughout this period. Elsewhere (Allen 1993) I have suggested that hunter-gatherer use of New Ireland may have changed about 18,000 to 20,000 years ago; this possibly involved the more intensive use of reef resources and the reduction in size and number of these *Turbo*.

Given that humans with lithic raw materials were moving around the Bismarck Archipelago in the late Pleistocene, evidence for plant and animal transfers of the same age should provide no real cause for surprise, especially as the Oceanic Pacific was colonized by canoes carrying exotic plants and animals. Most of these transfers were deliberate, but some, like the Pacific Rat (*Rattus exulans*), were accidental transfers of human commensals. The role of humans in distributing wild animals into island Melanesia has been raised on a number of occasions (Ziegler 1977; Menzies and Pernetta 1986; Flannery et al. 1988). The Pleistocene sites in the Bismarck Archipelago most commonly contain in their earliest levels the remains of endemic bats, rats, and reptiles as well as marine fauna, but additions to these faunal lists are found in the more recent levels of most of these sites. In Manus, at the site of Pamwak, Spriggs (pers. comm.) reports the presence of an introduced bandicoot, but whether this is Pleistocene or early Holocene is as yet unclear. In the New Ireland sites the phalanger *P. orientalis* is a late Pleistocene addition to the faunal lists of all the Pleistocene sequences there. The thylogale and a large rat, *Rattus praetor*, appear to be early Holocene introductions. While this movement is certainly earlier than the period of Lapita pottery in the Bismarcks, whether these animals might have arrived as part of a package that also included horticultural components is still open to question.

So far there is only a single mammal in the New Ireland sequences that seems to have gone locally extinct. *Rattus mordax* is a common element in the early levels of all sites analyzed but disappears during the Holocene and is unknown in New Ireland today. It seems likely that this animal suffered in competition with the introduced *R. praetor*. This local extinction may eventually not be the only one attributed to humans in New Ireland. Steadman (pers. comm.) has identified 32 bird species in the fauna from Balof Cave, of which eight are now extinct on New Ireland. Whenever these losses occur in the Balof sequence, they cannot represent a blitzkrieg by human colonists, because Balof was not occupied before about 15,000 B.P., at least 17,000 years after humans first reached New Ireland. The bird bones from Matenbek and Matenkupkum are currently being identified by Steadman, and a comparison between these sites and Balof should prove to be very instructive with respect to the avifauna.

Finally, evidence of late Pleistocene-early Holocene forest manipulations are very scanty in the Bismarck Archipelago prior to the appearance of Lapita. Wickler and Spriggs (1988) report artifacts from the Kilu site (ca. 29,000 B.P.) in the northern Solomons with residues suggesting they were used to cut root vegetables, and *Canarium* nuts (preserved as macroscopic charcoal) have been identified in sites like Pamwak and Kilu (Spriggs pers. comm.). In Panakiwuk, on New Ireland, a single specimen thought to be a wild form has been recovered from levels dating to ca. 8000 B.P. (Marshall and Allen 1991:88).

CONCLUSION

The various strands of evidence suggest that humans in Melanesia during the Pleistocene had a small but significant impact on the natural distribution of plants and animals. The record generally points to an increasing tempo in the earlier part of the Holocene, but whether this is related to the development of horticulture in Melanesia is still too early to say with any certainty. Yet it is quite likely, given the presence of horticulture in the New Guinea Highlands at 9,000 years ago (Golson 1988). Pig seems also to be present in the Highlands by at least 5,000 to 6,000 years ago, with some claims for a date of about 10,000 B.P. (White and O'Connell 1982:187-89). The single pig's tooth recently discovered in the 8000 B.P. levels of Matenbek is currently undergoing further ^{14}C dating by direct AMS technique. The question is whether horticulture was the impetus for the wider redistribution of wild animals and obsidian at this time. But however these wild animals were redistributed—under either a horticultural or a hunter-gatherer impetus—it does not seem necessary to argue that they were deliberately moved to stock empty landscapes.

This general lack of obvious intent is perhaps the conclusion to which this survey has led. At opposite ends of the Greater Australian continent we get a clear sense of difference between not only the kinds but also the degrees of impact Pleistocene colonists had on these empty landscapes, but in neither place can we detect the sorts of deliberate alterations that more intensive methods of food production necessarily impose. At the same time—and subtle though they may be—the traces of those impacts are nevertheless present, because hunter-gatherers, too, were a part of these past ecosystems. It also seems that in most parts of Greater Australia they were active agents of change rather than passive respondents. However, in the case of the major charge, implication in the demise of the Pleistocene megafauna, the jury is still out. If, as is now probable, there was considerable temporal overlap of humans and megafauna, at least in some parts of the Pleistocene continent, then blitzkrieg seems not to be the solution, and the argument devolves to questions of demography and ecology that the current archaeological data are not sufficiently robust to resolve. If we cannot overcome this problem, the jury may never return.

Note

I thank Nicola Stern, Brendan Marshall, and Richard Cosgrove for comments on a draft of this chapter.

3

From Horticulture to Agriculture in the New Guinea Highlands
A Case Study of People and Their Environments

Jack Golson

New Guinea is a long and relatively narrow island, almost totally within the humid tropics, where climatic variability is more a matter of altitude than of latitude. Its major topographic feature is a central cordillera stretching the length of the island, the largest mountains between the Himalayas and the Andes, rising in places above 4,000 m and commonly above 3,000 m.

Among these high ranges are a number of large basins and valleys, which, with the more elevated country between them, are the focus of our attention, typically between 1,300 and 2,500 m in altitude. They occupy two widely separated parts of the cordillera, a continuous block in the Highlands provinces of Papua New Guinea (Southern Highlands, Enga, Western Highlands, Simbu, Eastern Highlands) and an area in Irian Jaya comprising the Baliem Valley in the east and the Paniai Lakes in the west. In these regions large populations and their numerous pigs live in open landscapes at locally high densities on the produce of orderly plantations dominated by a food plant of tropical American origin, the sweet potato (*Ipomoea batatas*). These are the people I call Highlanders, and the characteristics I have enumerated for them contrast with those of the societies on their fringes. This chapter explores the development of Highlands landscapes and Highlands cultures thus defined.

THE PROPOSITIONS

The foundation was laid by the postglacial rise in temperature from about 15,000 B.P., which freed the Highlands valleys of their Pleistocene cloudiness and mist and allowed their colonization by the resource-rich mixed oak forest (Hope et al. 1983:40–41). Climatic amelioration had proceeded far enough by 9,000 years ago for an early form of horticulture to move up from lower altitudes, as argued from geomorphological and archaeological evidence provided by investigations at Kuk swamp at 1,550 m in the upper Wahgi Valley of the Western Highlands Province of Papua New Guinea (Golson 1991b and references). This event is seen as initiating a slow process of environmental change, through forest clearance under a system of shifting cultivation, which had important consequences in the socioeconomic domain, including the transformation of the agrarian system itself (Golson and Gardner 1990). Notable among the changes was the appearance of the tropical American sweet potato within the past few hundred years and its rapid acceptance as a staple due to its having greater edaphic and/or altitudinal tolerances than existing crops and its marked attractiveness as pig food. The major crops of the previous regime are the familiar ones of Pacific agriculture, yams (*Dioscorea* spp.), taro (*Colocasia esculenta*), and bananas (*Musa* spp.), the history of whose domestication is complex (cf. Golson 1991a and references) and need not be entered into here.

The environmental factors in this dynamic process are clear. On one hand, there are the benign conditions of Highlands valleys, which allow year-round cultivation in conditions of often remarkable landscape stability (see Hughes et al. 1991). Though climate was by no means stable through the Holocene (Brookfield 1989:308–10), the fluctuations are not likely to have affected the processes under review here in significant ways. On the other hand, there were factors affecting the stability of the vegetation cover that made it sensitive to disturbance, of which clearance for cultivation was now a permanent agent. These factors are partly rooted in the dynamic geological history of the island, whereby the continual creation of new environments provides "evolutionary opportunity [but] inhibit[s] the development of stability and give[s] rise to pronounced patchiness in the vegetation" (Walker and Hope 1982:266), partly in the incomplete adjustment of the forest to the late Pleistocene–early Holocene climatic change, which saw the treeline rise from a minimum level of 2,000 to 2,400 m at the glacial maximum of 18,000 to 15,000

B.P. to a point beyond its present limit of 3,800 to 4,000 m by 8500 B.P. (Walker and Hope 1982:266–69). Indeed, it has been doubted (Walker and Hope 1982:269–70) that the great intermontane valleys of New Guinea lying at or above 1,500 m, which as a result of frost and cold-air drainage were probably home to grassland extensions from higher altitudes for long periods during the lowered temperatures of the late Pleistocene, ever became completely forested during the general forest expansion of the terminal Pleistocene and earliest Holocene.

The impact of forest clearance for cultivation in these circumstances was exacerbated by a number of interacting features of the Highlands environment (Golson 1977:605–9; 1981:46–47). One feature concerns a tendency for slower forest regeneration after disturbance due to lowered temperatures, frequent low-level cloud, diminished amounts of photosynthetically active radiation, and a growth habit that puts production into thick and short-lived leaves at the expense of woody parts. Another is that the crops of the pre–sweet potato cultivation regime required fertile soils, which implies an extended scale of forest clearance because soil rehabilitation was delayed by slow forest regrowth. A third point concerns the laterally and altitudinally confined domain within which this form of cultivation took place: laterally because the Highlands cordillera is rarely more than 100 to 125 km wide, and often substantially less; altitudinally because the majority of the crops before the sweet potato do not flourish above 2,000 m, while their growth is inhibited by persistent cloud on the outer flanks of the mountains below 1,200 to 1,400 m.

A model incorporating these elements would propose the continuing replacement of older forest by increasingly degraded secondary growth, as cultivation expanded to its local limits and turned back on itself in a tightening cycle of more frequent clearances separated by ever-shortening fallow periods, driven by the pressures of production.

INVESTIGATING THE PROPOSITIONS

Vegetation

The story of forest alteration and destruction is told in a number of pollen diagrams resulting from research into Highlands vegetation history over the past 25 years. Though their coverage is markedly uneven, they extend from the Baliem Valley of Irian Jaya in the west to the upper Ramu Valley of Papua New Guinea in the east, with a concentration in the western and central parts of the Papua New Guinea Highlands zone. The majority of these pollen diagrams do not go back beyond the mid-Holocene, due to the character of the swamps from which the cores were taken, but by this date significant inroads into primary forest had been made, as a result, it is considered, of clearance for cultivation (see Powell 1982:218). At the Baliem Valley site of Kelela, however, both vegetation and landscape disturbance attributed to clearance are in evidence some time before 7000 B.P. (Haberle et al. 1991:31–37).

The parallel vegetation histories from different areas of the Highlands is significantly underscored by a consistent rise in the values for *Casuarina*, and sometimes other secondary taxa of quick growth, that appears in pollen diagrams about 1,000 to 1,200 years ago in the Wahgi Valley and neighboring areas, apparently around the same time in the Baliem and somewhat later in the upper Ramu (Haberle et al. 1991:38). Today *Casuarina* is widely planted for its timber and its capacity for soil rehabilitation through nitrogen fixation. The pollen diagrams register the appearance and expansion of this silvicultural technique, which can be seen as a response to local deforestation (Golson and Gardner 1990:399).

The ultimate result of deforestation, the establishment and expansion of grasslands, is less easily read in the pollen diagrams, for technical reasons (see Golson and Gardner 1990:402). However, in the far eastern Highlands, where there are less favorable conditions for forest regeneration after disturbance due to lower and highly seasonal rainfall and the exacerbating effects of fire, a diagram from the Norikori (Noreikora) swamp shows that little tree pollen was being trapped from 1,600 years ago, suggesting the presence after that date of the open grassed landscape of the area today.

Fauna

Given that the New Guinea fauna is characteristically forest adapted, though by no means wholly arboreal, the forest histories sketched above may be expected to have had considerable consequences in the faunal domain. Ecological studies show that loss of species occurs as a result not only of extensive forest clearance but also of the floristic and structural simplification of remaining forests through the activities of people and their pigs (Dwyer 1982:174–78).

There are a number of reasons why these effects have not been satisfactorily charted in the few available analyses of faunal assemblages from Highlands archaeological sites. Early efforts (e.g., White 1972) were hindered by lack of reference collections, especially of postcranial parts, and of information on the habits and ecology of New Guinea animals. Subsequently, as these deficiencies have been repaired, there have been continuing taxonomic revisions. In addition, for topographic reasons, Highlands assemblages of archaeological fauna contain animals from a number of different habitats within the range of human hunting.

Nevertheless, the faunal record from the Kamapuk Rockshelter on the fringes of the upper Wahgi Valley is interpreted as showing a marked degree of forest degradation from the beginning of its use some 4,000 years ago, with increased disturbance from 1,200 years ago as reflected in an overall reduction in the diversity of the lower montane fauna represented (Golson and Gardner 1990:402, citing unpublished work by Aplin). The rich Holocene faunas at Nombe Rockshelter, distributed between two archaeological levels dividing at about 5,000 years ago, show a significant increase in the proportion of terrestrial species in the upper level, with the Bush Wallaby (*Thylogale brunii*) marginally better represented than the Forest Wallaby (*Dorcopsulus vanheurni*) (Mountain 1991:9.15–16; for earlier comments on this site see Golson and Gardner 1990:402 and references).

The clearest case is provided by Aibura (White 1972:57–59), a cave site not far from the Norikori swamp whose pollen record, as we have seen, represents an extreme example of deforestation. The cave deposits span the past 4,000 years, and the faunal record registers a replacement of forest over this period by the grasslands that now characterize the region. In the lower levels forest animals like cuscus (*Phalanger* spp.) and other possums (mainly Pseudocheiridae) predominate and are overtaken in importance within the past 1,000 years by ground-dwelling species, mainly the Bush Wallaby. In the most recent level, however, these disappear, and small rodents dominate. In circumstances of less hunting pressure, it is certain that the wallaby would have colonized anthropogenic grasslands like those at Aibura, as it has areas of subalpine grassland above the forest limit (Flannery 1990:110; Menzies 1991:117, cf. Hope and Hope 1976:41). As it is, however, the faunal resources of the mid-altitude grasslands have been reduced to the smaller vertebrates and invertebrates (Bulmer 1968:306).

Pig Husbandry

A case has been argued that faunal impoverishment caused by forest disturbance and destruction set the scene for an intensification of pig husbandry, which had to be supported from increased agricultural production, because the availability of forage for pigs, as of forest resources for humans, was reduced by the vegetation changes we have been discussing (Golson and Gardner 1990:402–3 and references). Kelly (1988:166–68) suggests an alternative pathway, whereby the initial expansion of secondary forest through clearance for cultivation created favorable conditions for forage-based pig husbandry, but subsequent increases in human population upset the balance, so that fodder had to replace forage in order to maintain existing ratios of pigs to people (see Golson and Gardner 1990:406 for comment on this proposition). By both arguments the fully domesticated pig that now made its appearance represented a scale of investment that gained for it high cultural value, making it an appropriate indemnifying item in social transactions at many levels (cf. Kelly 1988:128–29).

Unfortunately, the history of the pig in the Highlands is poorly known, even as regards the date of its appearance (Golson and Gardner 1990:406 and references). Pig bone is only sparsely represented in the Highlands archaeological record before perhaps the past 1,000 years. In view of evidence that wild pigs do not seem to thrive in the mid-montane environment (Golson and Gardner 1990:406 and references; but cf. Flannery 1990a:32), it may be that this increased archaeological presence of pig registers a higher level of pig production made possible by controlled breeding under conditions of full domestication.

Exchange

There is limited archaeological evidence that systems of ceremonial exchange ethnographically characteristic of some Highlands regions, which fuel the demand for pigs, may have a general antiquity in the upper Wahgi Valley similar to that suggested for their full domestication. The case concerns ax stone from specialized quarries in the middle Wahgi and Jimi Valleys, which were supplying outside communities up to European contact. Ethnographically, highly institutionalized systems of ceremonial exchange like the Hagen *moka* are able to draw in valuables, including bush products no longer available in the denuded agricultural zone, items of localized natural occurrence like salt and superior ax stone, and true exotics from the lowlands and the coast (Golson 1982:130 and references). The predominance of middle Wahgi and Jimi stone in archaeological ax collections from the upper Wahgi suggests that the district had possessed this capacity for between 1,000 and 2,000 or more years (Golson and Gardner 1990:403–5, citing published and unpublished work by Burton).

From Horticulture to Agriculture

The central role of the upper Wahgi in a geographically extensive structure of exchange can be seen in the context of the agrarian history of the Highlands and its different regional outcomes. The archaeological record provides us with two contrasted instances that serve as exemplars of the whole.

The upper Wahgi. Investigations at Kuk swamp have provided evidence not only, as we have seen, for cultivation back to 9000 B.P. but also for subsequent developments in the agrarian domain. The discoveries in question concern episodes of swamp drainage for cultivation, taking place against a background of continuous dryland farming. There are six of these, forming two contrasting sets which reflect a significant change in the character of agrarian activities around 2000 B.P. (Golson and Gardner 1990:405–6 and references).

The first set of three (Phases 1 to 3 of the Kuk sequence) has approximate dates of 9000, 6000 to 5500, and 4000 B.P. to approaching 2000 B.P. Despite differences among the three, each is structurally diversified in ways suggesting that provision was being made for crops of different edaphic and hydrologic requirements growing side by side in the same plantings. In this sense the three earlier phases are horticultural.

The second set of three (Phases 4 to 6) has approximate dates of from something like 2000 to 1200, 400 to 250, and 250 to 100 B.P. All three exhibit standardized patterns based on long, straight ditches parallel and at right angles to each other that enclosed square to rectangular plots, thought to be planting areas for a single crop. In this sense the three later phases are agricultural.

The unintensive nature of the earlier swamp systems and the long periods of swamp disuse that separate them are compatible with a situation in which the focus was dryland cultivation of crops like taro under a forest-fallow regime. Such a regime became increasingly unviable, however; in the limiting conditions of topography, climate, forest, and crop type specified in our original model, its very operations deflected the vegetation succession to ever more degraded secondary growth and eventually, over wide areas, to grassland, as indicated in our review of vegetation and faunal histories. There is evidence from yields to show how unproductive dryland taro cultivation would have become in these conditions, except on ground of high and sustained fertility, including swamps (Golson and Gardner 1990:400–401 and references). The prolonged but ill-known Phase 3 at Kuk will no doubt prove to be a reflection of the long period of forest degradation during which shifting agriculture became increasingly unsuccessful.

Investigation of the later phases at Kuk has helped to identify the successive appearance of the different technologies that in combination today represent a successful response to the problem of productive cultivation in deforested environments (Golson and Gardner 1990:398–400 and references). First came the innovation of soil tillage, a remedy for the deficiencies of grassland soils, which is thought to be registered by a widespread stratigraphic change from clay particles to soil aggregates in the depositional record of the Kuk swamp about 2000 B.P., between Phases 3 and 4. Second was the innovation of silviculture, inferred from the *Casuarina* rise in the pollen diagrams, as already discussed. Finally came the innovation of raised-bed cultivation, a widespread and regionally varied practice in contemporary Highlands dryland agriculture, whose expression in the upper Wahgi is a checkerboard pattern of plots where the planting medium is the soil thrown up from the grid of intervening flat-bottomed ditches, a type of trenching recognizable in wetland and dryland contexts at Kuk during Phase 5.

These successful innovations in the technology of dryland agriculture may well have been factors in the periodicity of swamp cultivation at Kuk after 2000 B.P. Despite them, however, the swamp was under cultivation for substantial periods, and at a scale and uniformity of operation never seen before. The explanation must surely lie in the other consequences of deforestation previously proposed: the demands made on agricultural production by the increased importance and changed nature of pig husbandry following on faunal impoverishment and the retraction of foraging domains; and the impetus to greater production of pigs given by their likely importance in mediating transactions within and between groups, including exchanges that drew in valued items like middle Wahgi and Jimi axes. Kuk and other upper Wahgi swamps were well fitted for this role in view of their natural fertility (see Wood 1985:61) and their suitability both for taro cultivation and for pig foraging (on pig foraging, see Golson and Gardner 1990:406 and references). We may suppose that there were other land systems of similar capacity. However, there were also areas of greater sensitivity to human interference and less resilience in the face of it (the terms are those of Blaikie and Brookfield 1987a:10–11), which would be disadvantaged as a result of the environmental changes we have been considering. Circumstances such as these, at the local level, may have engendered inequalities between and within clans of the kind evident in the ethnography of the upper Wahgi (Golson 1982:130) and, at the regional level, been importantly implicated in some major differences between Highlands societies as seen at European contact (a thesis developed in detail by Feil 1987, for comments on which see Golson and Gardner 1990:409, 411).

The upper Ramu. With its lower rainfall of pronounced seasonality, the far eastern area of the Highlands is one of high sensitivity and low resilience. As we have seen, the pollen diagram from Norikori swamp shows rapid and comprehensive deforestation in its region, with grasslands dominant after 1600 B.P. Investigations of modest field systems on the benched landscape of the Arona basin 20 km to the northeast show that they were operating in increasingly treeless conditions around the same period and suggest that they served a moisture-regulating function in such circumstances (Golson and Gardner 1990:410–11 and references). Little is known about the duration of the systems in question or about their wider distribution, so that it is impossible to judge how successful they were in agricultural production. At Arona itself there is no evidence for any development in the system over time, and it seems to have gone out of effective operation by 250 years ago. The impression is the opposite of that given by the Kuk systems with their capacity for renewal.

The Argument Thus Far

We have reviewed the changing ecology of agrarian operations in the New Guinea Highlands over the millennia during which subsistence was based on a suite of tropical plants of Pacific affiliations with, overall, requirements for fertile soils and altitudinal limits on productive growth. A number of lines of evidence converge to suggest that in the course of interrelated developments under this regime the foundations were laid of modern Highlands landscapes and cultures and their regional differentiations, at least in country below 2,000 m. The chronologies of the separate lines of development in question do not fit each other as well as their claimed interrelatedness would require, but the general lack of resolution in the dates that are available could well account for this. However, the critical period when environment, agriculture, pig husbandry, and exchange assume features recognizable in the ethnographic record falls between 2,000 and 1,000 years ago.

AFTER THE SWEET POTATO

It is uncertain to what extent under the older regime agrarian settlement moved up from the great Highlands basins and valleys below 2,000 m for long periods, as distinct from seasonal or in other ways sporadic exploitation of high-altitude resources by communities permanently resident at lower elevations (see Hope and Hope 1976:39–43). In the pollen diagrams from the Sirunki plateau at 2,500 m toward the western margin of the Papua New Guinea Highlands zone there are indications of partial clearance of forest between about 4500 and 4000 B.P., and later—about 2000 B.P. of "more widespread and maintained replacement of forest by open land . . . which may have been the direct fore-runner of present conditions there" (Walker and Hope 1982:273). From a site at 2,240 m in the upper Kaugel Valley nearer the Wahgi a wooden spade at the base of a drainage ditch is dated to 4000 B.P. and may indicate possibly short-lived agrarian activity around that time (Golson and Steensberg 1985:369–70, 376).

At these high altitudes, however, sustained forest disturbance accompanied by rising values for *Casuarina* in pollen diagrams and notable increases in sedimentation rates in lakes is a feature of the past few hundred years and is thought to register the arrival of permanent agricultural settlement based on the sweet potato (Golson and Gardner 1990:400 and references).

Although the sweet potato has this greater capability at higher elevations than established crops, it is also more productive at the same altitudes and superior in poor and agriculturally degraded soils (yield data from various sources cited by Golson and Gardner 1990:400–401), the latter a significant product of agrarian exploitation over previous millennia. Changes in the character of swamp gardening at Kuk associated with Phase 6 of the drainage sequence beginning about 250 B.P. are attributed to the sweet potato (Golson and Gardner 1990:407 and references). The network of long straight parallel ditches intersecting at right angles characteristic of Phases 4 and 5 now becomes tighter and more gridlike, assuming the appearance and dimensions of the checkerboard of contemporary dryland sweet potato gardens. Significantly, the area under drained cultivation diminishes by two-thirds, reflecting the better performance of the new crop in the dryland sector.

These manifestations of the capacity of the sweet potato at higher altitudes and in depleted soils incidentally provide the best evidence for its late, post-Magellanic entry into New Guinea, against claims of an introduction from the Pacific around 1,200 years ago (cf. Gorecki 1986:164). The plant's ability to effect a spatial expansion of food production, of which it is evidence, is complemented by a corresponding temporal expansion, as its tolerance of poorer soils and its longer productivity (Yen 1974:72–73) allow it to assume the role of a follow-up crop in a rotational system (Clarke 1977:161). Add to all this its superiority as pig fodder, and there was the potential in sweet potato cultivation to confront some of the problems that we have identified in production systems under the previous regime and to ameliorate some of the inequalities to which they are likely to have given rise. There is little archaeological evidence to illustrate, let alone test, these propositions, but given the short interval between the appearance of the sweet potato and the beginning of European contact, it may be legitimate to use the data of the ethnographic record in its stead.

A striking feature of that record is the numerical size and extensive geographical spread of some speech communities of the central and western sections of the Papua New Guinea Highlands zone (see Wurm 1975:462–63, 467–69), each of them centered on valley systems where successful foci of population had long existed under the previous agrarian regime to extrapolate from the case of the Hageners of the upper Wahgi and neighboring valleys with whose history we are acquainted. Out of these, it is proposed (Golson and Gardner 1990:408 and references), historically recent expansion took place, made possible by the capabilities of the sweet potato over a wide range of environments, into elevated mountain country that had previously separated them and out of the Highlands proper into the upper reaches of lowland rivers. In the environmentally less favored region of the far eastern Highlands, heavily impacted (by the evidence of Norikori swamp and the Arona basin) long before the sweet potato made its appearance, agriculture with the new crop was able to support much more modest populations, grouped into several linguistically related speech communities (Wurm 1975:462–63, 467–68).

This linguistic picture goes hand in hand with the absence of developed institutions of intercommunity relations in the east (Feil 1987:238, "Kainantu" group of societies). Farther west, in contrast, expansion out of established population centers was accompanied by expansion of exchange networks. The origin of these, again from the evidence of the upper Wahgi Valley, has been sought in particular developments under the earlier agrarian regime, creating a need initially to intensify pig husbandry and subsequently to increase pig production, which could only effectively be satisfied on land systems of high productivity for taro. In principle, the sweet potato, as prime pig fodder that could be grown over a wider range of soils and environments, extended possibilities for successful pig keeping and offered an opportunity for more men, through the labor of their wives, to enter previously restricted systems of exchange in which pigs were a central item.

Commenting on the tendency of these exchanges to become inflationary as a result of competition between big men who put pressure on all members of their group to participate, Allen and Crittenden (1987:149, 151–52) suggest that the high levels of sustained production required could not be met without land degradation where they were attempted on land systems of high sensitivity and low resilience, in which they discern steeper slopes, inferior soils, and higher altitudes. The examples they quote (Allen and Crittenden 1987:152–55) include ascending grasslands as forest-edge gardens up to 2,600 m in altitude are cleared, cultivated, and abandoned; erosion on steep slopes with inferior soil at 1,900 m; and volcanic ash–derived soils over limestone at around 1,800 m, which have reverted to a cover of *Miscanthus* cane grass.

In the main valleys and basins, of which we have used the Wahgi as the exemplar, high levels of agricultural intensification have been achieved and sustained on soils developed on alluvium and volcanic ash (Allen and Crittenden 1987:152). In the western region of the Papua New Guinea Highlands zone in particular there have been periodic ashfalls during the Holocene, from volcanic sources outside the Highlands, which may have played an important role in cultivation by renewing the nutrient bank. In making this point, both

Haberle et al. (1991:30, 38) and Brookfield (1991:207) contrast the situation in the Irian Jaya section of the Highlands, where no volcanic ash is present and the land is more susceptible to degradation and erosion. Brookfield (1991:207) describes the intensively managed and densely peopled wetlands of the Baliem Valley and Paniai lakes as being separated by a wide expanse of sparsely populated degraded slopeland from agrarian settlements in tributary valleys and at the forest edge.

CONCLUSION

The massive transformation that the practice of shifting cultivation wrought in the ecology of Highlands New Guinea—from which we have derived some basic features of Highlands societies—as well as in agricultural practice, pig husbandry, and systems of exchange, took place over many millennia. The entry of the sweet potato was a fairly recent event, and it brought about changes in settlement and pig husbandry whose ramifications were still being worked out at the time of European contact. Today, in the Papua New Guinea Highlands the cash economy and rising populations are making new demands on natural and social systems and are requiring adjustments from them of a scale and urgency unparalleled in the past.

4

Extinctions of Polynesian Birds: Reciprocal Impacts of Birds and People

David W. Steadman

The loss of bird life is now well recognized as one of the major environmental consequences of the human colonization of Oceania. The decline of birds is related mainly to predation and to landscape changes wrought by prehistoric peoples to accommodate agriculture. For birds, perhaps the most influential landscape change has been the elimination and alteration of indigenous forests through cutting, burning, and the introduction of nonnative plants (Kirch 1983). These activities have eliminated natural habitats and have rendered the surviving birds even more vulnerable to predation from humans and nonnative mammals (rats, dogs, pigs). The erosion of topsoil associated with deforestation has removed large areas of nesting habitat for burrowing seabirds, such as shearwaters and petrels.

Ethnographic information gathered over the past two centuries is crucial for interpreting Polynesian uses of birds in prehistoric times. I shall concentrate as much as possible on the prehistoric and early historic relationships between birds and Polynesians, knowing that events of the post-European period have only exacerbated what already was a dismal situation from the birds' standpoint.

The geographic area considered here is all of Polynesia except the Hawaiian Islands and New Zealand. Although Hawaii (Olson and James 1982a, 1982b, 1991; James and Olson 1991; James et al. 1987) and New Zealand (Anderson 1984, 1989a; Cassells 1984; Trotter and McCulloch 1984; Holdaway 1989) both experienced major prehistoric losses of birds, the biotas of these two isolated archipelagoes are not closely related to those elsewhere in Polynesia.

The prehistoric record of birds is much better known in Polynesia than in Micronesia or Melanesia. Bones of birds from archaeological or paleontological contexts are now known from at least 20 Polynesian islands besides Hawaii and New Zealand. In Micronesia, studies have been completed thus far on only two islands. From Rota in the Mariana Islands, several small cave deposits have yielded bones of 13 species of extinct or extirpated birds, including shearwaters, terns, ducks, megapodes, rails, pigeons, parrots, swifts, and passerines (Steadman 1992a). Assemblages of archaeological bird bones excavated on Fais (Yap) represent species of indigenous resident birds, 12 of which probably or certainly no longer live on Fais (Steadman and Intoh 1994). In spite of the small amount of data currently available from Micronesia, I see no cultural, geographical, or biological reason why the extent of human-caused avian extinctions in this region of Oceania, when more fully studied, will differ in any major way from that of Polynesia.

The islands of Melanesia tend to be larger, and to support richer floras and faunas today, than the islands of Polynesia or Micronesia. The prehistoric record of Melanesian birds has been studied only on Fiji (Steadman 1989b), New Caledonia (Balouet and Olson 1989), New Ireland (Steadman, White, and Allen n.d.), and Mussau (Steadman n.d.). As in Polynesia and Micronesia, the limited record in Melanesia shows losses of a variety of both seabirds and landbirds, the latter dominated by hawks, megapodes, rails, pigeons, parrots, and owls but also including herons, ibises, buttonquails, snipe, owlet-nightjars, and crows. Preliminary data from the very large, high island of New Ireland indicate that even its relatively rich avifauna (currently about 108 species of resident landbirds) has lost about 25% of its species. A greater percentage of birds has survived on the larger islands of Melanesia than on typical Polynesian or Micronesian islands, or on small Melanesian islands. This seems to be due to the buffering effects that large island size, steep terrain, and diseases (such as malaria) have had on human impact. Common sense dictates that insular floras and faunas are easiest to deplete on low, small islands. In addition, many parts of island Melanesia have been occupied for tens of millennia longer than anywhere in Polynesia or Micronesia (Jones 1989; see Chapter 2, this volume).

HUMAN IMPACTS ON BIRDS

Background Extinction

The Holocene extinction of vertebrates on Oceanic islands has been calibrated by the fossil record of the Galápagos Islands (Steadman et al. 1991). Unlike any other group of islands in the tropical Pacific, the Galápagos never supported human populations before their discovery by Europeans in 1535 (Steadman 1986). As a result, human impact in the Galápagos is confined to the past five centuries, and it was relatively minor until about 1800. The Holocene fossil record of the Galápagos comprises about 500,000 bones, more than 90 percent of which predate the arrival of humans. These paleofaunas reveal the loss of only zero to three vertebrate populations in the 4,000 to 8,000 years before human arrival, compared to the loss of 21 to 24 populations in the past 150 to 300 years. Thus the rate of background (prehuman) extinction in the Galápagos was at least a hundred times less than the rate of human-related extinction. When undisturbed by humans, the natural processes of dispersal, colonization, and evolution may result in a very low rate of extinction for reptiles, birds, and mammals on tropical Oceanic islands.

The Polynesian Record

Moving west from the Galápagos to Oceania, we find that virtually all islands were inhabited at one time or another in prehistory and that a significant amount of prehistoric extinction has taken place. The prehistoric record of Polynesian birds is based on data collected from archaeological and paleontological sites in the Marquesas Islands (Steadman 1988, 1989a, 1991a, 1992b; Steadman et al. 1988; Steadman and Zarriello 1987; Dye and Steadman 1990), the Society Islands (Steadman 1989a, 1992b; Dye and Steadman 1990; Steadman and Pahlavan 1992), the Cook Islands (Steadman 1985, 1987, 1989a, 1991b, 1992b; Allen and Schubel 1990; Allen and Steadman 1990, Steadman and Kirch 1990; Kirch et al. 1991; Kirch et al. 1992), Henderson Island (Steadman and Olson 1985; Schubel and Steadman 1990), Easter Island (Steadman et al. 1994), Samoa (Steadman 1993b), Tonga (Steadman 1989a, 1989b, 1993a; Dye and Steadman 1990), and the Polynesian Outliers in Melanesia (Balouet and Olson 1987; Steadman et al. 1988; Steadman, Pahlavan, and Kirch 1990). I shall review this record briefly, using specific examples from several island groups.

The losses of birds in Polynesia fall into three categories: (1) extinction (loss of all populations of a species); (2) extirpation (loss of a species on an individual island, although one or more populations of the species survive elsewhere); and (3) reduced population (loss without replacement of individuals from a surviving population on an island). A single locality (Mangaia, the Cook Islands) can exemplify each category. The Conquered Lorikeet (*Vini vidivici*) is *extinct* because it has been exterminated on Mangaia as well as on every other island where it ever occurred. The Society Islands Pigeon (*Ducula aurorae*) is *extirpated* on Mangaia (and several other islands) but survives on Makatea (Tuamotus) and Tahiti. Audubon's Shearwater (*Puffinus lherminieri*), a pantropical seabird, has a *reduced population* on Mangaia today, surviving in numbers of less than 100, whereas archaeological and ethnographic evidence suggests that once it was common and widespread on the island.

The loss of seabirds in Polynesia has been particularly severe for petrels and shearwaters, although the ranges and numbers of many other kinds of tropical seabird (various albatrosses, storm-petrels, tropicbirds, frigatebirds, boobies, terns, gulls) have been reduced as well. The losses of landbirds in Polynesia have been greatest for rails, pigeons, doves, and parrots, although, as with the seabirds, no family of landbirds has been spared. In Western Polynesia, unlike Eastern Polynesia, the losses of landbirds also include herons, megapodes, hawks, shrikebills, whistlers, monarch flycatchers, thrushes, white-eyes, and honeyeaters, some of which represent taxa previously occurring only in Melanesia, not in Polynesia. By eliminating the "Melanesian" taxa, the anthropogenic extinction of birds in Tonga, for example, has artificially sharpened the biogeographic distinction between the avifaunas of Polynesia and Melanesia (Steadman 1993a).

Quite deservedly, the "biodiversity crisis" has received much popular and scholarly attention in the past decade. An understanding of the current biodiversity crisis requires proper historic perspective, which in turn tells us that few places on earth have escaped environmental change at the hands of prehistoric humans. From the standpoint of numbers of extinct species, the most dramatic story is that of flightless rails. Each island in the tropical Pacific with a thorough prehistoric record of birds has yielded the bones of one to three unique (endemic) species of flightless or nearly flightless rails. At least five genera are involved, *Porzana, Gallirallus, Nesoclopeus, Gallinula,* and *Porphyrio,* with *Gallirallus* providing the most species discovered thus far and one of the world's most exciting, albeit least understood, examples of adaptive evolutionary radiation. About 800 islands in Oceania are inhabitable by people and therefore, presumably, by flightless rails. Thus rails alone might account for as many as 2,000 species of birds that would exist today had people not colonized Oceania.

The first of several examples of depleted Polynesian avifaunas is from Ua Huka in the Marquesas Islands (Table 4.1), where the Hane site has yielded

Table 4.1 Resident Birds from Ua Huka, Marquesas Islands

	Bones from Hane Site	Exists today on Ua Huka
SEABIRDS		
Wedge-tailed Shearwater (*Puffinus pacificus*)	x	x
Christmas Island Shearwater (*Puffinus nativitatis*)	x	—
Audubon's Shearwater (*Puffinus lherminieri*)	x	—
Bulwer's Petrel (*Bulweria* cf. *bulwerii*)	x	x
Tahiti Petrel (*Pterodroma rostrata*)	x	—
Phoenix Petrel (*Pterodroma* cf. *alba*)	x	—
Unknown petrel (*Pterodroma* small sp.)	x	—
Polynesian Storm-Petrel (*Nesofregetta fuliginosa*)	x	x
White-bellied Storm-Petrel (*Fregetta grallaria*)	x	—
White-tailed Tropicbird (*Phaethon lepturus*)	x	x
Red-footed Booby (*Sula sula*)	x	x
Brown Booby (*Sula leucogaster*)	x	x
Masked Booby (*Sula dactylatra*)	x	—
Abbott's Booby (*Papasula abbottii costelloi*)	x	—
Great Frigatebird (*Fregata minor*)	x	x
Lesser Frigatebird (*Fregata ariel*)	x	x
Gray-backed Tern (*Sterna lunata*)	—	x
Sooty Tern (*Sterna fuscata*)	x	x
Brown Noddy (*Anous stolidus*)	x	x
Black Noddy (*Anous minutus*)	x	x
Blue-gray Noddy (*Procelsterna cerulea*)	—	x
Little Fairy Tern (*Gygis microrhyncha*)	x	x
LANDBIRDS		
Pacific Reef-Heron (*Egretta sacra*)	x	x
Tuamotu Sandpiper (*Prosobonia* cf. *cancellata*)	x	—
°Undescribed crake (*Porzana* new sp.)	x	—
°Undescribed rail (*Galliralus* new sp.)	x	—
Marquesas Ground-Dove (*Gallicolumba rubescens*)	x	—
°Giant Ground-Dove (*Gallicolumba nui*)	x	—
Red-moustached Fruit-Dove (*Ptilinopus mercierii*)	x	—
White-capped Fruit-Dove (*Ptilinopus dupetithouarsii*)	x	x
Nuku Hiva Pigeon (*Ducula galeata*)	x	—
°Marquesas Cuckoo-Dove (*Macropygia heana*)	x	—
Marquesas Lorikeet (*Vini ultramarina*)	x	—
°Conquered Lorikeet (*Vini vidivici*)	x	—
°Sinoto's Lorikeet (*Vini sinotoi*)	x	—
Marquesas Swiftlet (*Collocalia ocista*)	—	x
Marquesas Kingfisher (*Halcyon godeffroyi*)	x	—

Table 4.1 (continued)

	Bones from Hane Site	Exists today on Ua Huka
°Undescribed monarch (cf. *Myiagra* new sp.)	x	—
Iphis Monarch (*Pomarea iphis*)	—	x
Marquesas Reed-Warbler (*Acrocephalus mendanae*)	—	x
TOTALS		
Total species of seabirds	20	14
Combined total species of seabirds	22	
Total species of landbirds	15	5
Combined total species of Landbirds	18	

SOURCE: Modified from Steadman 1991a
° = extinct species

about 11,000 bird bones, the largest avian assemblage yet obtained from tropical Polynesia. Most of these bones represent seven species of shearwaters and petrels, easily obtained from their nesting burrows. While six of the 20 species of seabirds from the Hane site no longer occur on Ua Huka, this number alone does not represent the loss of seabirds. Most of the 14 species of seabirds listed in Table 4.1 as "exists today on Ua Huka" nest today not on the main island but only on tiny offshore islets. Even the surviving species of seabirds exist on Ua Huka or its islets in much reduced numbers.

Pigeons and doves (family Columbidae) are the most common landbirds at the Hane site, both in terms of number of species and number of bones. Only one of the six species of columbids in the Hane deposits survives today on Ua Huka. Rails and parrots are the next most common, each of these families having declined on Ua Huka from three to zero species. Although the exact chronology of the Hane site is in dispute (Sinoto 1979; Kirch 1986), the pattern of avian exploitation is clear (Dye and Steadman 1990:212): "The relative contribution of birds to the diet [more than half of all animal protein in the early period] is more than halved within the first 550 years and declines to insignificance in just over a thousand years."

The second example of a prehistoric avifauna is from the Fa'ahia site on Huahine in the Society Islands (Table 4.2). This assemblage is extremely rich in species, considering that it is based on only 336 identified bird bones. All but three of the 14 species of seabirds recorded from Fa'ahia occur at the Hane site as well. The species-level similarity among landbirds is much less striking, with only three species shared. Again, rails, pigeons, doves, and par-

Table 4.2 *Resident Birds from Huahine, Society Islands*

	Bones from Fa'ahia Site	Exists today on Huahine
SEABIRDS		
Wedge-tailed Shearwater (*Puffinus pacificus*)	x	—
Christmas Island Shearwater (*Puffinus nativitatis*)	x	—
Audubon's Shearwater (*Puffinus lherminieri*)	x	—
Tahiti Petrel (*Pterodroma rostrata*)	x	—
Phoenix Petrel (*Pterodroma alba*)	x	—
Herald Petrel (*Pterodroma arminjoniana*)	x	—
White-tailed Tropicbird (*Phaethon lepturus*)	x	x
Red-footed Booby (*Sula sula*)	x	—
Brown Booby (*Sula leucogaster*)	x	—
Great Frigatebird (*Fregata minor*)	x	—
Lesser Frigatebird (*Fregata ariel*)	x	—
°Undescribed gull (*Larus* new sp.)	x	—
Brown Noddy (*Anous stolidus*)	x	x
Black Noddy (*Anous minutus*)	—	x
Common Fairy Tern (*Gygis candida*)	x	x
LANDBIRDS		
Pacific Reef-Heron (*Egretta sacra*)	x	x
Mangrove Heron (*Ardeola striata*)	x	—
Gray Duck (*Anas superciliosa*)	—	x
Sooty Crake (*Porzana tabuensis*)	x	—
°Undescribed rail (*Gallirallus* new sp.)	x	—
Society Islands Ground-Dove (*Gallicolumba erythroptera*)	x	—
°Giant Ground-Dove (*Gallicolumba nui*)	x	—
Society Islands Fruit-Dove (*Ptilinopus purpuratus*)	x	x
Nuku Hiva Pigeon (*Ducula galeata*)	x	—
Society Islands Pigeon (*Ducula aurorae*)	x	—
°Huahine Cuckoo-Dove (*Macropygia arevarevauupa*)	x	—
Society Islands Lorikeet (*Vini peruviana*)	—	e
°Conquered Lorikeet (*Vini vidivici*)	x	—
°Sinoto's Lorikeet (*Vini sinotoi*)	x	—
Tahiti Swiftlet (*Collocalia leucophaea*)	—	e
Chattering Kingfisher (*Halcyon* cf. *tuta*)	x	x
Society Islands Reed-Warbler (*Acrocephalus caffer*)	x	e
Huahine Starling (*Aplonis diluvialis*)	x	e

Table 4.2 (continued)

	Bones from Fa'ahia Site	Exists today on Huahine
TOTALS		
Total species of seabirds	14	4
Combined total species of seabirds	15	
Total species of landbirds	15	4
Combined total species of landbirds	18	

SOURCE: Modified from Steadman and Pahlavan 1992
° = extinct species
e = recorded in nineteenth century but now extirpated

rots predominate in the prehistoric avifauna. Except for bone assemblages recently collected with screens of $\frac{1}{16}$ inch mesh on Aitutaki and Mangaia (Allen and Steadman 1990; Steadman and Kirch 1990), all of the Eastern Polynesian faunal assemblages are biased toward larger species of bird. (Screens of $\frac{1}{8}$ inch mesh remove most, but not all, of this bias.) Thus tiny species, such as swifts and many passerines, typically are absent or underrepresented. To the west, in Tonga, we now know through sampling with screens of $\frac{1}{16}$ inch that passerines underwent losses comparable to those of larger landbirds (Steadman 1993a).

The prehistoric exploitation of birds can be examined from a more detailed chronostratigraphic perspective at Tangatatau Rockshelter (site MAN-44) on Mangaia in the Cook Islands (Table 4.3). This stratified site ranges from about 900 (zone J) to 200 or 300 years old (zone A). During this period, the human consumption of vertebrates underwent dramatic change. Domesticates (the pig, *Sus scrofa,* and chicken, *Gallus gallus*) are rare in lower strata but increase as native landbirds decrease. Both pig and chicken decline, however, in the uppermost strata, perhaps because of overconsumption by a relatively large human population.

The number of bones from native birds—that is, all species except chickens—is high in zones I and J, low in the middle zones, and high again in zones B and C. The species composition, however, is different between the early and late zones. Extinct or extirpated species of landbirds dominate zones G to J, while seabirds, especially the Black-winged Petrel (*Pterodroma nigripennis*), account for the late increase in exploitation of birds. This differs from sites elsewhere in Polynesia, such as the Marquesas, the Society Islands,

Table 4.3 Vertebrate Faunal Summary by Analytic Zones (A to J), Main Trench (Squares C30 to G30), Tangatatau Rockshelter (MAN-44), Mangaia, Cook Islands, July to August 1989

Taxon	A	B	C	D	E	F	G	H	I	J	Total
Fish	51	667	973	937	2,464	1,206	875	735	1,379	80	9,367
Reptiles											
Sea turtle	1	—	—	1	—	—	—	—	—	—	2
Lizard	—	1	—	—	—	—	1	—	3	—	5
Birds											
All native species	7	27	22	4	7	1	9	13	49	47	186
Gallus gallus	1	—	2	3	13	3	6	—	3	—	31
Unidentifiable bird	5	27	20	10	23	4	11	13	36	52	201
Mammals											
Pteropus tonganus	—	—	—	—	—	—	1	2	15	13	31
Rattus exulans	19	125	83	50	39	14	61	34	40	1	466
Sus scrofa	—	—	3	10	13	2	1	1	1	—	31
Delphinidae sp.	—	—	—	—	—	—	1	—	—	—	1
Homo sapiens	1	1	—	—	—	—	—	—	—	—	2
Unidentifiable mammal	—	2	—	2	4	2	1	—	—	—	11
Unidentifiable bone	—	5	3	—	12	2	—	4	11	—	37
Totals											
All species	85	855	1,106	1,017	2,575	1,234	967	803	1,535	191	10,366
All nonfish	34	188	133	80	111	28	92	67	158	113	1,004
% nonfish	40.0	22.0	12.0	7.9	4.3	2.3	9.5	8.4	10.3	59.2	9.6
% fish	60.0	78.0	88.0	92.1	95.7	97.7	90.5	91.6	89.6	41.9	90.4

Source: From Steadman and Kirch 1990

Tonga, and Tikopia, where bones of seabirds are more abundant than those of landbirds in the oldest cultural levels of early sites (Steadman 1989a, 1989b; Dye and Steadman 1990; Steadman et al. 1990). A factor on Mangaia may be its precipitous, creviced limestone cliffs, which provided an extensive albeit narrow band of relatively rat-free nesting habitat for seabirds.

Based upon number of bones, predation on landbirds at MAN-44 was most intense in zones I and J and tapered off to practically nil by the end of zone G (Table 4.4). Based on relative abundance of bones (percentage of all vertebrates), predation on landbirds was by far most intense in zone J, after which most species survived into zones I, H, or G, but only in reduced numbers. No extinct or extirpated landbirds are recorded above zone G except the Cook Islands Fruit-Dove (*Ptilinopus rarotongensis*), which still survives on Rarotonga and Atiu and is known historically from Ma'uke and Aitutaki.

Conforming to the general Eastern Polynesian pattern, the eight species of extinct or extirpated landbirds from MAN-44 consist of rails, pigeons, doves, and parrots. Both of the extinct rails were flightless. Their obligatory existence on the ground for foraging and nesting must have facilitated predation by humans and rats. Pigeons, doves, and parrots are forest species favored as food by Polynesians. Parrots (and sometimes pigeons and doves) also were used for their brightly colored feathers, as I discuss below.

The five surviving species of landbirds recorded from MAN-44 are those that can withstand substantial forest clearance. Because they prefer marshes or dense growths of ferns, grasses, and sedges, the Gray Duck (*Anas superciliosa*) and the Sooty Crake (*Porzana tabuensis*) have probably benefited from anthropogenic landscape changes that promote irrigated taro cultivation. The Lesser Golden-Plover (*Pluvialis dominica*) is a migratory shorebird that nests only on the tundra of high northern latitudes. In Polynesia it occurs in open habitats like reefs and sand flats, grassy fields, and airstrips. The Mangaia Kingfisher (*Halcyon mangaia*) seems to tolerate moderate forest clearance but requires cavities in trees for nesting. Its current scarcity on Mangaia may be related to nest-site competition with the Common Myna (*Acridotheres tristis*), an abundant and aggressive Asian species introduced to Mangaia earlier this century. The Mangaia Reed-Warbler (*Acrocephalus kerearako*) occurs in forests, thickets, and shrublands of varying disturbance and maturity.

Hunting Methods

On many Polynesian islands today, birds are hunted with shotguns. Prehistoric Polynesians were highly skilled at catching birds by traditional methods, which are still used on some islands. Boys today (and presumably in the past) begin to hunt birds at a young age and often continue to hunt nearly throughout their lives. I have learned from personal experience that certain boys and young men can climb cliffs with amazing skill, speed, and fearlessness (to grab birds, especially tropicbirds and terns), can throw rocks at birds with deadly speed and accuracy, and may virtually run up a large tree to grab a nesting bird.

Perhaps the most common way that prehistoric Polynesians caught birds was simply by hand. The hand method of gathering birds, known as *tangotango* on Tikopia and *tango manu* on Pukapuka (Beaglehole and Beaglehole 1938:74), is rather straightforward: you simply "see a bird dozing in a tree, climb up behind it, grab it, and break its neck" (Feinberg 1981:34). Sticks are used to strike birds in trees, along shorelines, and on cliffs. Seabirds are particularly vunerable. Some species of tropicbirds, boobies, and terns, for example, nest on the ground, while most species of shearwaters and petrels nest in burrows or crevices. Any of these seabirds could (and still can) be plucked from their resting places just by reaching down or in and grabbing them. One such bird on Mangaia was the *titi*, probably a yet to be determined species of shearwater (*Puffinus* sp.) that no longer seems to exist on the island. In describing the eating habits of two late prehistoric fugitives living in the Mangaian forest, Gill (1894:26) stated: "The bird most easily caught by [the fugitives] Uriitepitokura and Temoaakaui was the *titi* (so called from its cry). In the month of December it leaves its burrowings in the red mountain soil, and comes to the rocks near the sea to fatten its young on small fish. By day it hides in holes, and sleeps. The hunter has only to call at the entrance to the dark cave, in a plaintive voice, *E titi e*, when the foolish bird, imagining it to be the voice of its mate, comes out of its secure hiding-place, and, dazzled by the unwelcome light, allows itself to be caught by hand." In a Mangaian dirge, a defenseless titi is compared to an unarmed man facing a man with a war club (Gill 1894:307).

Another example of predation on a defenseless seabird is Gill's description (1885:108–9) of the Herald Petrel (*Pterodroma [arminjoniana] heraldica*):

> A fishing-hawk, about the size of a pigeon, with black eyes and dark plumage, excepting underneath, where it is white, was formerly plentiful at Rarotonga [where it still survives; McCormack and Kunzle 1990b]. Its home is in the crannies of almost inaccessible rocks. The *koputu* (such is its name) lays but two eggs—exactly like those of a duck in size and color—in the season. It is considered by the natives to be excellent eating. A favorite pastime of young men in the olden time was to catch these birds in the breeding season at the risk of their lives. The plan was to lower a lad over the edge of a cliff with a stout coir rope round his waist, the upper end passed round the trunk of a tree and firmly grasped by a near relative [usually a brother, as distant relatives were not trusted]. A basket slung round the neck would soon be filled with dead birds and the lad hauled up again.

Table 4.4 Summary of Birds by Analytic Zones (A to J), Main Trench (Squares C30 to G30), Tangatatau Rockshelter (MAN-44), Mangaia, Cook Islands, July to August 1989

Taxon	A	B	C	D	E	F	G	H	I	J	Total
Seabirds											
Audubon's Shearwater (*Puffinus lherminieri*)	—	1	2	—	—	—	—	—	—	—	3
*Black-winged Petrel (*Pterodroma nigripennis*)	3	17	14	2	1	—	—	—	—	—	37
Unidentified petrel or shearwater (Procellariidae sp.)	—	—	—	—	1	—	—	—	—	—	1
Polynesian Storm-Petrel (*Nesofregetta fuliginosa*)	1	—	—	—	1	—	—	—	—	—	2
White-tailed Tropicbird (*Phaethon lepturus*)	—	1	1	—	—	—	1	—	1	1	5
Lesser Frigatebird (*Fregata ariel*)	—	2	3	2	—	—	—	—	—	—	7
Brown Noddy (*Anous stolidus*)	1	—	2	—	—	1	—	—	—	—	4
Blue-gray Noddy (*Procelsterna cerulea*)	—	—	—	—	—	—	—	—	1	—	1
Common Fairy Tern (*Gygis candida*)	—	—	—	—	1	—	—	1	3	1	6
*Little Fairy Tern (*Gygis microrhyncha*)	—	—	—	—	—	—	—	—	1	—	1
Landbirds											
Gray Duck (*Anas superciliosa*)	1	5	—	—	3	—	—	—	—	—	9
Lesser Golden-Plover (M) (*Pluvialis dominica*)	—	—	—	—	—	—	—	—	1	—	1
Chicken (I) (*Gallus gallus*)	1	—	2	3	13	3	6	—	3	—	31
**Ripley's Rail (*Gallirallus ripleyi*)	—	—	—	—	—	—	2	3	7	6	18
Sooty Crake (*Porzana tabuensis*)	—	—	—	—	—	—	1	—	—	—	1
**Mangaian Crake (*Porzana rua*)	—	—	—	—	—	—	—	2	13	11	26
*Society Islands Ground-Dove (*Gallicolumba erythroptera*)	—	—	—	—	—	—	—	—	4	3	7

Species												
°°Giant Ground-Dove (*Gallicolumba nui*)	—	—	—	—	—	—	—	—	—	—	1	1
°Cook Islands Fruit-Dove (*Ptilinopus rarotongensis*)	—	1	—	—	—	—	—	—	1	1	1	3
°Nuku Hiva Pigeon (*Ducula galeata*)	—	—	—	—	—	—	1	—	1	2	4
°Rimatara Lorikeet (*Vini kuhlii*)	—	—	—	—	—	—	—	2	3	5	10	20
°°Conquered Lorikeet (*Vini vidivici*)	—	—	—	—	—	—	—	2	3	5	9	19
°Rimatara or °°Conquered Lorikeet (*Vini kuhlii* or *V. vidivici*)	—	—	—	—	—	—	—	—	1	4	2	7
Mangaia Kingfisher (*Halcyon mangaia*)	1	—	—	—	—	—	—	—	—	—	—	1
Mangaia Reed-Warbler (*Acrocephalus kerearako*)	—	—	—	—	—	—	—	—	—	2	—	2
TOTALS												
All species	8	27	24	7	20	4	—	15	13	52	47	217
All native species	7	27	22	4	7	1	—	9	13	49	47	186
Seabirds	5	21	22	4	4	1	—	1	1	6	2	67
Native landbirds	2	6	—	—	3	—	—	8	12	43	45	119
Extinct or extirpated landbirds	0	1	—	—	—	—	—	7	12	40	45	105
Extinct or extirpated landbirds as % of all vertebrates	—	0.1	—	—	—	—	—	0.7	1.5	2.5	24	—
Extinct or extirpated landbirds as % of all nonfish	—	0.5	—	—	—	—	—	8	18	24	41	—
Extinct or extirpated landbirds as % of all birds	—	4	—	—	—	—	—	47	86	77	98	—

SOURCE: From Steadman and Kirch 1990
I: Introduced species
M: Migrant species
° Extant species extirpated on Mangaia
°° Extinct species

Gathering eggs of wild birds, particularly of seabirds, was another popular activity throughout Polynesia. By contrast, eggs of chickens seem to be shunned in many places, both today and in the past. Small, uninhabited islands were often visited to gather seabirds and their eggs. Among the many examples are Marquesans visiting Hatutu, Motu Iti, and Fatuhuku, Tahitians visiting Mopelia, Easter Islanders visiting Motu Nui, Mangaians and Aitutakians visiting Manuae, Atiuans visiting Takutea, Ma'ukeans visiting Maria, 'Euans visiting Kalau, and the Tikopia and Anuta visiting Fatutaka (see Dening 1963:121, 122). The relative abundance of seabirds on uninhabited islands provides an important clue to the role birds played in locating islands (discussed below).

On Mangaia, groups of men and boys often weed a taro field inwardly from all sides. Although volant, the Sooty Crake, or mo'o mo'o, is reluctant to fly. Thus several mo'o mo'o may be concentrated by the workers into a weedy patch in the center of the field that is only about 2 m in diameter, at which point the birds can be caught by hand or hit with machetes.

Nets were commonly used to catch birds. On Tikopia, a long-handled net known as *te kupenga veu* was and still is used to catch petrels, noddies, and swifts (Steadman, Pahlavan, and Kirch 1990). Positioning themselves strategically along cliffs, hillsides, and elsewhere, the Tikopia call to entice birds within reach of the nets, which have handles 3 to 4.5 m long.

Snares were an effective way to catch birds on the ground as well as in the trees. They were usually made of saplings and sennit cordage (see Buck 1944: Fig. 155). The slip noose of sennit cordage (Gill 1885:90–91) is a variety of snare often used in fruiting trees to catch members of feeding flocks of starlings, fruit-doves, or pigeons. The futility of an ensnared bird is revealed on Mangaia in "The War-Dirge of Tuopapa," which reads in part:

> Descendants of Tirango, destroy your foes!
>> They are but ensnared birds.
>> They fly for their lives.
> Whither indeed shall the vanquished fly,—
>> This vast host?
> Where can shelter be found, or life be safe?
> Can another stand be made? Is all hope gone? [Gill 1894:98]

Snares are still used occasionally. In 1988, I saw a Banded Rail (*Gallirallus philippensis*) snared by a 12-year-old boy on 'Eua, Tonga. Feral chickens, once commonly pursued with snares, are seldom hunted today because their flesh is less tender than that of tame chickens living in the yard. With the switch from subsistence to cash economies and the availability of electricity for freezers, even yard chickens are shunned if one can purchase and store the ultratender frozen chickens, already plucked and gutted, that are imported from New Zealand, Australia, or the United States, the same countries where people have begun to pay extra for "free-range" chickens and their eggs. Chickens need not cross the road to provide cross-cultural amusement.

Bird lime was an effective way to catch perching birds up to the size of pigeons (Gill 1885:90–91). This sticky substance was prepared by mixing the sap of breadfruit trees (*Artocarpus altilis*) with crushed, baked nuts of the candlenut tree (*Aleurites moluccana*). Bird lime was spread on limbs in places frequented by birds, whose feet would adhere to it. As with snares, trees heavily laden with fruit or flowers were targeted for bird lime.

Bows and arrows were used to kill birds in Samoa, including rails and pigeons (Muse and Muse 1982:83, 94). So far as I can determine, bows and arrows were of little or no importance for hunting birds elsewhere in Polynesia. Bub (1991: Fig. 173) illustrates a cage trap from Samoa that features a tethered Purple-capped Fruit-Dove (*Ptilinopus porphyraceus*) in a woven basket designed to catch more fruit-doves.

A specialized form of hunting the Pacific Imperial-Pigeon (*Ducula pacifica*) occurred in Tonga, Niue, and Samoa on large earthen "pigeon mounds" (called *sia heu lupe* in Tonga; *sia*—mound, *heu*—snare, *lupe*—pigeon; McKern 1929:19–21). Catching the pigeons involved nets of various sizes (or snares) and tethered pigeons (Loeb 1926:119; Bub 1991:207–8). It seems to have been more a sport for chiefs and nobles than a widespread method of hunting by commoners. The great importance of pigeons is indicated further by specialized words, such as the Samoan *mafu* (breast fat of pigeons), *matua'ie* (old and fat pigeon), and *tula* (decoy pigeon's perch; Milner 1966:119, 139, 285).

Finally, it should be noted that mammals introduced prehistorically to Polynesia preyed on native birds. Both dogs (*Canis familiaris*) and pigs must have killed ground-dwelling birds to some extent, although their impact may have been minor. Rats were far more important predators on the eggs, chicks, and adults of native seabirds and landbirds. Almost throughout Oceania, the Pacific or Polynesian Rat (*Rattus exulans*) was introduced prehistorically, either intentionally or as a commensal stowaway in voyaging canoes. In the past two centuries, the Black, Ship, or Roof Rat (*R. rattus*) and the Norway Rat (*R. norvegicus*), natives of Asia, have been spread unintentionally through much of Oceania. These aggressive species, particularly *R. rattus*, have displaced *R. exulans* on many islands. All three rats prey on island birds, although *R. exulans* does so less than the other two (Atkinson 1985). It is difficult to evaluate the prehistoric impact of *R. exulans* on native birds. Based in part on how abundantly their bones occur in carefully excavated archaeological sites, the prehistoric populations of *R. exulans* may have been large, and they are likely to have been important predators on a variety of birds.

Although my main concern here is prehistory, I should mention that the period of European influence has brought some other nonnative predators to parts of Oceania, such as the cat (*Felis catus*) and mongoose (*Herpestes* sp.). Among the various European-introduced herbivores to tropical Oceania, the goat (*Capra hircus*) has probably done the most damage to natural habitats through over-browsing native plants and dispersing the seeds of nonnative plants.

Understanding the Survivors

The human occupation of tropical islands has been detrimental to a wide variety of birds, leading to the extinction of many species and the elimination or reduction of innumerable populations. So far, it would seem that the impact of humans on Polynesian birds has been almost unfailingly negative. Most of the surviving species are those that prefer open or wetland habitats (herons, ducks, volant rails, and migrant shorebirds) or that tolerate some level of forest clearance (certain fruit-doves, kingfishers, and warblers). Even many of these species are declining, and some seem destined to extinction within decades. There are, however, opportunities to reverse some of the losses of Polynesian landbirds, including the forest-loving species.

One strategy would be to translocate species onto islands that they previously occupied, after it has been determined that adequate habitat now is suitable and that current human activities are compatible (Franklin and Steadman 1991). Uninhabited islands are especially well suited for such programs. As Polynesians move to major population centers, many relatively remote islands are being abandoned by humans or are decreasing in population. The chances of success for bird translocations would be enhanced by good background knowledge of the species involved. Unfortunately, we do not know much about the ecology and behavior of the landbirds that survive in Polynesia. With but a few exceptions, data are lacking on the precise distributions, current population sizes, habitat preferences, and vulnerabilities of Polynesian birds.

Surveys like those conducted in American Samoa by Amerson et al. (1982a, b) should be undertaken at regular intervals throughout Polynesia. Detailed studies of nesting and food habits are lacking for most Polynesian landbirds. Where such studies have been carried out, as for the endangered Rarotongan Monarch [Flycatcher] or *kakerori* (*Pomarea dimidiata*), an effective conservation program was implemented (McCormack and Kunzle 1990a). Many other similar programs could be undertaken if more scientists, and agencies that fund research, became seriously interested in Polynesian islands other than Hawaii and New Zealand.

Another topic for research would be to determine the potential impact on native birds of blood-borne parasites (hematozoa), especially those that cause avian malaria. These diseases are transmitted by dipterans (mosquitoes in the case of avian malaria). Steadman et al. (1990) found no hematozoa among blood smears from 79 native and introduced birds from the Cook Islands. While this absence would seem to be good news from a conservation standpoint, it may also indicate that native species of birds would have little resistance to blood-borne parasites should they be introduced to the islands. Atolls of the northern Cook Islands lacked mosquitoes in pre-European times (Gill 1885:199–200). Whether this was also true in the southern Cook Islands and elsewhere in Eastern Polynesia remains unknown.

Yet another subject for additional research and monitoring concerns the forests themselves. Most of the endangered species of Polynesian landbirds require forested or partly forested habitats, which continue to face threats from logging and the encroachment of nonnative species. Detailed descriptions and analyses of Polynesian forest vegetation, such as those of Stoddart (1975a), Merlin (1985, 1991), and Franklin and Merlin (1992), do not exist for most islands. The ecology of dispersal and pollination is virtually unknown for many species of Polynesian forest trees. I cannot imagine, however, that there have not already been serious long-term impacts on the dispersal and reproduction of forest trees caused by the loss of hundreds of populations of frugivorous, granivorous, and nectarivorous forest birds, such as megapodes, pigeons, doves, parrots, thrushes, starlings, and honeyeaters. Thus the loss of birds has undoubtedly had a negative effect on forests, which in turn has had a negative impact on the human populations that have depended on the forests for food, medicine, and building materials.

Flying foxes or fruit bats (Chiroptera: Pteropodidae) are also important pollinators and seed dispersers in the South Pacific (Cox et al. 1991; Fujita and Tuttle 1991). Like birds, flying foxes were hunted and eaten prehistorically, reducing their numbers and range (Steadman and Kirch 1990; Steadman 1991b; Koopman and Steadman 1995). These large bats continue to be popular food throughout much of the Pacific, resulting in dramatic further declines in recent decades (Wiles and Payne 1986; Wiles et al. 1989).

THE IMPACT OF BIRDS ON HUMANS

Prehistoric Polynesians had a major and negative impact on native birds. In this the Polynesians were not unique; I am aware of no group of humans, past or present, rich or poor, Westernized or traditional, that has not reduced populations of birds and other animals, mainly through habitat alteration and hunting.

We already know that, on island after island, bones from archaeological sites indicate that indigenous species of both seabirds and landbirds were extirpated within the first millennium of human occupation. And, as the various native birds became rare or extinct, the nonfish vertebrate diet of Polynesians depended more and more on domesticated and commensal species (chickens, dogs, pigs, and rats). Here, however, I want to explore the reciprocal relationship: What sort of impact, aside from the obvious one of providing food, did native birds have on the lives of prehistoric Polynesians? As we shall see, Polynesians were interested in birds for many reasons other than as food. Because the Polynesian ethnobiology of birds is such an extensive subject, my coverage cannot be comprehensive. Rather, for each category, I shall provide a few examples and some discussion.

Nothing is as effective as extinction to decrease the interest in, and uses for, a species of bird. For each of the categories to be discussed, the process of avifaunal depletion that followed the human colonization of an island continually narrowed the range of species that people could use, whether as navigational or fishing aids, in legends and imagery, for their feathers or bones, or as pets. Thus the importance of birds in Polynesian societies, which was substantial at European contact, must have been even much greater before many of the species were lost.

Human Colonization

The dispersal and colonization of humans in what now is known as Polynesia began about 3,500 years ago, when a pottery-making people characterized as "the Lapita cultural complex" arrived in Western Polynesia after a rapid migration through Melanesia (see various chapters in Kirch and Hunt, eds., 1988). Although the record of birds exploited by the Lapita people is incomplete, evidence from Mussau, New Caledonia, Tikopia, Anuta, Fiji, Tonga, Futuna, and Samoa indicate that a wide range of seabirds and landbirds were taken and that many species did not survive the first millennium of human occupation (Balouet and Olson 1987, 1989; Steadman 1989a, 1989b, 1993a, 1993b and 1995; Steadman, Pahlavan, and Kirch 1990).

In spite of uncertainties about the chronology and spatial pattern of human colonization in Eastern Polynesia (Kirch 1986), there is a consistent pattern of heavy exploitation of native birds early in the cultural sequence (which begins up to 2000 B.P.), followed by an increased dependence on domesticated and commensal species (Dye and Steadman 1990).

Because birds provided such an important and easily available source of protein to early human colonists throughout Polynesia, the seemingly intense search for undiscovered, pristine islands may have been stimulated, at least in part, by the abundance of tame birds that greeted the discoverers of each new island. Thus the pursuit of unexploited avifaunas may have been a factor in the remarkable rapidity with which early voyagers colonized much of the Pacific. The tameness of Polynesian birds at first human contact must have been remarkable, the various species having evolved in ecosystems free of mammalian predators. Obtaining birds in the first centuries of human occupation may have resembled our concept of gathering more than hunting.

The extinction and extirpation of birds in Polynesia are reminiscent of some aspects of the blitzkrieg model of vertebrate extinction (Mosimann and Martin 1975; Martin 1984, 1990): a rapidly dispersing people who hunted intensively wherever they went, wiped out many species, and then moved on to richer hunting grounds. Not everyone, of course, moved on. Polynesians had a rich and productive set of domesticated plants and animals; on many islands, some portion of the founding populations seems to have remained to establish a more agriculturally based economy.

Navigation and Fishing

Seabirds were involved in traditional methods of navigation in Oceania. An important aspect of detecting nearness to an island was to keep an eye out for certain seabirds that seldom stray more than 20 to 100 km from the islands where they roost at night (Heyen 1963:71; Hilder 1963:90; Dening 1963:114–16; Lewis 1964:364; Sharp 1964:40; Gladwin 1970:180, 181, 188, 195–200; Lewis 1972:162–73; Finney 1979:334). Useful in this regard were boobies (*Sula* spp.), the Blue-gray Noddy (*Procelsterna cerulea*), and especially Brown and Black Noddies (*Anous stolidus, A. minutus*) and the Common Fairy Tern (*Gygis candida*, = *G. alba*). Tropicbirds (*Phaethon* spp.), frigatebirds (*Fregata* spp.), and Sooty Terns (*Sterna fuscata*) also were helpful, although their wandering habits made them fallible as indicators of nearness to land. In addition to true seabirds, the melodic two-note whistle of a migrant shorebird, the Lesser Golden-Plover (*Pluvialis dominica*), could be heard both day and night, and indicated nearness of land.

Observing birds also assisted fishermen in finding fish. In the Caroline Islands, Gladwin (1970:30) described how nearshore surface feeding by noddies and fairy terns would trap "the little fish in their frenzy between enemies above and below. It is the birds which signal to the fishermen that the [larger] fish are running in a school. All the canoes turn and plunge toward the birds." On Mangaia, by contrast, fishermen today say that the *kakaia* (Common Fairy Tern) is a trustworthy bird whose feeding activities often reveal productive fishing grounds, whereas the *ngoio* (Brown Noddy) is a "cunning" bird that often will lead fishermen to sterile waters.

Inhabited islands tend to have far fewer seabirds than do uninhabited islands. Dening (1963:114) noted that "birds in great numbers became accepted in the Pacific by the explorers as the sign of an uninhabited island. In this we might find an explanation of why almost every uninhabited island in the Pacific gives signs of having been visited by the Polynesians. Lost voyagers would be easily attracted by the sign of birds." Fortunately for returning seafarers, noddies and fairy terns have been more resistant to over-exploitation than have other seabirds and thus have retained their usefulness in locating home islands in areas where sailing and deep-water fishing still occur.

The loss, however, of most other species of seabirds on most islands must have diminished the importance of seabirds as both navigational and fishing aids for prehistoric sailors. With few exceptions, the populations of shearwaters and petrels have been eliminated or depleted throughout Oceania. Especially out of the nesting season, most species of shearwaters and petrels are highly pelagic. The former role of these and other pelagic seabirds in blue-water navigation is uncertain (Hilder 1963:83, 84; Sharp 1964:42, 43, 47), although their feeding activities far offshore probably aided fishermen in locating pelagic species of fish and marine mammals. While I agree with Lewis (1972:172, 173) that the role of migratory landbirds in facilitating prehistoric long-distance voyaging is speculative and subject to criticism, those (like Sharp 1964:59, 61) who doubt the potential utility of pelagic seabirds in long ocean voyages should bear in mind that a respectable understanding of this matter is impossible today because: (1) seabirds are so drastically reduced in range and numbers (the total number of resident seabirds in the tropical Pacific today may be 100 to 1,000 times less than it was 3,000 years ago); and (2) the persons who study Polynesian navigation nowadays almost certainly do not understand the habits and field identification of Pacific seabirds as well as prehistoric sailors did.

Related to this is the possible role of seabirds as food to help sustain early Polynesian voyagers. Before their population declines occurred, seabirds may have been a significant food supplement, although methods of luring them at sea near enough to be captured are unrecorded. Ethnographic accounts of long-distance voyaging virtually lack mention of taking seabirds at sea; the only reports I have found are of a group of native missionaries from Aitutaki who killed a "few sea-birds" during five months at sea in the 1820s (Dening 1963:138) and of several Great Frigatebirds (*Fregata minor*) caught after landing in the rigging of a ship off Nassau in the Cook Islands in the mid-1800s (Gill 1885:31).

Feathers

Items made of feathers occurred throughout Polynesia. Although cloaks, headdresses, skirts, and fans are especially well known, feathers adorned many other items, such as wooden and tapa images and the tails of kites (Buck 1944:80–102, 258, 311, 318, 320, 327, 332–36, 345–47, etc.). The color red was associated with major deities, such as Tangaroa in the Cook Islands (Buck 1934:16) and Oro in the Marquesas, Tahiti, and Tonga (Parsonson 1963:29). Thus red feathers, when available, were particularly important and valuable. Red feathers were sometimes used to decorate canoes (Gill 1894:134). In the Cook Islands, the word *kura* was used variously to mean "red," "feather headdress," or "parrot." I believe that the parrot signified by kura is *Vini kuhlii* (the Rimatara Lorikeet), a predominantly red species that survives on Rimatara in the Austral Islands but has been found in archaeological contexts on Aitutaki, Atiu, and Mangaia and seems to have occurred into early historic times on Palmerston, Manuae, and Rarotonga. Words ranging from priceless and precious to sacred and beautiful were used to describe red parakeet feathers in Mangaian stories. After the Rimatara Lorikeet had been lost on Mangaia, its treasured feathers were imported in late prehistory (Gill 1894:235, 240–42, 255).

Although they are known only from bones, it is likely that the larger extinct parrots of Eastern Polynesia (*Vini vidivici* of the Marquesas, Society, and Cook Islands, and *V. sinotoi* of the Marquesas and Society Islands) also had significant amounts of red in their plumage. There is ethnographic evidence that red feathers were brought to Mangaia from as far away as Tahiti, where no red parrot exists today. (The Black-fronted Parakeet, *Cyanoramphus zealandicus*, and Raiatea Parakeet, *C. ulietanus*, which survived into the early or mid-nineteenth century in the Society Islands, had only minor amounts of red in their predominantly brown-and-green plumages.) In describing a late prehistoric conflict, Buck (1934:131) stated: "The father [Rongo-ariki], after handing his son [Rori] some precious red feathers from Tahiti, stayed behind to delay the enemy and so make good his son's escape." Ironically, Rori's own three sons were later killed in battle while wearing red feather headdresses (Buck 1934:206).

With the loss of red parrots virtually throughout Eastern Polynesia, the two long central tail feathers of the Red-tailed Tropicbird (*Phaethon rubricauda*) became the primary source of red feathers. Pursuit of these tail feathers took Polynesians to uninhabited islands, sometimes hundreds of kilometers from home. In the Fiji-Tonga-Samoa-Niue region, four species of parrots with red in their plumage survived into historic times in spite of human-caused range contractions. These species are the Red-throated Lorikeet (*Charmosyna amabilis*), Collared Lory (*Vini [Phigys] solitarius*), Blue-crowned Lorikeet (*V. australis*), and Red Shining-Parrot (*Prosopeia tabuensis*). A large extinct parrot, *Eclectus* new sp., was exterminated prehistorically on 'Eua (Steadman 1993a). Based upon the bright red plumage in the females of its only surviving close relative, the Eclectus Parrot (*E. roratus*) of Western Melanesia, it is probable that the extinct species of *Eclectus* from 'Eua was yet another source of red feathers for early Tongans.

Although red may have been preferred throughout Polynesia, feathers of virtually any other color were also used. Prehistoric Niueans wore girdles that included white, yellow, and green as well as red feathers of *V. australis* (Loeb 1926:93, 163). Gill (1894:26, 27) reported a cloak from Mangaia made of "the beautiful white, green, blue, and yellow feathers of the birds they had eaten." Feathers of frigatebirds and pigeons (*Ducula* spp.) were used on many islands. Chicken feathers were used for a variety of headdresses and other ornaments on Easter Island (Metraux 1940:220–28). The choice of chickens was an obvious one, as all native landbirds and most seabirds had been wiped out prehistorically on Easter Island itself. Certain seabirds could be obtained seasonally on two offshore islets, and F. E. Eyraud described "a sea bird whose carcass had been opened more or less carefully" and used as a headdress by an Easter Islander in 1866–67 (Metraux 1940:220).

Bones

Being light and hollow but very strong, bird bones were important raw materials for tools, toys, and ornaments. Because bones are readily preserved in archaeological contexts, our knowledge of them, unlike our knowledge of feathers, covers the entire period of prehistory.

Sewing needles were made of the ulna or, more often, the radius, by cutting one end perpendicular to the long axis, drilling a small hole in the shaft near the cut end, and filing the other end to a hollow point. This type of needle is commonly found in prehistoric contexts on Easter Island (Metraux 1940:213; Heyerdahl 1961:412), where it is called *ivi tia nua*. At Ahu Naunau, Anakena, Easter Island, two such needles were recovered in July 1991 from sediments dated to about 700–800 B.P. (Steadman, Vargas, and Cristino 1994). These needles had been made from the radii of Murphy's Petrel (*Pterodroma ultima*), a tropical seabird that no longer occurs on Easter Island.

Tattooing needles were made of bird bone on Easter Island (skeletal element and species unrecorded; Metraux 1940:237'38, 241), the Cook Islands (details unrecorded; Buck 1944:128), and Tikopia (humerus, ulna, and radius of frigatebirds; Steadman, Pahlavan, and Kirch 1990:147). Mammal bones also were used as tattooing needles in many places.

Whistles were made most often from the thin-walled bones of large seabirds, such as frigatebirds or boobies. From 'Anatu on 'Eua, I have excavated a still functional whistle (length 48 mm) made from the radius of a Lesser Frigatebird (*Fregata ariel*). From Hanatekua Shelter No. 2 site (MH-11) on Hiva Oa in the Marquesas, I have identified the cut proximal 44 cm of another radius of *F. ariel*. Presumably, the adjoining piece (not recovered) would have been the whistle. Certain skeletal elements of chickens and large native birds, especially the humerus, ulna, and tibiotarsus, were cut in short sections to serve as beads. From Tangatatau Rockshelter on Mangaia, P. V. Kirch and I recovered in 1991 the entire sternum of a chicken in which two holes had been drilled. This probable pendant came from strata dated to about 600 B.P.

Names, Imagery, and Legends

These three subcategories are too intertwined to separate clearly. Polynesian names for birds are significant from both a linguistic (Clark 1982) and an ornithological (Steadman 1985) standpoint. Many avian names reappear as cognates throughout much or all of Polynesia. One of many such examples is the proto-Polynesian *matuku* (variations include *matu'u, motuku, kotuku,* and others; Clark 1982), which refers to the widespread Pacific Reef-Heron (*Egretta sacra*). Occasionally, some rather drastic name changes reflect local onomatopoetic interpretations, such as *tanga'eo* for the Mangaia Kingfisher rather than *ngotare* or *kotare*, used for closely related Eastern Polynesian species. Other major name changes are difficult to explain, such as *iwa* for frigatebirds in Hawaii rather than the widespread name *kota'a* and its cognates.

In some areas of Polynesia, one can find names of birds in dictionaries that refer to species that no longer exist on a particular island. Such names generally have fallen into disuse. Dictionaries for Rarotonga (Savage 1980, actually compiled between about 1900 to 1940) and Tahiti (Andrews and Andrews 1944) are rich sources of such names. To mention only two examples, Savage (1980:117, 122) lists Rarotongan words that indicate the former presence there of a kingfisher (*kotare;* presumably the Chattering Kingfisher, *Halcyon tuta,* or closely related species) and a small red parrot (*kura;* presumably the Rimatara Lorikeet).

Bones from archaeological sites can corroborate the former existence of birds otherwise known only from stories or linguistic evidence. Clark (1982) deduced that a bird similar to the Purple Swamphen (*Porphyrio porphyrio*, proto-Polynesian name *kalae*) must have existed in the Marquesas or Society Islands, the presumed source areas of Hawaiian people, because the Hawaiian cognate *'alae* refers to similar large rails (Common Gallinule, or Moorhen, *Gallinula chloropus,* and American Coot, *Fulica americana*). No rails in the genera *Porphyrio, Gallinula,* and *Fulica,* however, had ever been found in Eastern Polynesia (Holyoak and Thibault 1984), the nearest occurrence being that of *P. porphyrio* in Tonga, Samoa, and Niue (Pratt et al. 1987). Raynal (1980–1981) noted that the Marquesan name *koau* refers to a flightless bird, with bluish purple plumage and yellow bill and feet, that existed earlier this century on Hiva Oa. Raynal proposed that the *koau* was related to the flightless swamphen of New Zealand, known as the *takahe* (*Porphyrio mantelli*). In 1986 and 1987, while examining bones from archaeological sites on Hiva Oa and Tahuata, I discovered 19 specimens that belonged to an undescribed species of swamphen, subsequently named *Porphyrio paepae* (Steadman 1988). While this adds support to the proposals of Clark (1982) and Raynal (1980–1981), it does not solve the linguistic discrepancy between *kalae* and *koau*. The bones from Hiva Oa are all more than 1,000 years old, while those from Tahuata are at least 700 to 800 years old. Raynal and Dethier (1990) have suggested that a "cryptozoological" search be made on Hiva Oa to see if *P. paepae* still exists.

A somewhat similar situation involves megapodes (*Megapodius* spp.). Clark (1982) pointed out that cognates of the Tongan and proto-Polynesian word for megapode, *malau,* are found in various places in Melanesia and Indonesia. Because megapodes are lacking in Fiji, through which proto-Polynesian speakers must have traveled on their way to Tonga, use of the word *malau* in Tonga suggests that megapodes must have existed at one time in Fiji. (In Tonga today, a single species of megapode, *M. pritchardii,* is confined to Niuafo'ou.) Archaeological bones have now shown that megapodes did exist two or three millennia ago not only in Fiji but also on Ofu (American Samoa) and on Lifuka and 'Eua in Tonga (Steadman 1989b, 1993a, 1993b). The most thorough record is in Tonga, where four species, three of them extinct (°), have been recorded (°*M. molistructor,* °*M. alimentum,* and *M. pritchardii* on Lifuka and Foa, °*M. alimentum, M. pritchardii,* and °*M.* new sp. on 'Eua). As megapodes became increasingly localized in the Fiji-Tonga-Samoa region, their eggs became one of the prestigious trade items of a well-organized long-distance exchange network (Steadman 1991c; see Dening 1963 and Kirch 1984:238–42, 1988:257–60 for details of this network).

Clark (1982) also used cognates of the proto-Polynesian *lulu* (Common Barn Owl, *Tyto alba*) to propose that some sort of owl once existed in Eastern Polynesia. This suggestion has not been substantiated by bones. Lastly, Clark (1982) interpreted cognates of the proto-Polynesian *siwili* and *kula* to suggest that a larger parrot once existed in Fiji or Tonga, a suggestion now supported by discovery on 'Eua of the large extinct parrot, *Eclectus* new sp. (Steadman 1993a).

The Polynesian names of birds were sometimes used to name other things. Te Ana o Kakaia (the Cave of the Fairy Tern) is a large cave in the Ivirua district of Mangaia that served as a place of refuge for Ruanae and his clan after they were defeated in a battle said to have occurred in 1718 (Gill 1885:74, 1894:167). True to the cave's name, each of the six bird bones I recovered from Te Ana Kakaia in 1984 were those of the *kakaia* (*Gygis candida*). The lack of bones from extinct or extirpated species suggests that the cave was inhabited only late in prehistory.

Two other well-known Polynesian caves have names involving birds. Ana Taketake (var. Ana Takitaki) on Atiu is named after the sound made by echolocating Atiu Swiftlets (*Collocalia sawtelli;* Holyoak 1974, 1980; Holyoak and Thibault 1978; Steadman 1991b). Ana Kena on Easter Island is named after the Masked Booby (*Sula dactylatra*). Ana Kena, originally the name of a cave only, has been merged into the single word *Anakena* to refer to the surrounding region, which consists of a valley mouth and protected bay, a site of much archaeological and paleoecological importance (Kirch, Christensen, and Steadman n.d.; Steadman, Vargas, and Cristino 1994).

Two vanquished leaders in Mangaian prehistory were named after birds (Buck 1934:35): Mokora (Gray Duck, *Anas superciliosa*) and Kota'a (frigatebird; Polynesians seldom distinguish between the two local species, *F. minor* and *F. ariel*). The bird's nest fern (*Asplenium nidus*) of Cook Island forests is called *rau kota'a* (Gill 1894:224) or, more often today, simply *kota'a* (Wilder 1931:9; pers. observation), in reference to its leaves, which resemble in shape the long, narrow wings of frigatebirds. The graceful kites once made in the Cook Islands were called *manu tukutuku* ("letting go a bird"), *manu— aka-rere* ("flying bird"), or simply *manu* ("bird"; Buck 1944:257; Savage 1980:139).

The words *rupe* and *lupe* refer to the Pacific Imperial-Pigeon in the Cook Islands and Western Polynesia, respectively. On Mangaia, where no species of pigeons survive, there is a valley named *rupetau* (Gill 1894:209). *Tau* means "to alight, to come to rest" (Savage 1980:363), thus *rupetau* seems to be named after a former pigeon roost.

All sorts of birds were involved in Polynesian stories and sayings, which often were based in part on the habits of a bird. On Niue, for example, many *pekapeka* (White-rumped Swiftlets, *Collocalia spodiopygia*) were said to leave their caves and "dance" in the sky before a rain (Loeb 1926:179). Thus the swiftlets' period of heavy aerial feeding on insects helped Niueans to predict rain. Although Polynesian stories often reveal an appreciation and understanding of birds, they indicate a "close harmony with the natural world" (as purported by Muse and Muse 1982:ix) only if such harmony includes predator-prey relationships. Many Polynesian stories include or imply the death of the bird, particularly if it is a highly edible species, such as a shearwater or pigeon.

Birds were often associated with Polynesian deities. Such an association may have been accompanied by a taboo on killing that particular species. The fact that bones of "deified" species are frequently recovered in archaeological sites may reflect local or chronological differences in their protective status. For example, a common migratory shorebird, the Lesser Golden-Plover, is regarded as sacred by modern Tikopia, who also associate with deities the resident Pacific Imperial-Pigeon and migratory Long-tailed Cuckoo (*Eudynamis taitensis*). Bones of all three species were found on Tikopia in archaeological contexts (Steadman, Pahlavan, and Kirch 1990).

The endemic Mangaian Reed-Warbler (*Acrocephalus kerearako*), onomatopoeically called *kerearako* on Mangaia, was regarded as the incarnation of the god Tane (Buck 1934:163, 171). The Bristle-thighed Curlew (*Numenius tahitiensis*) is a large migratory shorebird variously called *kiu* or *kau'a*, again because of its melodic voice. Prehistoric Mangaians regarded both the curlew and warbler as "mouthpieces" of the god Tane (Gill 1885:116–17, 1894:332; Buck 1934:172). The *kau'a* was "considered to be good eating by the natives, the tribe of Tane excepted" (Gill 1885:117). Another vocal bird, the Mangaia Kingfisher, or *tanga'eo,* was associated with the gods Utakea and Tekura'aki (Buck 1934:166, 171).

According to legend, a Pacific Imperial-Pigeon belonging to the god Tangaroa mated with "a female shadow of great beauty" and became the father of the first man on Atiu (Gill 1894:262–63). On Mangaia, the god Motoro was "proudly called *te io ora* = 'living-god,' as his worshippers were *not* eligible for sacrifice" (Gill 1894:332). Motoro was represented in sennit work and carvings from the *oronga* (a small tree, *Pipturus argenteus*), which used the *mo'o* or *mo'o mo'o* (Sooty Crake, *Porzana tabuensis*) as his incarnation (Buck 1934:166, 171). Association of the mo'o mo'o with life may have been because this bird inhabits taro swamps, which for centuries have been of utmost important to Mangaian subsistence.

Easter Island is well known for wooden carvings and petroglyphs of stylized birds and birdmen, the latter consisting of the head of a bird (often the Great Frigatebird) attached to a human body (Metraux 1940:256–59, 270–72; Lee 1986). Bird glyphs also appear in the famous *rongorongo* wooden tablets of Easter Island (Metraux 1940:389–411). The "bird cult" of Easter Island involves the small population of Sooty Terns that nests opposite Orongo on the offshore islet of Motu Nui (Metraux 1940:331–41). This cult may have increased in religious significance late in prehistory because of the rarity (and therefore value) of seabirds and their eggs. The seabirds and landbirds on Easter Island itself had already been exterminated.

Birds also appeared in Polynesian riddles, such as "Who with a black skin is ever clothed in purest white? A species of tern [Common Fairy Tern], common in the Pacific" (Gill 1885:268). Following the arrival of missionaries in the nineteenth century, birds continued to have roles in stories that were Christianized (Gill 1885:92, 108, 110, 136–37; 1894:372–73).

Pets

Polynesians enjoy keeping birds as pets, a habit that may be decreasing in popularity. Generally the birds are obtained as nestlings and are hand-reared, increasing their tameness. On Mangaia I have seen a Brown Noddy (obtained locally) and a Red-footed Booby (*Sula sula*, obtained on Penrhyn) kept as tame pets in sheds behind houses. The noddy is also sometimes kept on Tikopia. Frigatebirds were kept as pets in Samoa (Armstrong 1932:17) and Tuvalu (Gill 1885:17–18), as were White-tailed Tropicbirds (*Phaethon lepturus*) in Tonga (Lewis 1972:169). The Pacific Imperial-Pigeon and Crimson-crowned Fruit-Dove were commonly kept in Samoa (Armstrong 1932:1, 58). The Red Shining-Parrot is indigenous to Fiji (Layard 1876; Rinke 1989) but was kept as a pet in Samoa (Armstrong 1932:91) and still is held captive (and exported?) on 'Eua, Tonga (pers. observation). The Tahiti Lorikeet (*Vini peruviana*) was noted in captivity on Aitutaki (Townsend and Wetmore 1919).

CONCLUSION

It is not surprising that birds were extremely important to early Polynesians, who arrived on one pristine island after another, each inhabited by a variety of birds but few if any species of reptiles or mammals. The importance of birds, however, did not save them. The net effect of the human occupation of Polynesia has been the elimination of much of the original bird life. Also gone is most of the natural habitat that supports those species of birds that have survived.

We know from archaeological sites that birds were killed and consumed regularly by early Polynesians. Nevertheless, it is difficult from this evidence to assess the relative contribution to avian extinction from direct human predation versus other prehistoric anthropogenic factors, such as habitat loss, disease, or predation from rats, dogs, and pigs. The importance of each factor probably varied from island to island. Regardless of the specifics of any individual case, human presence on tropical islands has been called an environmental catastrophe (Olson 1989). Although this statement may be controversial from an anthropological standpoint, you would find few birds that would argue against it. I say this without judgment; I am not criticizing any individual person, living or deceased, who has killed birds (I have killed some myself, in the name of science) or who has destroyed forest (which we all do, at least indirectly, for food, lumber, paper, and so forth).

I have mentioned some of the conservation programs that have been or could be undertaken to improve the lot of Polynesian birds. Although conservation agencies can perhaps save more species per dollar on the mainland tropics than on islands, there is something undeniably special about islands and their unique biotas. Programs to conserve Polynesian birds should be considered for the sake of the birds themselves, not to mention preserving an important part of the human heritage of Polynesia. These programs are important for the future of science as well. It is no coincidence that Charles Darwin and Alfred Russel Wallace came up with many of their most brilliant insights while studying island faunas, as did other great names of zoology, such as Ernst Mayr, David Lack, Robert MacArthur, Edward Wilson, and Jared Diamond. In anthropology, such prominent figures as Bronislaw Malinowski, Margaret Mead, Sir Raymond Firth, and Marshall Sahlins were inspired to new heights of creativity through studies of island peoples.

Any attempts to preserve the remaining fragments of Polynesian bird life must consider the importance of involving local people (Hay 1986). Interest in nature seems to be waning as cash economies replace subsistence agriculture. If environmental protection is to succeed in Polynesia, this interest in nature must be kept alive in young Polynesians as they inherit their island environments. More than 50 years ago, Aldo Leopold wrote: "Conservationists have, I fear, adopted the pedagogical method of the prophets; we mutter darkly about impending doom if people don't mend their ways. The doom is impending, all right; no one can be an ecologist, even an amateur one, without seeing it. But do people mend their ways for fear of calamity? I doubt it. They are more likely to do it out of pure curiosity and interest" (1991:99, 101). In the next few decades it will be fascinating to participate in the interplay of science, culture, economy, and environment in the South Pacific.

Note

I thank T. L. Hunt and P. V. Kirch for the opportunity to make a contribution to their symposium at the seventeenth Pacific Science Congress, upon which this chapter is based. The research was supported in part by the National Geographic Society (grant

4001-89), the National Science Foundation (grant BSR-8607535), and the Smithsonian Institution. For research permits, logistics support, and other cooperation, I thank the governments of Easter Island (Chile), French Polynesia, the Cook Islands, and Tonga. T. L. Hunt, H. F. James, P. V. Kirch, P. S. Martin, N. G. Miller, and S. L. Olson kindly commented on the draft manuscript.

5

Landscape Catastrophe and Landscape Enhancement: Are Either or Both True in the Pacific?

Matthew Spriggs

That significant changes have occurred in the vegetation and landscape of the Pacific Islands during the course of human occupation is no longer doubted and has been amply demonstrated by publications during the past ten to fifteen years.[1] What caused these changes has occasioned debate, however. Three commonly considered possibilities (not necessarily mutually exclusive) are often canvassed (cf. Butzer 1974): (1) climate change affecting vegetation and other aspects of the ecosystem; (2) catastrophic events or sequences of events such as volcanic eruptions, earthquakes or hurricanes; and (3) human interference with the vegetation cover by clearance and/or fire.

There is at present no evidence that the islands of Remote Oceania (Pawley and Green 1973), those south and east of the main Solomons chain, were occupied before 3200 B.P. (see Spriggs in press, b, for discussion). For Eastern Polynesia, the time depth of human occupation is even shorter: less than 2,000 years in Hawaii, 1,000 years in New Zealand (Aotearoa). There is no evidence for major climatic shifts in the region during the past 3,000 years, and it is not clear how minor climatic shifts, as argued by some researchers (e.g., Nunn 1990, 1991:17–28), could have led to the vegetation and landscape changes that have been documented. Higher relative sea level between about 5000 B.P. and 2000 B.P. (Nunn 1991) may well have had an important role in shaping coastal environments, but where the sediment came from that subsequently filled former marine embayments still needs to be considered and is not adequately explained by Nunn's appeal to river-channel response or coincident uplift (Nunn 1991:12).

Catastrophic events or natural hazards have been important in shaping the island environments of the Pacific (Kirch 1984: Chapter 6), but on their own they cannot be the full explanation for the changes observed. This leaves human impact as an extremely important factor—though certainly not the only one—in vegetation and landscape changes in most nonvolcanically active areas of the region. Nunn's claim that "most writers have linked post-settlement environmental changes in the Pacific islands solely to human activities" (1991:16, cf. 8, 28) is absurd. Recent researchers have sought to give human impact its due place among the causal agents, following a long period when such impact was scarcely acknowledged. It should be remembered, too, that natural hazards can be particularly damaging "when ongoing social and economic conditions are such as to expose the production system and the land to abnormal harm from such events" (Blaikie and Brookfield 1987:142).

All human groups have an impact on their environments, whether hunter-gatherers (for Australia see Flood [1983:213–15] and Kimber [1983]) or, as in the case of the settlers of Remote Oceania, agriculturalists. Clearing of vegetation from hillsides to create gardens leads to higher erosion rates, particularly on tropical soils. Any gardening of the usual tropical crops will inevitably lead to higher than natural erosion rates.

Sustaining human life in Remote Oceania without an agricultural economy based on introduced plants may well have been impossible prehistorically (Spriggs in press, b). Few edible plants were naturally available, and the only edible nonmarine fauna of significance were fruit bats (which never reached Hawaii), birds, and, on some islands, lizards (Dye and Steadman 1990). The islands were certainly not "Paradise" when they were first settled.

Agriculture requires gardens, and gardening on forested islands requires forest clearance, usually by fire. Windward forests tend to regenerate quickly, while the more diverse, leeward dry forests regenerate much more slowly after clearing and are more susceptible to further human disturbance and alteration.

CATASTROPHE OR ENHANCEMENT: DOES IT DEPEND ON WHERE YOU STAND?

The scale of forest clearance for agriculture on some islands was massive: in the main Hawaiian Islands, for instance, many thousands of square kilometers of forest had been cleared before European contact. The leeward dry forest had virtually disappeared by A.D. 1778, replaced by grassland or agricultural field systems. On Hawaii Island three of the leeward field systems have been partially investigated and are known to have covered at least 216 km². The full extent of these, and of other large agricultural field systems on the island that have yet to be examined in any detail, have never been defined, and so the total area cleared of forest that was there before contact may be at least an order of magnitude higher.

The forest was replaced by a managed, highly productive environment capable of supporting dense human populations (Kirch 1985). Integral to this environment were the "barren" grasslands described by early visitors to Hawaii. These grasslands produced crop mulch, which was necessary in the drier garden areas, and *pili* grass, which was the usual thatching material for the roofs and walls of all Hawaiian structures.

One effect of this massive vegetation alteration was the removal of the habitat of many species of birds (Olson and James 1984) and of land snails, such that more than half of the species of each group became extinct. Today we see this as an ecological catastrophe but there is *no* evidence from oral tradition that this process was even remembered by later Hawaiians.[2] As Steadman et al. have noted of Polynesia in general, "Even if these hunters became aware of the scarcity of certain species after decades or centuries of exploitation, they could do little to prevent predation by rats and dogs, or possible avian pathogens introduced with chickens. Moreover, these people were unlikely to alter their agricultural practices for the sake of preserving forest habitats for birds" (1990:148).

Another concomitant of forest clearance that might be seen as disastrous is greatly accelerated erosion. Today soil scientists often denounce practices such as slash-and-burn agriculture that lead to increased runoff and erosion from hill slopes. These same soil scientists, however, also describe the alluvial soils in the valleys that are the products of this erosion as some of the prime agricultural lands of the region. These alluvial plains were previously viewed as the products of slow natural erosion processes, whereas they can now be seen to be extremely recent creations and the direct result of humanly accelerated hill-slope erosion (Spriggs 1985).

It might be argued, however, that all the benefit is to the valleys, leaving the hills stripped of soil and barren. In some cases this is clearly the case, as "badlands" and "fern desert" areas of various Pacific islands testify (see Chapter 8). But one might argue that they are no great loss to agricultural systems, often representing areas of naturally thin and not very productive soils on steep slopes that would require extensive terracing to make them fit for sustained cultivation.

Clearly critical, too, are erosion and soil-formation rates on various slopes and bedrock types. Accelerated erosion is not always of a catastrophic kind, and not every gardened hillside is inevitably destined to lose all of its soil cover. Soil development may keep pace with soil loss. Drier lower-slope areas in leeward Hawaii have benefited from the products of erosion from higher, wetter areas in the past 500 years. With the development of gardening on the upper slopes, accelerated erosion has transported sediments to lower areas of previously barren lava flows, allowing gardening in places where it was not previously possible (Schilt 1984:270, 274–76). There is no evidence that the upslope areas were losing soil at such a rate that gardening in them was impaired by this process.

Sudden influxes of large amounts of sediments into the valley bottoms during extreme rainfall periods would have led to flooding and destruction and burying of agricultural systems, creating at least short-term catastrophe. One such flood on Aneityum in Southern Vanuatu, "the greatest known on the island," was described by the Reverend Gunn in the early years of this century:

> Though the wind sometimes reached hurricane force, there was no cyclone; but the rain poured incessantly for 28 hours. Next morning the harbor—ordinarily one of the finest sights in the islands—was one great mud pond, with earth carried down by the river. As news came in from other districts about the rain, I learned that all the valleys were under water when the flood was at its height, and natives seeking places of safety had to wade breast-high, or swim. Many taro plantations were buried in mud. . . . But in one narrow, deep valley, formerly a death-trap in times of flood, the height and force of the water surpassed that in any other part of the island. Owing to a land-slip, the water rose 60 feet high, then, breaking the barrier swept seaward in resistless torrents, carrying huge boulders, taro patches, houses, the school, lately built, coconut trees, etc. Fortunately few people were in the valley at that time, or there would have been a large death roll. As it was, a man and a boy were drowned. [Gunn 1911:7]

The interplay between human interference with the vegetation and natural catastrophic events can be seen at work in such situations (see Spriggs 1981:96–102). But would the occasional catastrophic flooding of valley floors, clearly reflected in their alluvial sequences, have rendered them unfit for agriculture for significant periods? In considering this, we come to some methodological problems to do with the identification of garden soils and the interpretation of radiocarbon dates.

THE IDENTIFICATION OF GARDEN SOILS

On Aneityum, soils of former furrow-irrigation gardens on Pleistocene alluvial terraces were examined by the geomorphologist Marc Latham and myself in 1979 (see Spriggs 1981). The only features of these soils that distinguished them from natural topsoils were homogeneity of the humic horizon, a marked difference in structure and firmness between the upper (cultivated) and lower horizons, with a sharp break between them, and the presence of manganese staining at depth. Of these features, only the manganese staining could possibly be said to distinguish these soils from dryland agricultural soils. On the recent alluvial plains, at least where they are adjacent to the rivers and are particularly well drained, even manganese staining is absent, so that irrigation—and indeed cultivation of any kind—can only be established on the basis of associated remains of stone-lined plot boundaries or channels.

In other areas of the Pacific, soils associated with different techniques of irrigation may be more distinctive in structure. Thus, as described by Kirch (1977:253-55), pondfield soils are generally distinct from naturally formed soil types. The effect of waterlogging creates an eluviation, reduction state in the upper A horizon and an illuviation, oxidation state in the lower B horizon. Kirch noted that this effect on soils under pondfield irrigation has been reported from Hawaii, (East) Futuna, Japan, Thailand, and Malaya but not from Ifugao pondfields in the Philippines. Such conditions do not occur in all soils under pondfield conditions, and they are at least in part a function of the length of time a pondfield has been in use.

Kirch (1977:254) notes another physical feature of irrigated (specifically pondfield) soils to be the presence of limonite concretions, "hydrated iron-oxide tubes which apparently formed around *Colocasia* roots under aqueous conditions. . . . It is probable that the limonite concretions resulted from the oxidation of the surrounding sediment by plant roots." There are, however, a variety of ways in which iron concretions can form in the soil, and the reported association with the pondfield soil could be fortuitous (Phillip Hughes, pers. comm.). Although (as noted by Kirch) pyrolusite concretions around sugarcane roots in Hawaiian soils have been described, similar tubes forming around taro roots have not been reported in field studies of the crop.

Kirch (1977) suggested that additional criteria, such as textural difference, pH, or total organic content, should be investigated to try to identify pondfield soils in archaeological contexts. As I mentioned, textural difference is one criterion used on Aneityum to distinguish cultivated soils, but with the more sandy soils on recent alluvium the difference is not distinctive. Total organic content is generally higher in the upper rather than the lower horizons of natural soils and so again is unlikely to be distinctive of irrigated or dryland gardening. On Aneityum, pH content did not distinguish garden soils from other sediments revealed in section.

Thus when a sediment is found in an excavation or exposed river section, it can only be firmly identified as a garden soil when it is associated with structural remains (stone-lined channels, plot boundaries, terraces, and so on) that can be identified as features related to agricultural exploitation. Where such evidence is lacking, identification as a garden soil can only be made very tentatively. For further discussion of these issues see Allen et al. (1987:36-37).

INTERPRETING RADIOCARBON DATES

Radiocarbon dates from within irrigated or other garden soils may be misleading for a number of reasons. In discussing charcoal found within pondfield soils, Kirch (1975a:306) notes three possible sources: (1) initial clearing and burning of the site before pondfield construction; (2) burning of fallow growth during the period the field was in use; and (3) an upstream source with charcoal carried in by the irrigation water. Thus dating material from a pondfield soil does not necessarily date first use of the pondfield. A similar range of sources for the charcoal found in the furrow irrigated soils of Aneityum Island can also be postulated. Similarly, with tilled soils (dryland or irrigated), charcoal present within them may only relate to the latest phases of use when the soil was last turned over. This must always be borne in mind when interpreting dates on charcoal within garden soils.

In cases where charcoal has been fluvially transported into the garden area, there is a possibility that it will give a date far older than that of the time of its incorporation into the soil, because of charcoal storage in sites in the catchment. This has been found to be the case under certain circumstances in Australia (Blong and Gillespie 1978). The opportunities for such storage on most Pacific islands, however, would appear to be much less than in the very much larger sandstone catchments studied by Blong and Gillespie. A sequence of radiocarbon dates from a stratigraphic section should always be preferred to single dates when evaluating the possibility of charcoal storage, and all dates on dispersed charcoal found within sediments should be treated with caution.

It is debatable whether dispersed charcoal found within garden soils can be attributed to any particular source among the possibilities I have mentioned. The most secure dates are those obtained for in situ hearths or ovens, but even here the possibility of old wood being used for fuel means such dates can only be used as maxima (see Gillespie and Swadling 1979). Initial agricultural use may well have involved the clearing and burning of old-growth forest, with trees perhaps several hundred years old entering the archaeological record as charcoal. Early dates for initial clearance may in part reflect this old-wood effect.

A further obvious problem is that working even with calibrated dates a single age range will often span more than 100 calendar years at one standard deviation (s.d.) and 300 years at two s.d.[3] The best we can do in assessing whether there is a "significant" time difference between two layers representing garden soils is to see whether they overlap at two s.d. (a 95% confidence level). If such an overlap does not occur, it might be suggested (but note my caveats on charcoal sources) that the period of abandonment between them was of considerable duration.

ALLUVIAL AND COLLUVIAL SEQUENCES

There are few studies of alluvial deposition in areas of prehistoric Pacific agriculture that give more than single dates from within river sections or excavations. The examples I shall use come from (East) Futuna in Western Polynesia and appear in the work of Kirch (1975a, 1976, 1981, 1994) and of Frimigacci and his colleagues (Di Piazza 1990; Di Piazza and Frimigacci 1991; Frimigacci 1990), from Luluku in windward O'ahu (Allen et al. 1987), from Makaha in leeward O'ahu (Yen et al. 1972), and from my own research on Aneityum in southern Vanuatu (Spriggs 1981, 1985, 1986).

An analysis of the dates from sites on these islands reveals two important findings. First, in several cases there is a significant gap of several hundred years between initial human presence at, and use of, a site and evidence for a second (usually gardening) use. Second, at sites used within the past 1,000 years no significant gaps in use between periods of alluvial or colluvial deposition can be established. This applies whether we are talking of initial human use of a site within the past 1,000 years or second and subsequent uses after a significant gap. A single depositional sequence is often illustrative of both findings.

Futuna, Western Polynesia

Kirch's excavation at the Tavai site (FU-11) on Futuna revealed 10 layers (Fig. 5.1), including an occupation layer (Layer IX) representing a village site on a

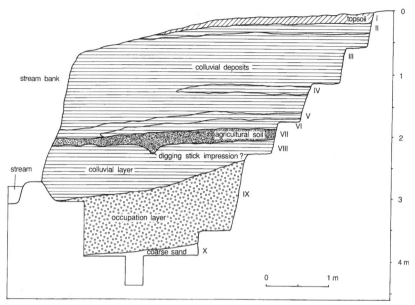

Figure 5.1 Stratigraphic section at Tavai (FU-11, now SI-09), Futuna, Western Polynesia (after Kirch 1981)

gravelly clay coastal plain a few meters from the shoreline (Kirch 1981). This occupation is buried by a clay sediment derived by erosion from upslope (Layer VIII), and Layer VII above this is interpreted as an agricultural soil on the basis of abundance of charcoal, a concentration of charcoal in the northwest corner of the excavation suggestive of a swidden burn pile, and a probable digging-stick impression. The soil is "a compacted horizon of clay-sand which stands out clearly within the section" (Kirch 1981:129). The upper six layers are interpreted as having been transported to the site through sheet wash, debris flow, and slumping. Although some charcoal flecking was noted near the base of Layer VIII, these upper layers appear to have been devoid of charcoal. Whether any of them might also represent gardened soils cannot be established. The important point, however, is the significant gap in time between initial use of the site for settlement at 2303 to 1998 B.P. and its subsequent garden use dated to 1382 to 1060 B.P.

In later research on Futuna, Frimigacci and his colleagues (Di Piazza 1990; Di Piazza and Frimigacci 1991; Frimigacci 1990) investigated the site of Asipani (Fig. 5.2). Here an early Lapita and Plainware occupation is buried by alluvial deposits, on top of which is a taro pondfield soil (Layer 6). Above this are two more topsoils (Layers 4 and 2), separated by further alluvial deposits.

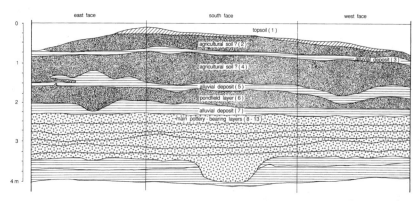

Figure 5.2 Stratigraphic section at Asipani (SI-001), Futuna, Western Polynesia (after Di Piazza and Frimigacci 1991)

Areal excavation of the Layer 6 garden soil revealed individual planting holes and a drain (Di Piazza 1990). The two upper topsoils are of the same texture as the pondfield horizon and have also been interpreted as agricultural, although not necessarily irrigated. There is an area of pondfields currently in cultivation adjacent to the Asipani site. The latest dates for the ceramic occupation of the site are 2054 to 1873 B.P. and 2704 to 1860 B.P. The Layer 6 pondfield has been dated to 1122 to 958 B.P. As with the Tavai site, there is a possible 1,000-year gap in human use between the initial and the second utilization of the site. The upper two garden soils are undated, but use and rebuilding of three agricultural systems within the past 1,000 years or so certainly betokens a quickening level of activity at the site compared to the previous millennium.

Another site investigated by Frimigacci's team is at Moasa, an upland site situated on the edge of the *toafa*, or fern desert, in the interior of the Vailala Valley (Frimigacci 1990:167–68; Di Piazza and Frimigacci 1991). Excavation revealed four humic clay horizons, interpreted as swidden agricultural soils separated by alluvial and/or colluvial deposits (Fig. 5.3). The lowest of them (L. 13) has been dated to 1235 to 970 B.P., the middle two (L. 11, 10) combined to 434 to 0 B.P., and the upper garden layer (L. 7) to 490 to 314 B.P. The gap between first postulated garden use and secondary use is significant, on the order of 500 years, whereas the upper dated garden soils are not separable by radiocarbon dating. As with the Asipani site, we find higher rates of deposition and greater evidence of human use over time. A note of caution must be raised here, however. These layers have been identified as agricultural on general textural properties and the presence of charcoal. Their identification

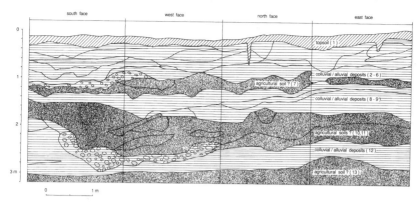

Figure 5.3 Stratigraphic section at Moasa (SI-013), Futuna, Western Polynesia (after Di Piazza and Frimigacci 1991)

as gardened soils, though plausible, cannot be said to have been firmly established by these criteria.

An excavation by Kirch in an abandoned pondfield system between the Nuku and Leava areas revealed a pondfield horizon buried beneath the surface system (Kirch 1976:47–49). The buried horizon has been dated to 304 to 0 B.P., revealing rebuilding of the system in the recent past at a time indistinguishable by radiocarbon from the present. Other buried pondfield horizons on the island remain undated (Kirch 1976:47–49; Di Piazza 1990:160). As Di Piazza notes, land suitable for irrigation is scarce on Futuna, which is why after floods and cyclones these systems are reestablished in the same place. Overbank flooding, such as occurred on Futuna in 1986 during cyclone Radja, may necessitate the temporary abandonment of some pondfields but may also create conditions suitable for the creation of new ones (Di Piazza 1990:161).

O'ahu Island, Hawaii

Extensive agricultural excavations have taken place in windward O'ahu at Luluku (Allen et al. 1987). Trench 3 in site G5-85 (Features 34, 35, and 38) is particularly instructive (Allen et al. 1987:76–87). Layer VIII has been interpreted as a buried pondfield topsoil remnant that had developed on the Layer IX alluvial deposits (Fig. 5.4). It was truncated by a colluvial deposit (L. VIIb) and the fill behind a buried pondfield-terrace wall. There are three radiocarbon dates for this layer, although one of them is clearly anomalous (see discussion by Allen et al. 1987:174, 177). If this is excluded from consideration then Layer VIII dates to 1550 to 1340 B.P. and 1350 to 1070 B.P., a combined range of 1505 to 1305 B.P. Such an age would represent one of the earliest dates for hu-

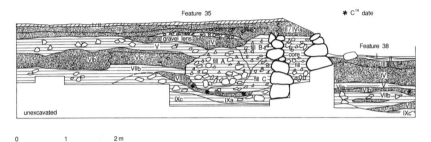

Figure 5.4 Stratigraphic section of parts of Features 35 and 38, Trench 3, north face, Luluku (G5-85), O'ahu, Hawaiian Islands (after Allen 1987)

man occupation in the Hawaiian Islands. Incorporation of old wood seems likely as part of initial forest clearance, and the dates should be used only as maxima.

Layer VI, above the colluvium that covers and partially truncates Layer VIII, represents one or more pondfield horizons associated with some of the fill phases and terrace-facing construction. Combining the two dates from this layer in different pondfields gives a generalized date of 675 to 551 B.P. It is again worth noticing a significant gap between initial use of the area and second use, even allowing for the "old-wood factor" in the earlier dates. Layer V represents another episode of erosion and colluvial deposition, and Layers III and IV represent pondfield A and B horizons that developed on this deposit, relating to the surface terrace facing. Layer III dates to 540 to 327 B.P. at one s.d. but overlaps comfortably with Layer VI when taken at two s.d. (650 to 290 B.P. for L. III, 700 to 520 B.P. for L. VI).

Other depositional sequences at Luluku do not go back as far as Layer VIII in Trench 3, but they do illustrate the point about rapid response to catastrophic flooding and deposition in the irrigated system. Test Pit 2 (Feature 9) contains multiple pondfield layers separated by colluvial and alluvial episodes of deposition (Allen et al. 1987:71–74). Layer VI is a pondfield horizon dating to 305 to 0 B.P. Colluvial sediments (L.V) cover this, and Layer IVb above is interpreted as a probable pondfield topsoil. Above this is a fill deposit, part of a platform for stream retention. A buried stone facing may have served the same function in relation to earlier pondfield use. Layer III is a pondfield layer dating to 445 to 293 B.P., and the current topsoil also relates to former pondfield use. Up to four separate pondfield uses are thus indicated for the past 300 to 450 years.[4]

In nearby Trench 1 (Features 7–10) the same picture is revealed (Allen et al. 1987:65–71), with up to five separate pondfield episodes (Fig. 5.5). Layer

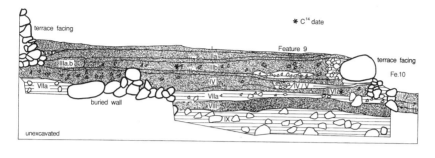

Figure 5.5 Stratigraphic section of part of Features 9 and 10, Trench 1, north face, Luluku (G5-85), O'ahu, Hawaiian Islands (after Allen 1987)

VIII is the lowest of these and is undated.[5] It is separated from another pondfield topsoil (L.VI) by a colluvial deposit. The Layer VI pondfield dates to 670 to 520 B.P. Between this and the next undated pondfield level (L. IIIc/IV) in some places is a further colluvial deposit. Immediately above this pondfield episode is another, represented by Layers IIIa and IIIb, and the current surface layers are also former pondfield deposits. Layer IIIb has produced a date of 476 to 290 B.P. This sequence suggests four separate pondfield uses in the past 600 or so years, and an earlier undated use.

Also on O'ahu, the pioneering excavations of Yen and his colleagues in the upper Makaha Valley in 1970 (Yen et al. 1972) revealed two separate pondfield layers in a series of excavations within an extant agricultural system (site C4-286). One radiocarbon sample related to the initial construction of the earlier system at 660 to 510 B.P., while two further samples related to the period of use of that system at 544 to 441 B.P. and 526 to 319 B.P. The upper pondfield use dated to 313 to 0 B.P., overlapping at two s.d. with the dates for the earlier system. The two periods of use were separated by an episode of alluvial flooding and destruction.

Comparable studies on a Kaua'i Island pondfield system by Schilt (1980) and Athens (1982) have produced somewhat confusing results but also point to an early initial use of the site, a significant period of abandonment, and then at least two phases of pondfield use, probably in the past 300 years.[6]

Aneityum Island, Vanuatu

Four sites on Aneityum (Fig. 5.6) offer somewhat comparable information (see Spriggs 1981 for details): Imkalau (AT37), Lelcei River (AT555), Aname (AT196), and Anetcho River (AT188).

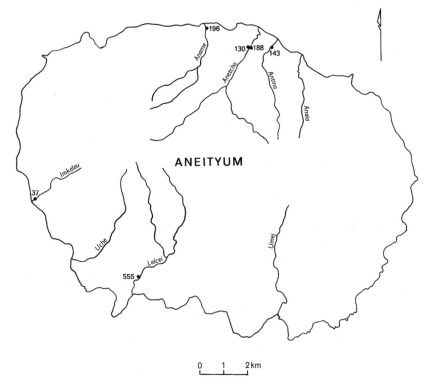

Figure 5.6 Map of Aneityum, Vanuatu, showing sites mentioned in the text

At Imkalau the evidence of initial human use of the area is in the form of charcoal-rich alluvial deposits covering a coral reef platform that is now some 175 m inland (Figs. 5.7, 5.8). These deposits were laid down between 2328 and 2065 B.P. and between 1863 and 1617 B.P. There was then a hiatus in alluvial deposition, and marine sand or beach deposits built up. A burial near the base of these dates to 1248 to 1142 B.P. On top of this marine deposit is an occupation site providing a range of dates from 1056 to 929 B.P. down to 669 to 540 B.P. This occupation site is buried by a further marine-sand deposit, on top of which developed a garden soil associated with structural remains of a terraced dryland garden system. This soil gives a date of 537 to 463 B.P. and is buried by recent alluvium. The Imkalau site is another example where there is a significant time gap, more than 500 years, between initial use and secondary use of a site. Admittedly the evidence for the initial use is indirect, showing significant erosion within the catchment as a result of the burning of vegetation cover, presumably in gardening.

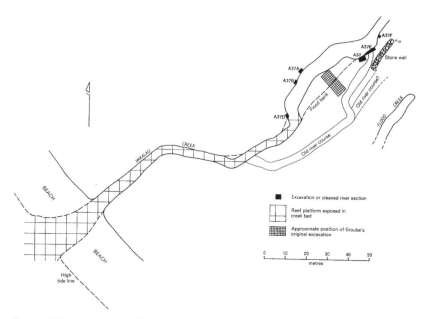

Figure 5.7 Location map for Imkalau Creek excavations (AT37), Aneityum

In the Lelcei Valley a river exposure (AT555) revealed three buried A horizons. As at Imkalau, the lowest horizon revealed only indirect evidence of gardening, in the form of charcoal within the sediment. A concentration in Layer III.2 just above the river level gave a date of 1682 to 1320 B.P. An earth oven was associated with Layer II above this and dated to 642 to 515 B.P. It is the first direct evidence of use of the valley floor. In other sections exposed along the Lelcei River three or four main horizons have been found, but evidence of agricultural exploitation, in the form of stone plot boundaries and walls of dryland gardens, is only seen in the upper two horizons. A similar time gap between first use and second use of the area is evident in Imkalau and Lelcei.

Moving to the north side of Aneityum we find a similar pattern in exposures along the Aname River at site AT196 (Fig. 5.9). Three hundred meters behind the present shoreline a reef platform and overlying beach deposits were revealed in the riverbed, beneath nearly 2.5 m of alluvial deposits (River Section 2). The uppermost part of the beach deposit contains charcoal flecking as well as shell and has been dated to 1880 to 1290 B.P. (charcoal) and 2148 to 2013 B.P. (shell). These dates overlap at two s.d. The alluvial deposits immediately above this correlate with a horizon from another river section approximately 50 m upstream that dates to 490 to 0 B.P., with a probability of 0.69 that

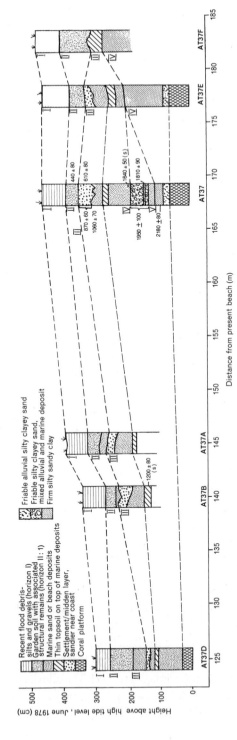

Figure 5.8 Imkalau Creek stratigraphic sections (AT37), Aneityum. S after a radiocarbon date denotes a marine-shell sample

Landscape Catastrophe and Landscape Enhancement   95

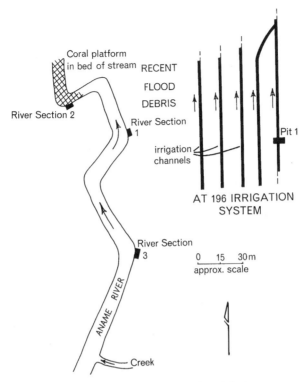

Figure 5.9 Location map for Aname River sections (AT196), Aneityum

the date is between 480 and 260 B.P. (L.II.3 of River Section 1). A buried topsoil (L.III.1) below this dated horizon provides an age from a charcoal concentration of 644 to 513 B.P. The two dates overlap at two s.d. A lower buried topsoil (L.IV) at this exposure has not been dated but correlates with a layer dated at a farther exposure, River Section 3, about 85 m upstream, dated at 663 to 519 B.P.

Adjacent to the river is a large furrow-irrigated garden system. A test pit within it (Test Pit 1) showed a similar general stratigraphy, with the two phases of irrigation use postdating a layer equivalent to that at River Section 1 dated to the last 490 years B.P. (Fig. 5.10). Using the strict criteria I outlined near the beginning of the chapter it is not possible to establish that the topsoils buried below the irrigation system were garden soils. There is evidence, however, for at least four cycles of soil formation and subsequent alluvial deposition within the past 650 or so years. Erosion and deposition rates seem to have increased considerably in the catchment at some point between about 1550 and 550 B.P. given the beach and reef deposits exposed at River Section 2.

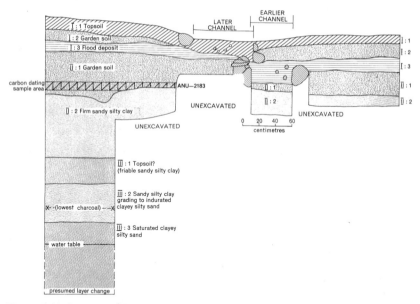

Figure 5.10 Stratigraphic section, Test Pit 1, AT196 Irrigation System, Aname, Aneityum

Similar rapid deposition within the past 950 years has occurred in the Anetcho catchment to the east, and probably in the next river east again, the Antina. In the Anetcho River a deep section was revealed, with up to five possible topsoil horizons exposed (site AT188). Charcoal flecking occurs throughout the exposure. The lower horizon is undated, but an earth oven in Layer IV.2 at 3 m below the top of the riverbank has provided an age of 990 to 720 B.P. This overlaps at one s.d. with a date on dispersed charcoal from Layer III.1 between 1.35 and 1.5 m below the ground surface of 1060 to 790 B.P. Associated with horizon II is a stone-lined creek or large storm drain. Stone-lined storm drains for dryland gardens are exposed in a river section 50 m upstream associated with horizons II and III at site AT130 (Fig. 5.11). There are further structural remains associated with the top near-surface horizon I.

The close network of storm drains found in horizon III is not found in the surface agricultural systems recorded during the 1978-80 archaeological survey of the northern half of the island. It may represent an initial attempt to control flooding and allow gardening during a period when the valley floor was much wetter than it is today and therefore more liable to inundation.

A sequence similar to that at AT130 was recorded in the Antina River to the east at site AT143 (Fig. 5.12). Parallel stone-lined storm drains were revealed at the base of the riverbank in the lowest of three agricultural horizons. The

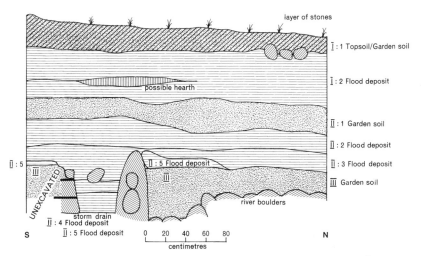

Figure 5.11 Stratigraphic section, River Section 1, Ivanauad (AT130), Anetcho River, Aneityum

lowest storm drains both here and at AT130 are also "stylistically" similar in that the boulders and cobbles generally used in their construction are smaller than those in surface examples of storm drains noted during the archaeological survey. Dates from horizons I and II associated with dryland-garden remains all came out as "modern," whereas no date was obtained for horizon III. By analogy with AT130 and AT188 horizon III, this horizon would appear to be of equivalent age, about 900 B.P. Downstream of this site structural remains associated with horizon I appear to relate to an irrigation system (AT389) that seems to be a recent feature relating to the latest prehistoric or early historic (post-1830) period.

Nunn (1990:130), in reviewing the Aneityum data, suggests that "tectonic change may have had a significant effect; rapid coseismic uplift may have been responsible for the unexplained breaks in alluviation at some sites." He suggests a similar explanation for the Futuna data. While seismic effects are certainly felt on Aneityum and have been a contributing factor to landscape change (Spriggs 1981:102, Appendix 1), the pattern of significant breaks in alluviation occurring *only* in the early parts of the valley-fill sequences comparatively soon after human arrival would not seem to be convincingly explained by such effects. If the uplift rates proposed by Nunn had the effects he suggests, then a major change in the tectonic regime in the past millennium or so, as opposed to the previous one, would have to postulated.

I interpret the sequences from these different island groups to mean that

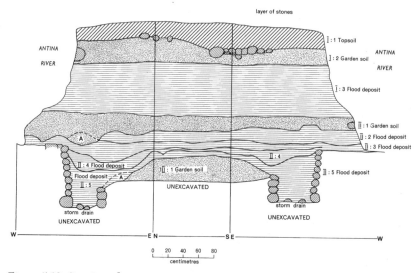

Figure 5.12 Stratigraphic sections, Pit 1, AT143, Antina River, Aneityum

early occupation by small groups was less intensive, as land was not in short supply. These groups could respond to problems of land degradation and catastrophic flooding by moving elsewhere. Later on, with greater subsistence and social demands (see Brookfield 1972) in evolving chiefdoms, this option was no longer possible. In addition, a greater labor force was available to repair and/or rebuild agricultural systems damaged by human-accelerated erosion and deposition. The upper parts of alluvial sequences often reveal several rebuildings of irrigation and other garden systems, at intervals not distinguishable from each other by radiocarbon dating. It was these intensive valley agricultural systems that often formed the economic base for evolving chiefdoms in the region (Spriggs 1986, 1990).

DESTRUCTION OF THE INSHORE ENVIRONMENT

Perhaps more important than the denuding of hillsides in the creation of the valley-floor agricultural systems were the effects on the near-shore and reef environments. The most direct result was the physical covering of former embayments or reef flats. On small islands this would have had an important effect on the availability of reef resources. On Tikopia, for instance, the result of human-induced erosion combined with natural inputs was a reduction in total reef area by 41% (Kirch and Yen 1982). Suspended sediment from runoff would have had a deleterious effect on reef growth in inshore waters as well. One would need to have an idea of reliance on marine resources by the inhabitants of particular islands before and after phases of erosion in order to put an economic value on the loss of marine resources. A feedback seems quite possible between degradation of the marine resource base and development of the terrestrial economic base: a decrease in the marine resource as a result of erosion caused by agriculture would necessitate a further intensification of agriculture, leading to further effects on the marine environment.

Late shifts toward greater exploitation of pelagic fishing grounds after early reliance primarily on inshore reef fishing have been noted in some island groups. Green (1986) points to an almost exclusive reliance on inshore fish at Lapita-period sites, in contrast to more varied fishing strategies and fishing for pelagic species later in the archaeological sequences. Kirch and Yen (1982:289–90) note that coral reefs have a substantially greater biomass than benthic and pelagic communities. The quest for deep-sea fish may have been at least in part necessitated by a deterioration of the inshore reef environment. The development of fishponds and aquaculture in Hawaii (Kikuchi 1976) may also have some connection to such factors.

CATASTROPHE AND ENHANCEMENT

Here and elsewhere I have argued that what has previously been argued as "landscape degradation" in the Pacific can often be viewed as "landscape enhancement," in terms of an island's capability for feeding its human population. Many of the issues that allow for judgment as to whether land degradation has occurred are detailed in Blaikie and Brookfield's book on the subject (eds., 1987). Starting from a definition of degradation as "a reduction in the capability of land to satisfy a particular use" (1987a:6), they note that this is context specific. A shift from hunting and gathering to agriculture brings into relevance a different set of intrinsic qualities in an area of land that may be more capable or less capable in the new context.

Blaikie and Brookfield also discuss the issues of time lags and temporal scale (1987b:66–68). I earlier raised the question of whether what we can now see as long-term benefit, such as the creation of alluvial coastal plains suitable for intensive agriculture, would be perceived in the short term of a human life as land degradation: erosion of hillsides, dumping of flood-borne sediment on valley floors, and so on. At this scale radiocarbon dating does not help, except to point out that several episodes of flooding and rebuilding of agricultural systems can be identified on various islands over periods of a few hundred years. Long-term deleterious effects on the agricultural system do not appear to have resulted from flood episodes.

At the ethnographic, short-term scale, events such as landslips whose initial effect can be to destroy crops and trees may also be viewed in a positive light. They potentially add to the total gardened area, especially where there is a slumping of cliffs as described by Kirch and Yen: "Various economic trees were buried under debris in two affected areas, estimated at 3 and 2 ha. An informant—while acknowledging that it was fortunate that one of the landslides stopped just short of the highest house at Paepaevaru village—was more concerned with plans to try plantings of *taro* or *ango* (turmeric) on the 'new' land" (1982:148). Kirch and Yen (1982:43) also note that the two largest taro gardens they observed in 1977 were on old landslips of some 15 to 25 years earlier.

There are also cases where catastrophic erosion is not simply taken advantage of but is purposefully induced. "Hydraulicking," the shifting of soil and gravel by the release of impounded water, is a commonly used technique in irrigation-terrace construction and repair in the Ifugao area of Luzon in the Philippines. Conklin describes how "hundreds of meters of temporary canals and ditches may be dug across house terraces, hillsides, and even through other pond fields so that artificial streams can transport rock and gravel fill as well as earth and soil to new terrace levels" (1980:16–17). Similar techniques are known from the Americas and are reported as part of stone-ax quarrying techniques in the New Guinea Highlands (Spores 1969:563; Vial 1940:159). Such techniques would have been eminently suitable for the construction of terrace systems in the Pacific, requiring only a mimicking of processes easily observed in nature. They are not, however, remembered or used in Pacific communities today so far as I am aware.

I conclude that changes to the landscape, although sometimes dramatic at the level of individual events, such as landslips or flood-borne deposition, were expectable, controllable in their effects, and to an extent channeled toward particular outcomes. Blaikie and Brookfield note that not all human impact increases degradation, and they view it as "an equation in which both human and natural forces find a place" (1987:74). Their formula for assessment is: net degradation = (natural degrading processes + human interference) − (natural reproduction + restorative management). The numerous examples that one can cite of terracing, storm drains around and in gardens, stone lining of rivers to control their courses, planting of trees to stabilize dunes and shorelines, and so on, speak to the many responses of populations in the Pacific to potential degradation. In all but a few cases, Easter Island being the most notable (Flenley and Bahn 1992), the landscape changes that humans induced created conditions for continued growth in agricultural production rather than putting such growth at risk.[7] Given a large labor force, the floods and landslips that on occasion buried their gardens may have been seen as opportunities for extension of the gardened areas, with the soil refertilized by the influx of sediment. There is at least no evidence from the archaeological sequences of any long-term disruptions caused by such events.

POSTSCRIPT: THE WEST DISCOVERS THE ENVIRONMENT

Until comparatively recently it was often assumed that small-scale non-Western societies in regions like the Pacific were so much a part of their natural surroundings that they were very like the other fauna and the flora of the region, and their presence did not alter the natural equilibrium or balance of the islands (see Chapter 1). These populations were seen as representing Rousseau's *homme naturel,* man in a state of nature, "noble savages" in harmony with the environment, in contrast with "civilized man," who had somehow fallen from grace and so was out of step with the environment. It is a romantic notion, a viewpoint informed by the biblical view of the Fall, and it also allows for considerable paternalism and exploitation with regard to the supposedly childlike indigenous populations.

I find it significant that in independent Pacific nations people find it easy to live with the idea that their ancestors, and frequently they themselves, actively altered their island environments. Although these Pacific Islanders often manage their resources wisely, it seems inappropriate to attribute to them a traditional, self-conscious conservation ethic of the kind that has developed strongly in the West since the 1960s and 1970s. In terminally colonized Pacific countries that have been heavily affected environmentally by immigrant populations, people often talk of a golden age before the arrival of the "white man," when their ancestors lived in harmony with the environment. They profess that their ancestors did indeed have a strongly developed conservation ethic, which the West is only now catching up on.[8] In such places—Hawaii and New Zealand (Aotearoa), for example—if evidence is presented that the indigenous inhabitants had in fact had an impact on the environment, the news is greeted with angry denial by native activists and a certain degree of glee by the media. The implication of reporting in the media is often that although the environment has been altered (!) by Westerners, this is acceptable because the natives had already ruined it, thus weakening claims for indigenous land rights and the self-respect of the native community, or at least respect by nonnatives.

Although it was clearly the case previously that because they were "innocent" the indigenous populations could justifiably be dispossessed of their land, the new message is perhaps that because they are not so innocent after all, they deserve all they got. Thus "liberal" guilt is assuaged.

An example is press coverage of a dispute in Australia over the proposed mining of Coronation Hill in the Northern Territory, opposed by Aborigines, who are said to believe that impact on an area they hold sacred will disturb a Dreamtime creator-being and bring about a catastrophe. A report on the Aborigines' claim and in particular on the findings of anthropological consultants has been summarized and quoted by the *Melbourne Age* (22 April 1991): "It [the report] says that according to the mythology of conservationists, hunting and gathering people such as the Aborigines had a special association with the wilderness. 'But there is little substance to this myth . . . in fact, Aborigines played a major role in altering the Australian environment,' the report says. 'They probably contributed to the extinction of many species of large animals.'" The report also describes Aborigines as Australia's first miners, "extracting flint and ocher from deep underground mines."

I do not mean here to criticize the quoted report by Ron Brunton of the Institute of Public Affairs (Brunton 1991) but to make the point that, in demonstrating the human element in environmental change in the Pacific during the past several thousand years, we must be aware of the political content and implications of our findings. Unless presented sensitively, they are likely to be seized upon by those with political agendas we may not have considered and may well not want to subscribe to. We surely owe the descendants of the people we study some consideration in guarding against misrepresentations of our findings that seek to deny indigenous people their dignity and their land.

Notes

I thank Patrick Kirch and Terry Hunt for soliciting a contribution to their session at the Pacific Science Congress. My attendance at the congress was partially funded by the Australian National University, and this chapter was typed in the Department of Prehistory (now the Division of Archaeology and Natural History) of that institution. The figures were drawn by Winifred Mumford and Ian Faulkner of the Australian National University, Canberra.

1. Some notable works, by no means an exhaustive list, include Allen-Wheeler 1981; Allen 1987; Beggerly 1990; Brookfield, ed., 1979, 1980; Flenley and King 1984; Golson 1977; Hope and Spriggs 1982; Hughes 1985; Kirch 1975, 1976, 1982, 1983, 1988; Kirch and Kelly 1975; Kirch and Yen 1982; McGlone 1983; Powell 1970; Schilt 1984; Spriggs 1981, 1985; Yen et al. 1972.

2. In this discussion I shall, like Blaikie and Brookfield (1987:26), duck "the difficult environmental-ethical question such as the extinction of endangered species, or conflicts between national parks and other human uses of the biome, where ethical judgements assume greater importance." Like them again, I have chosen in this chapter to see land degradation "in terms of the altered benefits and costs that accrue to people at that time and in the future."
3. Calibrated radiocarbon ages are presented in this paper using the CALIB computer program (version 2.0) of Stuiver and Reimer (1986). Unless otherwise indicated, dates are given as the calibrated range at one s.d.
4. Allen's statement that "while the ^{14}C date from Layer III (A.D. 1435–1665) overlaps the A.D. 1135–1435 glass date from Layer IVb, the A.D. 1490–1950 date from Layer VI is either in error due to the small size of the accelerator-processed sample or reflects disturbance" (1987:74) is a misrepresentation of the radiocarbon results. They overlap at one s.d. and so are essentially indistinguishable. There is thus no need to postulate any error in the Layer VI determination. Little reliance can be placed on volcanic-glass age determinations from Hawaii, given problems of source-specific hydration rates, effect of storage temperature, and lack of inter-laboratory comparability in hydration measurement.
5. A date for a horizon supposedly equivalent in Trench 1A gave an unacceptably late date of 290 to 0 B.P. Allen notes that this layer is "a very sandy soil affected by groundwater" (1987:76, cf. 177).
6. Schilt's original report of excavations at site D10-12 suggested that the lower of two pondfield horizons (her Layer III) was related to a radiocarbon date that calibrates to 1297 to 1094 B.P. Stone waste flakes and bifacially flaked tools were found in this level. Layer II represented a flood deposit. From dispersed charcoal within a possibly equivalent layer came a date of 514 to 314 B.P., said to be associated with habitation activities. This and the alluvial Layer II deposit were overlain by an upper pondfield horizon, which was dated to late prehistoric and early historic periods on the basis of some stone artifacts found within it and historical records of the use of this area for irrigated gardening (Schilt 1980:60–61). Further excavations by Athens (1982) failed to confirm the early date for agriculture. Two pondfield horizons were again encountered in Athens's Trenches A, B, and C (Layers V and II in his more detailed stratigraphy). Three Layer V dates were all within the range of 302 to 0 B.P. Layer V and the upper part of Layer VI were associated with a stone-tool assemblage and three possible post molds, suggesting to Athens an initial precultivation use of the site for habitation, which had been disturbed by later pondfield use. It is likely that Schilt's early date was in fact associated with this disturbed-habitation component. If so, there is a considerable hiatus between first (habitation) and second (pondfield) use of the site. In Trench B a second date was obtained from Layer III.2, in between the two pondfield horizons. Athens rejected it as anomalously early, but it does in fact overlap at two s.d. with the Layer V date from Trench B. Schilt's Layer II date came from another part of the site, where only a single pondfield horizon (her Layer I) was present. It can be interpreted as a terminus

ante quem for pondfield agriculture, supporting its dating to the past 300 or so years.
7. The refutation of an oft-quoted similar but more extreme scenario for Easter Island in the case of Kaho'olawe in the Hawaiian Islands is detailed elsewhere (Spriggs 1991).
8. I base these remarks on personal observation of reactions to the proposition that indigenous Pacific peoples had altered their environments in conversation with villagers in Vanuatu and Papua New Guinea, compared to reactions among generally urban-dwelling Hawaiians and Maori. I claim no statistical validity for my sample, but I have witnessed the enjoyment of these villagers in setting large conflagrations in and outside their garden areas. Whether Hawaiians had such a traditional "conservation ethic" is a topic worthy of further research, particularly given the somewhat unique (for the Pacific) level of social stratification recorded at European contact. Does a particular kind of environmental consciousness follow from this, different to that in less-stratified parts of the region?

6
The Historical Ecology of Ofu Island, American Samoa, 3000 B.P. to the Present

Terry L. Hunt and Patrick V. Kirch

Thirty-five years ago Raymond Fosberg (1963:5) wrote, "It is clear that the arrival of man has invariably increased, to some extent, the degree of instability in these [island] systems. With the advent of modern man this increase has frequently assumed catastrophic proportions." At the same symposium in 1961, Cumberland (1963:191) pointed to early "Moa-hunters" in New Zealand as responsible for massive disturbance and modification: hunting and the widespread use of fire had driven several species of birds to extinction. In contrast, Cumberland (1963:193) argued, the Maori—then believed to be descendants of a second Polynesian migration to New Zealand—were conservationists and nowhere caused wholesale transformation of the environment or disastrous disturbance of the ecosystem. These and similar views expressed at the symposium pointed to an emerging paradox: Polynesians were seen as conservationists, yet island environments had been greatly transformed.

Archaeologists and natural scientists have learned a great deal in the Pacific through interdisciplinary research since the Fosberg symposium. Their studies document biotic and landscape transformations resulting in some cases from natural causes, in others from human agency. Recently, Nunn (1991, 1994) has stressed the significance of natural environmental changes. In contrast, numerous studies (Christensen and Kirch 1986; Flenley et al. 1991; Hughes 1985; Kirch 1982c, 1983; Kirch et al. 1992; Kirch and Yen 1982; McGlone 1989; Olsen and James 1984; Steadman 1989a) have demonstrated the fragility of island environments and the significant role of direct and indirect human-induced changes, especially to island biota. We have begun to attain a more sophisticated understanding of the relative importance of changes in island environments induced by natural agencies, prehistoric human populations, and post–European contact populations. This success owes a great deal to the collaborative, interdisciplinary efforts of archaeologists, geologists, and biologists.

In this chapter we present the results of interdisciplinary research at the To'aga site on Ofu Island, part of the Manu'a Group of American Samoa (Fig. 6.1). Our archaeological work, which included studies of coastal geomorphology and sediments, marine and mammalian fauna, avifauna, and land snails, points to the dynamic interplay of natural and human-induced change on Ofu Island. We suggest that this change is paradigmatic of many other cases in the Pacific. As the chapters in this volume illustrate, documenting change in island environments provides an essential historical context for research in such fields as ecology, evolutionary biology, biogeography, and anthropology.

THE OFU ISLAND ENVIRONMENT

The Manu'a Group, an intervisible cluster of three small, relatively young islands—Ta'u, Olosega, and Ofu—lies 100 km west of the larger island of Tutuila (Fig. 6.1). Ofu and Olosega are closely adjacent, separated by a narrow and shallow strait, now spanned by a concrete causeway. The weathered volcanic cone of Ofu is only 3.4 km^2, reaching a maximum elevation of 638 m; approximately 91% of Ofu's land is comprised of slopes steeper than 30 degrees. Ofu's geologic youth (averaging 0.3 million years, McDougall 1985:318) is attested by its precipitous topography, shaped by faulting and landslide events that have formed sheer cliffs around much of the island. Erosion by intermittent streams has only slightly altered the island's morphology. Coastal terraces range from only 50 to 150 m in width and are found along the western and southeastern shorelines. These coastal terraces are especially important in the island's human history, and today are where the island's roughly 250 people live.

Samoa lies within the humid tropics, with near constant temperature and humidity and abundant rainfall (Buxton 1930:17). Average temperatures vary

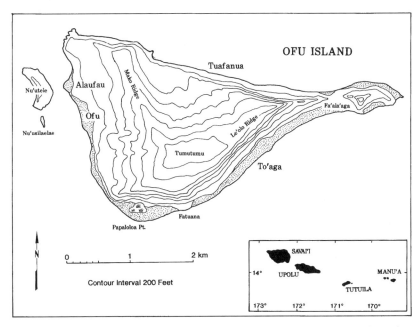

Figure 6.1 Topographic map of Ofu Island; inset map of the Samoan archipelago; the stippled area represents the coastal terraces

from only 25.7 to 26.2°C, with humidity ranging between 80 and 86%. The total annual average rainfall (recorded from Tutuila Airport) is 3,100 mm (Nakamura 1984, Table 1). Rainfall is variable, and there are extended dry seasons, with droughts that can significantly affect agricultural production. The wet season (October to March) can bring torrential rains and damaging floods—some associated with tropical storms and hurricanes (Nakamura 1984:3). Visher (1925:27, Table 6) indicates an annual frequency of two to three hurricanes in the area of Samoa, some hitting the islands directly and bringing devastation to crops, houses, and human life. The torrential rains cause landslides, and storm surge can reshape the shoreline and configuration of the coastal terrace in a matter of hours (e.g., see Bayliss-Smith 1988).

Soils on Ofu are young and in consequence poorly developed (Nakamura 1984). Most of the steep interior is covered with "Ofu silty clay," a deep, well-drained soil formed in volcanic materials. In the steepest areas (slopes 40 to 70%) this soil is covered in forest. On the gentler slopes (15 to 40%) these silty clays are under shifting cultivation. These interior gardening zones are confined to the western slopes and to the northern slopes of Tumutumu

Mountain. Very steep slopes and areas of talus, such as on the interior edge of To'aga, have a soil that Nakamura (1984:11) describes as a very steep rock-outcrop association. On the interior edge of the coastal terrace is "Aua very stony silty clay loam" (Nakamura 1984:10), characterized as very deep, well drained soil formed in colluvium derived dominantly from basic igneous rock. The zone of stony colluvium is used for subsistence gardening, primarily cultivation of bananas and breadfruit. Finally, the coastal terrace that skirts most of the south and west shores of Ofu has a "somewhat excessively drained soil . . . derived from coral and sea shells" (Ngedebus mucky sand [Nakamura 1984:15]). This zone supports extensive subsistence gardening, including cultivation of traditional root crops and arboriculture (Kirch 1993a).

Contemporary vegetation communities on Ofu attest to several millennia of human land use. The vegetation of the lower elevations and coastal terrace of the island is strongly anthropogenic, with a mosaic of coconut stands, breadfruit and banana orchards, and aroid gardens interspersed with second growth, a "transported landscape" in the sense of Anderson (1952). (See Kirch 1993a:19–20 for descriptions of the floral dominants along vegetation transects across the coastal terrace at To'aga.) Only on the higher, steeper slopes does rainforest persist. Yuncker (1945:4) lists 421 taxa, including mosses, pteridophytes, and flowering plants, in his botanical survey of the three islands of Manu'a.

The vertebrate terrestrial fauna of Ofu is quite restricted, although somewhat richer in invertebrates (land snails and insects). The only indigenous mammal is the Fruit Bat, (*Pteropus samoensis*), found in abundance and still taken for food. The Pacific Rat (*Rattus exulans*), domestic pigs (*Sus scrofa*), and dogs (*Canis familiaris*) were introduced prehistorically by Polynesians. The modern avifauna on Ofu includes native landbirds and nesting seabirds (Watling 1982). The higher elevation forests of Ofu are inhabited by the *lupe*, or Pacific Imperial-Pigeon (*Ducula pacifica*), and the *manutagi*, or Crimson-crowned Fruit-Dove (*Ptilinopus porphyraceus*), both occasionally taken for food. White-collared Kingfishers, or *ti'otala*, (*Halcyon chloris manuae*, a Manu'a subspecies), Banded Rails, or *ve'a* (*Gallirallus philippensis*), the Polynesian Starling, or *mitivao* (*Aplonis tabuensis*), and the *iao*, or Wattled Honeyeater (*Foulehaio carunculata*), are relatively common in the coastal bush and gardens. A variety of seabirds and some migratory birds also nest on Ofu, including the *tava'e*, or White-tailed Tropicbird (*Phaethon lepturus*), and the seasonal migrant, *tuli*, or Lesser Golden-Plover (*Pluvialis dominica fulva*). The only other vertebrates on Ofu are small lizards in the families Geckonidae and Scincidae.

Of particular significance to paleoenvironmental reconstruction are the terrestrial mollusks often preserved by and recovered from archaeological contexts. These include endemic, indigenous, and synanthropic taxa introduced by prehistoric Polynesians or after European contact. Kirch (1993b) has analyzed land-snail assemblages from the To'aga site, and we draw on these results here.

Ofu is surrounded by a fringing reef, providing microhabitats for abundant fish, shellfish, and other food resources. The fringing reef is widest and most sheltered on the western side of the island. Here a diverse array of mollusks inhabit the algal crest and reef flat, including bivalves like *Periglypta reticulata*, *Tridacna maxima*, *Hippopus hippopus*, and *Asaphis violascens*, as well as gastropods like *Trochus maculatus*, *Turbo setosus*, *Nerita* spp., *Cypraea* spp., *Drupa* spp., *Thais armigera*, and *Conus* spp. Many of these shellfish have been important sources of protein, as shown by the analysis of the archaeological faunal remains (Nagaoka 1993). Also present on the reef are spiny lobsters (*Panulirus* sp.), sea slugs (holothurians), sea urchins (echinoderms), octopus, and a variety of edible seaweeds. Approximately 800 taxa of inshore fishes occur in Ofu's waters (Jordan and Seale 1906), and they form a significant part of the traditional Samoan diet. Among the fish commonly taken are jacks (*Caranx* spp.), parrot fish (*Scarus* spp.), wrasses (Labridae), and acanthurids. The open sea beyond the reef provides rich pelagic resources, such as the prized tuna (Scombridae). Marine turtles (*Chelonia mydas* and *Eretmochelys imbricata*) are not commonly seen in Ofu waters today, but they were undoubtedly once more common, as archaeofaunal evidence attests to their heavy exploitation in prehistoric times.

In sum, the contemporary environment of Ofu, as of many other Pacific islands, reflects its natural geological and biotic history, as well as the consequences of human settlement and some three millennia of cultural manipulation. Precisely because Ofu is a rather small island it may be ideal for modeling the dynamic interplay of natural and human-induced factors. In spite of its diminutive size, Ofu supports a variety of habitats, including the persistence of comparatively "pristine" rainforest. However, the island also has a substantial anthropogenic component, particularly on the coastal terraces and areas of low elevation where the vegetation has been transformed into an economic landscape. Ofu is thus a microcosm of larger high islands, where the human hand has affected microenvironments differentially. Ofu, unlike some other small islands (e.g., Mangaia, Kirch et al. 1992; Kirch, Chapter 8), has not seen massive environmental degradation. Rather, the island's historical ecology reflects the delicate balance between human impacts and natural resilience and recovery. Understanding Ofu's environmental history has implications in comparative context for many other Pacific islands.

THE TO'AGA ARCHAEOLOGICAL PROJECT

Archaeological investigations of the To'aga coastal terrace on Ofu Island were carried out from 1986 to 1989 (Kirch and Hunt, eds., 1993). Numerous surface stone structures and pottery-bearing deposits at least 2,000 years old (based on comparative evidence from Western Samoa [Green 1974]) were discovered in 1986. In 1987 and 1989 we conducted excavations totaling 27 m^2, systematically sampled along 17 transects designed to crosscut the geomorphic variability of the coastal terrace (Fig. 6.2). Details of the fieldwork and results from several specialized analyses are reported in the monograph we edited (Kirch and Hunt, eds., 1993).

A major objective at To'aga was to reconstruct the island's occupational chronology in relation to the geomorphological and paleoenvironmental history of the coastal lands. For this reason, we paid special attention to details of coastal-terrace stratigraphy and correlation across units. Geoarchaeological analysis of sediment samples and radiocarbon dating were used to reconstruct the depositional history and chronology of the coastal terrace. Our excavations also recovered extensive faunal remains (marine and terrestrial vertebrate remains, marine mollusks, and terrestrial gastropods), which provided critical data on the human-exploited biota over the past three millennia.

The chronology of human occupation and of coastal-terrace formation along Ofu's southern coast is indicated by a suite of 14 radiocarbon age determinations, spanning the period from about 3500 cal B.P. to 1000 cal B.P. (Kirch 1993c, Fig. 6.1). Polynesian populations had colonized Ofu by at least 3257 to 2879 cal B.P. (Beta-35601), and there are indications from two older samples that colonization might have occurred slightly earlier. These first settlers produced earthenware pottery for at least 1,000 years (Hunt and Erkelens 1993), then abandoned ceramics for reasons that remain unclear. Other artifacts recovered in our excavations include stone adzes, shell ornaments and fishhooks, hammerstones, and coral and sea urchin spine abraders. There is evidence for interisland transport, perhaps of ceramics and, more definitively, of fine-grained basalt adzes (Weisler 1993).

We now turn to a review of the evidence from the To'aga site for changes in the geomorphology and biota of the Ofu coastal terraces, evidence from which the historical ecology of the island of the past 3,000 years may be reconstructed. The full data sets on which our summary is based have been publish-

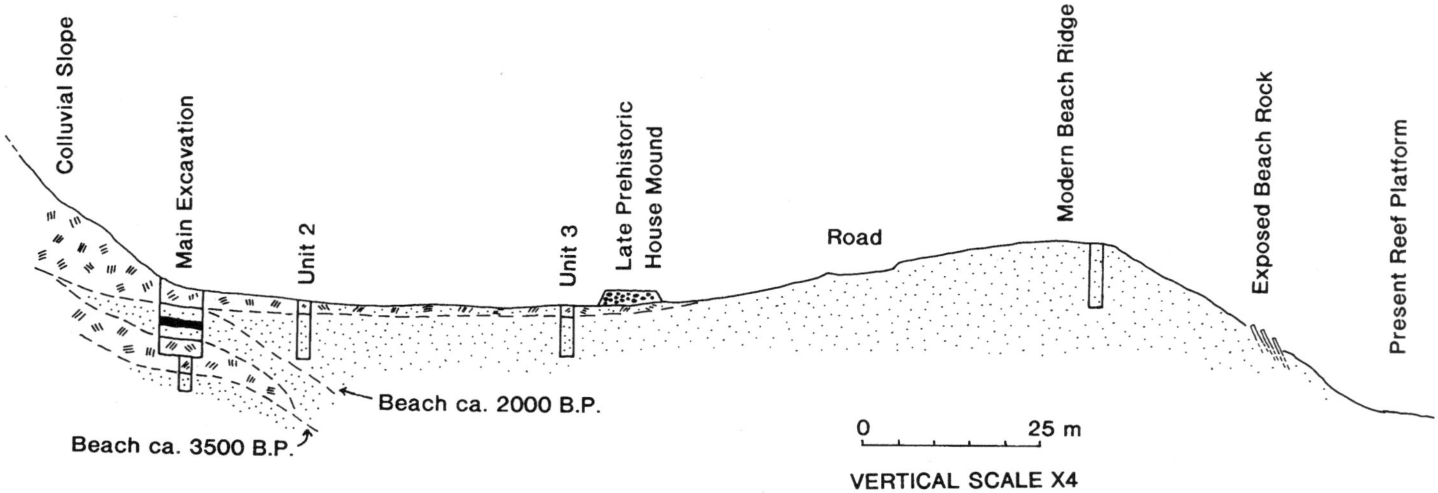

Figure 6.2 Schematic profile along the 1987 excavation transect at To'aga, showing the main geomorphic features and stratigraphic zones

ed elsewhere (Kirch and Hunts, eds., 1993), and we shall draw only on selected examples to illustrate our key points.

LANDSCAPE CHANGE

During excavation and mapping of the To'aga coastal plain we noted that archaeological deposits containing pottery—known in Samoa to be approximately 2,000 to 3,000 years old—were restricted to a zone along the interior of the coastal terrace. This interior edge of the coastal terrace is defined by 500 m high cliffs, large boulder talus slopes, and gravelly clay-loam soils. Moving only a few meters seaward, terrace stratigraphy changed to interbedded strata of terrigenous colluvium (derived from mass wasting and slope wash) and calcareous sands of biogenic origin. Near the present shoreline, deposits consisted of unconsolidated calcareous sands with only recent cultural materials. Exposed beach rock and the undercutting of large shoreline trees indicated active coastal erosion. These initial observations suggested that the coastal terrace along Ofu's southern shore was geomorphically dynamic, that a sequence of shoreline progradation had occurred during prehistoric occupation of the area, and that a new phase of erosion was under way. Any understanding of the chronology of occupation and deposition of cultural remains—or "site formation," as it is known to archaeologists—would have to account for the formation of the coastal terrace. Careful stratigraphic analysis, radiocarbon dates, and geologic comparisons in the region provided the elements of a model for environmental change at To'aga. The long-term geomorphic sequence at To'aga includes interrelated factors of sea-level change, island subsidence, and sedimentary budget. Kirch (1993b) has reconstructed and modeled this sequence, testing it against our field geomorphological and archaeological data (Kirch and Hunt 1993).

The stratigraphic profiles revealed by several transect excavations revealed a consistent depositional history for the To'aga coastal terrace. The reconstructed sequence for Transect 5 provides an example of landscape evolution of the coastal terrace:

> *Stage 1* (>3200–2800 cal B.P.): A narrow coastal bench at the base of the steep talus was formed. The active shoreline at this time would have been in the vicinity of Units 16–17, considerably inland of the modern shoreline. The "salt-and-pepper" lithology of the beach ridge sediments [mixed basaltic and calcareous grains] indicates exposure of volcanic headlands along the coastline, providing a source of volcanic lithic grains. In addition, the presence of larger clastics (coral cobbles, branch coral fingers) indicates a fairly high energy shoreline.

Stage 2 (ca. 3200–2800 cal B.P.): Humans began to occupy the narrow bench formed during Stage 1, resulting in non-concentrated midden deposits that contain thin, fine-tempered, orange- or red-slipped pottery in Units 28 and 15/29/30. The main area of occupation was probably farther inland from Unit 28, and thus is now deeply buried under talus rockfall and colluvium. The deposits exposed in Units 28 and 15/29/30 appear to represent the seaward periphery of such an occupation, down the slope of the former beach ridge toward the old shoreline. Archaeological exposure of the putative main occupation zone would require the use of heavy machinery, since as much as 5–15 m overburden of boulder talus and colluvium would probably have to moved.

Stage 3 (ca. 2800–2000 cal B.P.?): Deposition of calcareous sands onto the beach ridge continued, with significant seaward progradation of the shoreline occurring late in this phase....

Stage 4 (ca. 2000–1600 cal B.P.?): A stabilized land surface formed during these four centuries over the now wider and prograded coastal terrace, marked by a paleosol horizon (Layer IIIA-1 in Units 15/29/30; Layer IC in Unit 16; Layer IVB in Unit 17).

Stage 5A (ca. 1600–1400 cal B.P.): The stabilized coastal terrace in the vicinity of Units 16 and 15/29/30 was occupied during this terminal phase of ceramic manufacture and use on Ofu Island....

Stage 5B (ca. 1300–1000 cal B.P.): Aceramic occupation on the coastal terrace in the vicinity of Unit 17 resulted in the construction of a low house mound formed by several successive gravel (*'ili'ili*) pavements.

Stage 6 (< 1000 cal B.P.): A tongue of clay-silt colluvium was deposited out onto the coastal terrace, probably due to increased up-slope forest clearance, agricultural activity, and subsequent erosion. At this time the coastal terrace was used for tree-cropping and shifting cultivation, continuing into the present era. [Kirch and Hunt 1993:67–68]

Accounting for the depositional history revealed at Transect 5 and other localities—with a rapid coastal progradation commencing after about 2000 B.P.—requires a morphodynamic model that takes account of regional eustatic sea-level variations, local tectonics, and changing local sediment budgets. A rapid eustatic rise in global sea level associated with the terminal Pleistocene is well established (e.g., Fairbridge 1961; Sheppard 1963). More complex are mid-to-late Holocene changes, varying dramatically with local conditions and processes. Using global Holocene data, Bloom (1980, 1983) argued for a 1–2 m higher sea level for the South Pacific. Substantial evidence from several South Pacific islands now supports Bloom's model for 1 to 2 m higher levels during the period between 4000 and 2000 B.P. (see Kirch 1993b, Fig. 4.4, for a summary). Nunn's work on Fijian shorelines, for example, points to a 1 to 2 m higher stand of the sea (Nunn 1990, 1994). In the central Pacific, comparable evidence has come from recent research in Tonga (e.g., Kirch 1988; Nunn 1994), Western and American Samoa (e.g., Nunn 1991, 1994), the Cook Islands (Stoddart et al. 1985; see also Chapter 7), and French Polynesia (e.g., Pirazzoli and Montaggioni 1986). The general pattern provides strong evidence for 1–2 m higher sea level from about 5000 B.P. until sometime between 1000 and 2000 B.P., when sea levels fell to modern positions.

A second factor in explaining the depositional history at To'aga is local island subsidence. As with most volcanic archipelagoes of the central Pacific, the Samoan Islands formed on a "hot spot" on the Pacific Plate. As the plate migrates, islands subside as they age. Subsidence also occurs with point loading, where the volcanic mass of a young island causes crustal deformation. While there is no direct geological documentation of subsidence of Ofu Island, evidence from Mulifanua in Western Samoa suggests how rapidly the Samoan Islands may be sinking. Mulifanua is a ceramic-bearing archaeological deposit now situated 1.5 m below mean sea level, capped by 75 cm of reef rock (Leach and Green 1989). As the site represents an occupation on a former coastal terrace, about 2.6 to 3 m of subsidence is indicated. This points to a rate of approximately 1m/kyr subsidence for Upolu in Western Samoa. Given Ofu's younger geologic age—and hence high point loading—its rate of subsidence could be more rapid.

As we have described (Kirch and Hunt 1993:68), the stratigraphic sequence from Transect 5 and its correlation to present sea level, provide critical evidence for testing a morphodynamic model for landscape evolution (Kirch 1993b). We draw on the evidence of Transect 5 again where the earliest cultural deposit has been dated to 3257 to 2879 cal B.P. (Beta-35601, see Kirch 1993c:88). This deposit correlates to the modern mean high tide at 1.8 m above the reef flat. This elevation correlation provides solid evidence of local tectonic subsidence: "The *in situ* cultural materials in Units 15 and 28 were clearly deposited on a narrow terrace or beach ridge that must have been at least 1 m, and more likely 2 m, above the sea level at 3 kyr B.P. Given a +1–2 m high sea level at 3 kyr B.P., this means that the To'aga site has undergone between 2–3 m of tectonic subsidence over the past three thousand years, as suggested by the morphodynamic model. . . . The alternative hypothesis—that there was no tectonic subsidence—would require the deposition of the occupation deposit under water, a physical impossibility given the sedimentological evidence" (Kirch and Hunt 1993:68). Corroborating evidence was also obtained from Transect 9 (Figure 6.3).

Sea-level change and subsidence alone do not account for the shoreline progradation at To'aga. As Chappell (1982) points out, the sedimentary

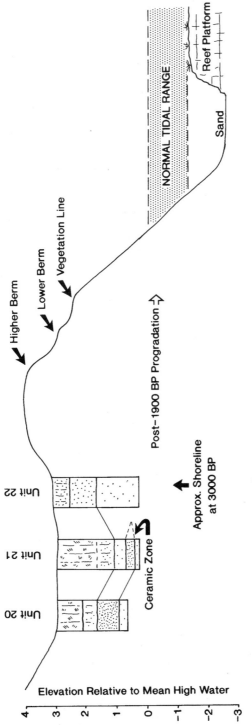

Figure 6.3 Schematic profile along Transect 9 at To'aga, showing the relation of buried occupation deposits to the modern shoreline and sea level

budget is an important factor in the progradation process. While some sediment was provided by weathering of the volcanic cliffs above the site, most of the coastal terrace is composed of marine biogenic sediment—calcareous sands and larger clastics of coral or reef conglomerate. Sediment is formed by wave erosion and biologic processes, such as generation of sand by parrot fishes that rasp and grind coral to extract algae. High-energy storm surf and cyclones are also important means of generating rapid accumulation of sediment, transforming many coastal and atoll landforms in a matter of days (e.g., Bayliss-Smith 1988). Under normal Holocene conditions, during periods of rapid sea-level rise when corals are actively growing below mean sea level, the generation of sediment would be reduced. When sea level dropped or was stable, coral growth would have caught up, allowing erosion and an increased sediment budget. Kirch (1993b, Fig. 5.5) has modeled these factors in temporal context for the morphodynamics of the Ofu coastal terrace.

Terrigenous sediment also contributed to the buildup of the coastal terrace, through the processes of rockfall, mass wasting, and sheet erosion. The presence of people on Ofu from nearly 3000 B.P. accelerated the erosion of terrigenous sediment when land was cleared by slash-and-burn techniques for agriculture. That humans played a role in erosion and landscape change is shown by the presence of charcoal flecks in the earliest colluvial deposits at To'aga (Kirch and Hunt 1993; see also Kirch and Yen 1982 for discussion of similar evidence for Tikopia Island). In the Ofu case, however, human-induced erosion played only a relatively minor role, in concert with other dynamic processes in the dramatic development of the coastal terrace.

In sum, Holocene glacio-eustatic rise in sea level reached a maximum at +1–2 m in the South Pacific between 4 to 2 kyr B.P. Before about 3200 B.P., the Holocene sea-level maximum would have worked its erosive forces on the cliff that now defines the interior of the To'aga coastal terrace; indeed, this feature has been described as a remnant sea cliff by Stice and McCoy (1968). Evidence from To'aga indicates that sea level assumed its modern position about 2000 B.P. after which time the coastal terrace prograded primarily through rapid deposition of biogenic sediment. With tectonic subsidence of Ofu, the local relative sea level was stable or fell slightly, a period when the biogenic sediment budget would have increased significantly (Kirch 1993b). Over the past 1,000 years, as sea level stabilized and subsidence continued, the sediment budget declined and shoreline erosion recommenced. Human activities also contributed to the formation processes of the coastal terrace. Forest clearance for swidden cultivation and the associated use of fire would, by inference, have meant destruction of natural habitats for native plants, birds, and other fauna. The model for dynamic changes to the Ofu coastal terrace is graphically summarized in Figure 6.4.

CHANGES IN ISLAND BIOTA

Vertebrate and invertebrate fauna deposited as food remains, and in some cases as a natural component in the sedimentary process, are well preserved in the calcareous deposits of the To'aga site. Indeed, the To'aga excavations produced one of the largest and best preserved faunal assemblages yet recovered from Western Polynesian archaeological contexts. We examine three sources of evidence for biotic changes from To'aga: (1) the fish, shellfish, and mammalian remains (analyzed by Lisa Nagaoka 1993); (2) the avifaunal remains (analyzed by David Steadman 1993); and (3) the subfossil land snails (analyzed by Kirch 1993d). These independent lines of evidence provide a complementary view that forms a reliable measure of biotic change on the island.

One of the impacts wrought by prehistoric human populations on Pacific islands was the introduction of larger vertebrate animals, especially the domestic pig and dog, and the fowl. Chicken (*Gallus gallus*)—though not abundant—is the most common of the Polynesian-introduced domesticates in the To'aga faunal assemblage (16 NISP) and the most common bird species found in the avifaunal assemblage. It is present in an early dated context, about 2800 to 2300 cal B.P. indicating that chicken was introduced to the island by the initial human colonizers. Pig, however, is only unambiguously present in later contexts, and we are therefore uncertain as to the date of its first introduction to the island. Some of the unidentifiable medium mammal bone in early stratigraphic contexts may well be pig, dog, or both, so that the absence of these domesticates in early contexts is not certain. But, perhaps significantly, despite the large and well-preserved faunal sample, pig, dog, and chicken are not well represented, suggesting that they may not have been present in large numbers in prehistoric times. The Pacific Rat (*Rattus exulans*), a synanthropic species transported either inadvertently or intentionally by Polynesians, is present in the earliest cultural deposits at To'aga and through the rest of the sequence (total 380 NISP). As on many other Pacific islands, *R. exulans* appears to have arrived with the earliest human colonizers of Ofu.

Remains of marine animals dominate the To'aga cultural deposits, with some 2,196 identified fish bones, 56 marine-turtle bones, and large quantities of mollusks and other invertebrates. Nagaoka's (1993) analysis of fish bones shows that the composition of the To'aga marine fauna changed little over time, suggesting that the level of exploitation was not severe enough for extir-

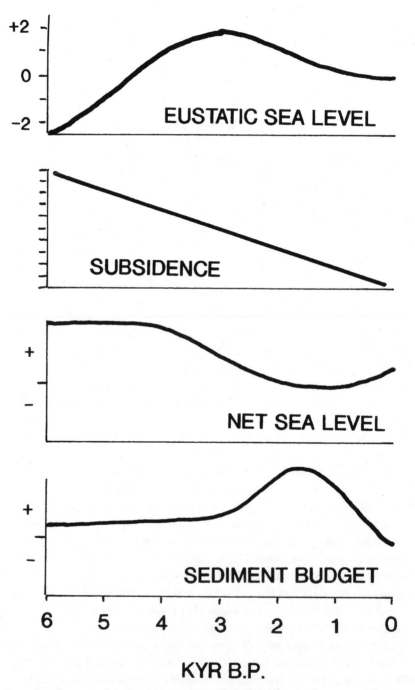

Figure 6.4 Time trends in four key variables affecting the morphodynamics of the To'aga coastal terrace

pation or extinction of taxa. The rank-order structure of exploitation shows emphasis on a small number of taxa from the marine environment. Most fish are from inshore-reef or reef-edge habitats, taken with a variety of strategies, including netting, spearing, poisoning, and angling. When Diodontidae (42.1%) is removed from the analysis (it may bias estimations because identifiable parts are robust and distinctive), fishes from the Acanthuridae (18.2%), Serranidae (17.7%), and Holocentridae (13.7%) families are most common (Nagaoka 1993:211). Other fishes, represented in lower rank-order abundances, include Scaridae (parrot fish), Carangidae (jacks), Labridae (wrasses), Lutjanidae (snappers), Muraenidae (moray eels), Balistidae (triggerfish), and Ostraciidae (boxfish). These, like the high-ranked taxa, are inshore-reef or reef-edge fishes. Also represented, however, albeit in small numbers, are members of the Scombridae (tunas and mackerels). These fish come from the offshore, pelagic zone and indicate the strategy of trolling.

The invertebrate fauna displays a comparable rank-order structural pattern. More than 165 kg of shellfish remains were recovered in the To'aga excavations, with more than 40 families represented. Nagaoka (1993:197) shows that more than 76% of the identified shell comes from just three families: Turbinidae, Trochidae, and Tridacnidae. The large gastropod *Turbo setosus* dominates the assemblage, ranging from 29 to 40% of the total shellfish based on weight in samples from To'aga.

A relative shift from the exploitation of wild resources—such as bird, turtle, or fish—to a greater reliance on domestic resources—such as pig—has been postulated for certain island sequences (e.g., Dye and Steadman 1990). In the To'aga case, however, no clear quantitative shift is evident (Nagaoka 1993:207). While some wild resources (bird and turtle) are represented early in the sequence, but not later, they never comprised a predominant component of faunal record. Instead, a subsistence economy focused largely on marine resources, at least from the perspective of the faunal record, appears on present evidence to have been stable, in spite of the dramatic landscape changes occurring on Ofu. The stability in marine subsistence, different from many cases documented in the Pacific (e.g., Kirch and Yen 1982), may well reflect the high productivity of a living reef. Unlike other reef environments, the fringing reef of Ofu continued to grow as the island subsided (Kirch 1993b). Continuous coral growth would support a rich, diverse fish and molluscan fauna. The high productivity of such reef environments (Wiens 1962) could apparently sustain continued human predation without a discernible impact on the archaeofaunal record. Thus, we should anticipate variability in subsistence composition, structure, and temporal trajectory, given the unique historical and environmental factors from islands across the Pacific.

While the marine fauna shows no evidence of substantial human impact, the terrestrial avifauna from the To'aga site does reveal significant human-induced changes. The bird bones (139 NISP) from To'aga have been analyzed by David Steadman (1993), whose analysis provides a striking example of human impacts on the biota of Ofu. Among the indigenous, resident species recorded from Ofu, five of 10 seabirds and one of three landbirds are extirpated on the island today. Two bones from a *Megapodius* sp. in the earliest dated deposits at To'aga indicate the presence of megapodes farther east than previously known. This bird—not found in later archaeological contexts at To'aga—must have been exploited to the point of extinction early in the island's prehistory. The majority of bird bones from To'aga (62%) are of some five species of petrels or shearwaters, none of which nests on Ofu today. Indeed, Steadman (1993:226) notes a pattern of systematic butchering among the shearwaters and petrels suggested by both ends of the humerus, ulna, and tibiotarsus being broken off, implying direct predation of these birds for food. Only two of the extirpated species, Audubon's Shearwater (*Puffinus lherminieri*) and the Tahiti Petrel (*Pterodroma rostrata*), are known to nest today in American Samoa. Two of the surviving species (Sterninae sp., *Gallicolumba stairii*) exist on Ofu only in small, threatened populations. As Steadman (1993:226) concludes, "Should these [two] species be lost from Ofu, the proportion of bones of extirpated species at To'aga would increase from 85% to 93%."

The Ofu case thus joins a number of archaeologically and paleontologically documented instances of severe avifaunal depletions on Polynesian islands following prehistoric human colonization (Steadman 1989a; Chapter 4). The reduction in bird populations and species diversity on Ofu and elsewhere can be attributed in large part to habitat alteration—habitat destruction for many birds—as well as direct predation of the avifauna. In many cases, birds offered a valuable, easily procured source of protein and fat, and they were widely hunted in Polynesia (see Chapter 4). Removal of large tracts of native vegetation through the creation of agricultural landscapes would also threaten birds. This sequence is one that appears to have occurred throughout the Pacific—with New Zealand one of the most dramatic examples from the region (Anderson 1989a, 1989b).

Finally, the sequence of terrestrial gastropods recovered from the To'aga deposits also informs us of the nature and degree of human-induced changes on Ofu (Kirch 1993d). These land snails are small (1–3 mm), and their presence in archaeological deposits is due not to human deposition but to their presence as natural components of the depositional environment. Land-snail assemblages thus reflect vegetative conditions in the immediate vicinity at the time of deposition. Analysis of land snails has a comparatively long history in

British and European archaeology (e.g., Evans 1972) but was neglected in Pacific archaeology until the work of Christensen and Kirch (e.g., 1981, 1986). They have shown the value of land-snail studies for documenting local environmental changes (especially in vegetation) and the introduction of exotic biota by prehistoric humans. Exotic introductions are indicated by synanthropic or anthropophilic snails closely associated with human cultigens, gardens, and habitations.

Kirch (1993d: Tables 8.1, 8.2) found that the To'aga land-snail assemblages were dominated, in early deposits, by anthropophilic rather than indigenous or endemic land-snail taxa. In the earliest sample represented, at about 2500 cal B.P., five human-introduced, anthropophilic land-snail taxa (*Assiminea* cf. *nitida, Lamellidea pusilla, Gastrocopta pediculus, Liardetia samoensis,* and *Lamellaxis gracilis*) had already become well established on the coastal terrace. Two indigenous snail taxa, *Pleuropoma* sp. and *Sinployea* sp., are represented, but already in very small quantities. Kirch (1993d:118) also notes a striking trend in the distribution of snail taxa across strata: "The oldest sample contains only one [adventive] species. This species is joined by *Lamellidea pusilla* and *Gastrocopta pediculus* in Sample 2, and by *Lamellaxis gracilis* and *Liardetia samoensis* in Sample 3. Thus, by the time of occupation represented in Layer IIB, ca. 2500 cal B.P., five species had been introduced to, and become established on, the coastal terrace of Ofu Island. Because *Lamellaxis gracilis* is particularly associated with human gardening sites, its appearance by 2500 cal B.P. is important in terms of the economic prehistory of the To'aga site."

In sum, the dominance of anthropophilic land snails supports the view of the To'aga coastal terrace as a largely modified, anthropogenic habitat occupied by gardens and human settlements. This transformation apparently occurred within the first 500 years of occupation at To'aga, even as the land area grew through shoreline progradation that formed the coastal terrace.

Regrettably absent from our paleoenvironmental analyses on Ofu is a sequence of vegetation change based on such evidence as pollen spectra, plant macrofossils, or wood charcoal identifications. Unfortunately, no palynological work has been done on Ofu. Taxonomic identification of wood charcoal from the archaeological deposits is possible, and the technology in the Pacific has developed dramatically in the past few years. We suspect that the vegetation history of the lower elevations of Ofu would parallel changes indicated in the land-snail and bird evidence. Probably early in the colonization of the island a large proportion of coastal land and some upland slopes were converted from native vegetation to gardens. As other Pacific Island sequences attest, Polynesians transported—that is, quickly established economic landscapes—on islands that otherwise had little to offer in the way of plant foods.

CONCLUSION

Dramatic physical and biotic changes occurred on Ofu Island over the course of its nearly 3,000-year human occupation. Some changes, such as the progradation of the To'aga coastal terrace, cannot be attributed to human-induced changes alone. Although people have played a notable part in the physical transformation of the island, natural processes were primarily responsible. Changes in sea level, island subsidence, and related sedimentary budgets have dramatically altered Holocene shorelines. In the past, some researchers in the Pacific have conflated many of these natural and human-induced changes, placing emphasis on either the natural or the human-induced causes at the expense of an integrated perspective. We believe To'aga is instructive in this regard, because it shows that it is critical to tease apart natural and human-induced processes in reconstructing paleoenvironmental change.

We have reviewed the major classes of evidence from To'aga that record dramatic landscape and biotic change, and have reconstructed the significance of the morphodynamic history of the coastal terrace. This history includes geologic processes of eustatic and net sea-level changes, island subsidence, and related sedimentation changes. Geomorphological change on the coastal terrace was also affected by human activities. Slash-and-burn forest clearance of the steep, narrow slopes above To'aga accelerated erosion of terrigenous sediment. Agricultural activities have also indelibly changed the pattern of vegetation on the island, transforming a lowland native flora into an adventive, economic one. The land-snail record reflects these transformations dramatically. The abundance of synanthropic taxa found early in the sequence illustrates how quickly change must have occurred on the Ofu coastal terraces. Similarly, the extirpation and extinction of avifauna indicate the degree of human impact through predation and habitat modification (Steadman 1993). The nonavian vertebrate faunal record, however, suggests relative stability in the structure of subsistence. In contrast, the human population of Ofu had little significant impact on the marine environment and its abundant resources. These natural and human-induced changes, documented for the past three millennia, have transformed Ofu's ecosystem from a natural landscape to an economic (or "transported") landscape where human populations can flourish.

Finally, our work in To'aga, as elsewhere in the Pacific, underscores the importance of interdisciplinary collaboration among archaeologists, geologists, geomorphologists, paleontologists, and other natural scientists. Only through such continued interdisciplinary collaboration will we come to understand the relative contributions of natural and human agency in specific Pacific Island environmental histories.

Note

Research at To'aga was supported by grants from the Historic Preservation Office of the Government of American Samoa, to which we would like to express our appreciation. We also thank the chiefs and people of Ofu Island for their support and gracious hospitality.

7
Coastal Morphogenesis, Climatic Trends, and Cook Islands Prehistory

Melinda S. Allen

Over the past decade, the magnitude of Holocene geomorphic changes on Pacific islands has become increasingly apparent (Hughes et al. 1979; McLean 1980; Spriggs 1981; Kirch and Yen 1982; Kirch 1983, 1988). Long-term tectonic processes, sea-level fluctuations, stochastic cyclical events, and human-induced perturbations have all played their part in reshaping island landscapes. The attendant geomorphic processes had several implications for human settlement, subsistence, and technology. As island shores emerged or subsided, the availability of dry land and reef flats was altered, as were associated biotic resources. Inland areas, too, have been dynamic throughout the Holocene, with recent sea-level regression accelerating valley incision, slope instability, and river sediment loads (Nunn 1991).

Human settlement also contributed to landscape change, in particular the character and configuration of coastal areas. Removal of natural forest cover increased slope erosion and deposition of terrigenous sediments in lowland areas (e.g., Spriggs 1981, 1986; Kelly and Clark 1980; Kirch and Yen 1982). These two processes, aided by sea-level fall and changing climatic conditions (Nunn 1991), have in several documented cases transformed shallow marine bays into marshes (e.g., Kelly and Clark 1980; Hunt and Kirch 1988) and wetlands into alluvial plains (e.g., Spriggs 1981; Kirch and Yen 1982) over the past few thousand years.

Geomorphic conditions also affect the prospects of archaeological recovery on Pacific islands, particularly for occupations dating to the early human-settlement period of the region. This situation is exacerbated by the generally *coastal* location of initial occupations, the most dynamic of all sedimentary contexts (see McLean 1980; Reinecke and Singh 1980). In many cases, the discovery of early settlements has been fortuitous, as with the 3000 B.P. Mulifanua site in Western Samoa (discovered through dredging [Leach and Green 1989; Jennings 1974]) or the To'aga site in American Samoa (initially exposed during landfill excavation [Hunt and Kirch 1988]). Elsewhere, as on Tikopia (Kirch and Yen 1982) and Ta'u in American Samoa (Hunt and Kirch 1988), geologically informed research has identified colonizing occupations in the coastal zone.

This chapter considers how climatic trends, sea-level change, and local sediment budgets have affected human populations on the "almost-atoll" of Aitutaki. Many of the issues raised here have relevance for other Pacific islands, particular small islands and atolls. Drawing on a combination of regional evidence and recent archaeological research in the southern Cook Islands, I provide a model for environmental conditions on Aitutaki Island over the past 3,000 years, and I relate questions of initial human occupation in the southern Cook Islands to issues of geomorphic change. I also consider the effects of climatic trends and human-induced landscape change on biotic resources.

GEOGRAPHIC SETTING

The southern Cook Islands lie on the western boundary of the Eastern Polynesian region (Fig. 7.1). Defined on biological, linguistic, and archaeological criteria, Eastern Polynesia is geographically delimited by a 1,500 km water gap between the Cook Islands and islands of Western Polynesia. This stretch of open water between Tonga and Rarotonga, interrupted only by isolated Niue Island, has been a deterrent to colonization of more eastern islands by landbirds (Steadman 1989a), reptiles (Pregill and Dye 1989), and possibly people. There has been much debate over when human populations settled Eastern Polynesia, with some arguing for settlement continuous with 3000 B.P. Lapita colonizations in the west (Kirch 1986; Irwin 1981; Kirch and Ellison 1994) and others suggesting settlement much later (e.g., Jennings 1979; Spriggs and Anderson 1993).

126　Melinda S. Allen

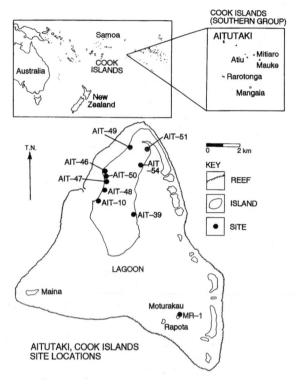

Figure 7.1 Map of the southern Cook Islands and Aitutaki, with primary excavation localities identified

Geologically, the southern Cook Islands are an exceptionally diverse archipelago. Rarotonga is a high volcanic island; Manuae, Palmerston, and Suwarrow are atolls; Mangaia, Atiu, Mitiaro, and Ma'uke are *makatea* islands; Aitutaki is an almost-atoll. As a consequence of this geological variability, the southern Cook Islands also vary significantly in natural resources. Aitutaki, with its volcanic mainland and coral-ringed lagoon, is a blend of atoll and high-island characteristics. The volcanic soils and low rolling hills offered excellent gardening opportunities, while the lagoon and encircling reef provided abundant marine resources. Streams and wetlands suitable for wet taro (*Colocasia esculenta*) cultivation, in contrast, are limited on Aitutaki. Dense, fine-grained basalts appropriate for adz manufacture occur in moderate quantities, both on the Aitutaki mainland and on two volcanic offshore islets. The technologically important pearl shell (*Pinctada margaritifera*) is found in Aitutaki's lagoon and at Manuae Island but is rare elsewhere in the southern Cook Islands. The availability and distribution of these and other resources, however, have varied through time, as attested in the archaeological record (Allen 1992a, 1992b).

Aitutaki's main volcanic island measures 2.75 km from east to west by 7.5 km from north to south, with a total land area of roughly 16 km² (Fig. 7.1). Another 2.2 km² of land is partitioned among the offshore islets (Stoddart 1975a, 1975c). The encircling reef, which encloses a large triangular lagoon, is about 45 km in circumference. The reef closely hugs the main volcanic island at its northern end, flaring away to maximum distance of 8.5 km at the island's southern extremity. Fourteen of the named offshore islets are coralline in composition; Maina is a sand cay, while the other 13 are coral detrital islets. Two volcanic islets, Rapota and Moturakau, are found within the lagoon near the southern end. The main island is characterized by low weathered volcanic hills that slope gently toward the lagoon on the eastern side and drop much more steeply to a sandy coastal plain on the western coast. A craggy exposure of volcanic tuff known as Maungapu at 124 m is the highest point on the mainland (Survey Department 1983).

The volcanic foundation of Aitutaki is largely Tertiary in age, dating to 6.6 to 8.4 million years ago (Turner and Jarrard 1982). Most of the island's basalt formations occurred in this early period. Following a significant hiatus, volcanic activity resumed about 0.7 to 1.9 million years ago but was volumetrically minor (Turner and Jarrard 1982). The Moturakau bedded-tuff breccia formation, which includes blocks of dense fine-grained basalt, dates to this later period of volcanism. This late volcanism could have resulted in local point-loading and subsidence, which I discuss below.

RECENT ARCHAEOLOGICAL STUDIES ON AITUTAKI

Archaeological studies on Aitutaki were initiated by Peter Bellwood (1978b) in the 1970s, and his work identified the 1000 B.P. coastal site at Ureia (AIT-10). David Steadman and I returned to Aitutaki in 1987 (Allen and Steadman 1990) to continue excavations at Ureia as part of Steadman's study of Cook Islands avian biogeography. In 1989, archaeological work expanded to include the whole of the central western coastal plain, the offshore islet of Moturakau, and limited areas elsewhere on the main island (Allen 1992a; Fig. 7.1, this volume). The research goals were to locate occupations related to initial settlement of Aitutaki and secure faunal assemblages from these and later components of the sequence. The central western coast was selected for study based on the presence of a major reef passage (commonly associated with Lapita settlements), the prior identification of a relatively early occupation at

Ureia (SIT-10 [Bellwood 1978b]), and alkaline, as opposed to acidic, soil conditions that would favor faunal preservation.

Augering was carried out at intervals of 50 m or less along six transects placed perpendicular to the coast and spaced 250 m apart. Additionally, 30 shovel trenches 1 by 2 m wide were opened to expose subsurface stratigraphy and to aid in geomorphic interpretations. Based on the results of this initial work, eight localities with cultural occupations were selected for stratigraphic excavations using trowels, fine ($\frac{1}{4}''$ and $\frac{1}{8}''$) mesh screening, three-dimensional plotting of in situ finds, and detailed recording.

In addition to this work on the Aitutaki mainland, the peripheries of 11 of the 15 offshore islets were surveyed, test excavations were carried out on two islets, and 16 m^2 were excavated in two contiguous rockshelters (MR-1) on Moturakau Islet. These shelters, initially tested in 1987, yielded an exceptional record of fish exploitation, shell-fishhook manufacture, and basalt adz production (Allen and Schubel 1990; Allen 1992a, 1992b).

CHRONOLOGY OF HUMAN SETTLEMENT

The antiquity of human settlement in the southern Cook Islands remains an open issue. At present the earliest occupations date to ca. 1000 B.P. and are found at Ureia (site AIT-10) on Aitutaki (Allen 1994; Allen and Steadman 1990) (Beta-25250 and NZ-1252, Table 7.1) and at Tangatatau Rockshelter (site MAN-44) on Mangaia (Kirch et al. 1991). At Ureia the evidence suggests an environment already significantly modified by human activities at 1000 B.P. (Allen 1992a). The land-snail assemblages are dominated by anthropophilic taxa commonly associated with garden areas, including *Lamellaxis gracilis, Lamellidea oblonga,* and *Gastrocopta pediculus.* Wood-charcoal assemblages are species poor and include at least one extirpated taxon (cf. *Planchonella grayana*). The use of economically important species for firewood, such as *Artocarpus altilis* (breadfruit) and *Calophyllum inophyllum* (*tamanu*), suggests that other fuel resources may have been on the decline as early as 1000 B.P. Equally important is Aitutaki's impoverished record of indigenous birds, which, though represented by fewer than 150 specimens, includes both extirpated and extinct species (Allen 1992a; Steadman 1991b). The small size of Aitutaki Island probably contributed to rapid environmental change and significant loss of biodiversity soon after human settlement. Given the extent of human impact recorded at 1000 B.P., however, human colonization *at least* a few hundred years prior is considered quite likely.

The earliest evidence from Aitutaki contrasts in certain respects with that from Tangatatau on Mangaia Island, where a nearly synchronous sequence is

Table 7.1 Aitutaki Radiocarbon Dates

Lab No.	Site	Provenience	Conventional age B.P.	Calibrated age range B.P. (2σ)
Beta-41061[a]	Amuri	Trench 6	3540 ± 90[b]	3599–3159
Beta-27439	AIT-10	Zone J	790 ± 70	912–572
Beta-40759	AIT-10	Zone I	1120 ± 60[b]	730–530
Beta-25250	AIT-10	Zone G	1040 ± 80	1171–763
NZ-1252	AIT-10	Layer 7	969 ± 83[c]	1058–694
Beta-25767	MR-1	Zone K	840 ± 80	948–670

[a] Noncultural association (see text for full discussion)
[b] Shell dates calibrated using delta factor of 45 ± 30 from the Society Islands
[c] From Bellwood 1978; sample not adjusted for $^{13}C/^{12}C$ ratios

found (Kirch et al. 1991; Chapter 8). Unlike the situation on Aitutaki, extinct birds are well represented at 1000 B.P. in the Tangatatau Shelter (Steadman and Kirch 1990). The more impoverished bird fauna on Aitutaki at 1000 B.P. probably reflects the smaller size of this island and the lack of relatively inaccessible zones, like the Mangaian makatea, where native species might have found refuge from human predation. Palynological (Lamont 1990; Ellison 1993) and sedimentological studies (Dawson 1990) from the nearby Lake Tiriara and other localities augment the Mangaian archaeological data and point to a significant human influence in the Mangaian interior region as early as 2500 B.P. (Kirch et al. 1992; Kirch and Ellison 1994). At this time, there is evidence of a decline in native tree species, an increase in weedy ferns, a sudden appearance of charcoal particles, and changes in the sedimentological character of the lake deposits. The complementary and contemporaneous cultural record of the island, however, has not yet been located.

Other human settlements in the southern Cook group date to a few centuries later (e.g., Allen 1994; Bellwood 1978b; Sinoto et al. 1988; Walter 1990). Elsewhere in Eastern Polynesia, specifically the northern Cook Islands (Chikamori 1987) and the Marquesas (Ottino 1985; see also Kirch 1986), occupations dating to between 2300 and 2000 B.P. have been recovered, strengthening the possibility of contemporaneous settlements in the southern Cook group. Based on the Mangaian pollen sequence, Kirch et al. (1991) and Kirch and Ellison (1994) take the view that initial human settlement in the southern Cook Islands most probably dates to as early as 2500 B.P. I have argued elsewhere (Allen 1992a) that Eastern Polynesian colonization continuous with Lapita settlement of Western Polynesia also remains a possibility on other grounds.

CURRENT CLIMATIC CONDITIONS AND HOLOCENE TRENDS

Both short-term climatic patterns and long-term climatic trends have affected the morphogenesis of Aitutaki's coastal areas. Overall, the climate of the past hundred years has been fairly mild and moderately predictable. Seasonal variation in rainfall is the most striking climatic feature, with intense but short periods of rainfall (Thompson 1986). Much of this precipitation is brought by tropical storms, more than half of which originate in the west; the daily trades, in contrast, come from the east. Major storms pass through the islands on average every three to five years (Thompson 1986; Stoddart 1975a; Visher 1925), and they have been primary depositional agents in coastal areas, as is well attested in the archaeological record (Fig. 7.2).

Over the past few thousand years, the southern Cook Islands may have been affected by two, possibly three, second-order climatic trends. Stoddart (1975b:50) raises the possibility of an increase in storminess ca. 2500 to 3000 B.P. based on cemented rubble ramparts found in several Pacific localities. The widespread distribution of conglomerate platforms of roughly comparable dates could reflect a slight transgression, a change in the rate of transgression that aided sediment accumulations, or an increase in storminess (Stoddart 1975b).

A second climatic trend is registered in varied paleoclimatic monitors from the Pacific Basin at large (Nunn 1991). The Little Climatic Optimum was characterized by an average temperature increase of 0.5 to 1.0°C between 1200 and 650 B.P. During this period, much of the tropical Pacific would have been drier than at present, and erosion and landslides on hillslopes may have been more frequent (Nunn 1991:18). An associated reduction in storminess and tropical cyclones, coupled with changing wind conditions, has also been hypothesized (Nunn 1991; Bridgman 1983). Nunn (1991:17) argues that these changes are not well recorded in existing pollen records from Pacific islands because of a variety of ecological and methodological factors. Ice fluctuations in the New Guinea Highlands and New Zealand, however, indicate a temperature change during this period.

The Pacific-wide evidence further suggests that this warmer, drier period may have been followed by a 1.5°C temperature decrease, the Little Ice Age, which had a maximum time range of 650 to 50 B.P. (Nunn 1991). Notably, there is considerable variability in the dating and temperature history of the hypothesized Little Ice Age, and no *direct* evidence for its effects on tropical Pacific islands (Nunn 1991:19–22). Once again the best evidence for this climatic trend comes from the high-altitude areas of New Guinea and New Zealand. There is also limited evidence from the Marquesas (Sabels 1966) and

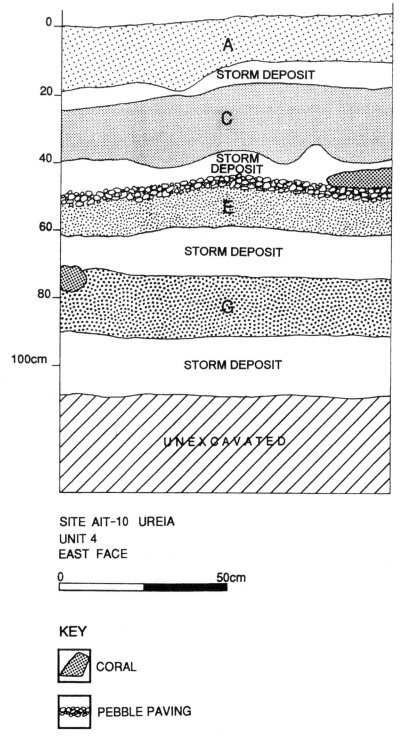

Figure 7.2 Typical stratigraphic sequence from Aitutaki's central western coast

the Galápagos (Sanchez and Kutzbach 1974), as well as other Pacific-rim areas (Nunn 1991), for cooler, wetter conditions. The onset of the Little Ice Age was most likely traumatic, as landforms in equilibrium with warmer, drier conditions were exposed to comparatively wetter, cooler conditions and increased storm activity.

The implications of these climatic trends for Polynesian peoples are varied. Increased storminess at 2500 to 3000 B.P. and 650 to 50 B.P. would have negatively affected the success of open ocean travel (fishing, interisland voyages, and long-distance migrations). The great success of Lapita voyages between 3500 and 2500 B.P. are noteworthy in this respect and suggests that the geomorphic evidence should be evaluated carefully. Other potential changes arising from increased cyclonic activity relate to settlement patterns, with localities less proximate to the immediate coast being favored. Biotic resources would have suffered from frequent storms, with crops being destroyed and marine fauna temporarily depleted. The expansion of certain inshore-fishing techniques and elaboration of carbohydrate pit storage may also correlate with environmental perturbations such as extended periods of storminess. On Tikopia, for example, pit storage facilities appear at about 650 B.P. (Kirch and Yen 1982), and their acceptance and persistence could be tied to environmental conditions.

On Aitutaki, significant storm activity is registered throughout the archaeological sequence, with no clear indications of periods of abatement or intensification. Although settlement locations remained relatively constant, layers of sterile sands indicate periodic disruption of human occupations along the western coast over the past 1,000 years (Fig. 7.2). In the protohistoric to early historic periods, more inland coastal localities were settled, and surface occupations with European goods are found at several localities near the base of the inland slopes. Augering and trenches indicate that this more inland portion of the coastal plain had not been occupied at earlier times. During the protohistoric to early historic periods, localities closer to the coast also continued to be occupied, as, for example, the Ureia site (AIT-10). The relatively recent settlement of inland coastal areas could relate to a variety of factors, including—but not limited to—increased storminess.

Late in the sixteenth century, other kinds of cultural trends may reflect deteriorating climatic conditions. On Aitutaki, the quantity of imported fine-grained basalt declines, as does the number of sources from which materials are being imported. On the southern Cook Island of Ma'uke, Walter (1990) found that imported goods were prominent in the fourteenth and fifteenth centuries at the Anai'o site but lacking from sites dating to later periods. Increased storminess, associated with the Little Ice Age, may have played a role in lowering the frequency of successful interisland voyaging. However, changes in the availability of certain exchange goods could also have been a contributing factor, as I discuss below.

RELATIVE SEA LEVEL: TECTONIC ACTIVITY AND HOLOCENE SEA-LEVEL CHANGE

Over the past 3,000 years, the combined effects of local tectonics and eustatic fluctuations (that is, relative sea level) have had important implications for the possibility of human settlement in the southern Cook Islands. The Holocene tectonic history of Aitutaki is currently not well known and may be rather complex. McNutt and Menard (1978) argue that Pleistocene volcanics on Rarotonga, Aitutaki, and Manuae were responsible for local point-loading and uplift of Ma'uke, Atiu, Mitiaro, and Mangaia. Jarrard and Turner (1979), however, observed that the role of Aitutaki in this uplift may have been relatively minor, given the small scale of Pleistocene volcanism on that island. Thus, local volcanism may have created some point-loading on Aitutaki, while activity on Rarotonga and Manuae may have resulted in some uplift. McNutt and Menard (1979) specifically note that Aitutaki lies on Rarotonga's lithospheric flexural bulge, and that Rarotonga may have lifted it up. The net effect, therefore, would have been relatively little shoreline displacement. This contrasts with a more general model proposed by Kirch and Hunt (1988), who suggest that the failure to identify Lapita-age deposits in Eastern Polynesia may be related to island subsidence.

The nature of Holocene eustatic sea-level fluctuations is more clear-cut than the tectonic evidence and comes from several South Pacific localities, including observations made locally in the southern Cook Islands. Of the main southern Cook Islands, only Aitutaki has not been studied in detail. Twenty-five years ago Schofield (1970) first noted raised reef formations on Rarotonga that he thought reflected a 1 m higher sea-level stand. This formation was dated to 2030 ± 60 B.P. but Schofield cautioned that recrystallization may have resulted in a date too young by several hundred years. Wood and Hay (1970:76) observed raised platforms and notches on Aitutaki, Rarotonga, and Mangaia, which they took to indicate a Holocene shoreline 1 m higher. Limited work by Yonekura et al. (1988) on Aitutaki dated an emerged microatoll to 1530 ± 210 B.P. suggesting a 0.4 to 0.5 m higher stand for this island.

More systematic work on Mangaia by Yonekura et al. (1986, 1988) has identified emerged notches, benches, and microatolls consistent with a maximum sea-level stand of 1.7 m above present levels around 4000 to 3400 B.P. The Mangaian evidence further suggests that sea-level fall from this high was relatively rapid, probably within the succeeding 500-year period. Recent surveys of Ma'uke, Mitiaro, and Atiu have also identified raised notches and benches generally consistent with the foregoing evidence, although the elevations of these features are somewhat greater than in the Mangaian case (Stoddart et al. 1990).

Corroborative evidence for higher Holocene sea-level stands in Eastern Polynesia comes from the Societies (Pirazzoli et al. 1985) and Tuamotus (Montaggioni and Pirazzoli 1984; Pirazzoli and Montaggioni 1986). Based on cementation patterns in emerged coral conglomerates, as well as other reef formations, the evidence suggests a 1 m higher sea-level stand between 5,000 and 1,500 years ago. In contrast to the Mangaian evidence, the French studies suggest that sea-level fall was gradual (Pirazzoli et al. 1988). Athens and Ward (1993) observe that coastal peat marshes appear rather abruptly in the Hawaiian record at approximately 2100 B.P., presumably related to sea-level regression after the mid-Holocene high stand. Research in other South Pacific localities provides evidence that is generally consistent with the magnitude of the regression indicated in the Eastern Polynesian cases, but it offers no further insights as to the timing or rate of sea-level fall (e.g., Clark and Lingle 1979; Bloom 1983; Thom and Roy 1985; Ash 1987).

Radiocarbon determinations from Aitutaki's offshore islets also fit well with the evidence for a higher sea level in the recent past. *Tridacna maxima* valves from clastic conglomerate platforms of Akaiami and Muritapua islets yielded dates of 2040 ± 90 (GaK-3496) and 160 ± 80 (GaK-3500) B.P. (dates uncorrected for $^{13}C/^{12}C$ ratios), respectively (Stoddart 1975b:42–43). These dates indicate the recentness of the materials composing these platforms. They further suggest that some or all of these islets may not have emerged or stabilized until after 2000 B.P., and thus were not available for human occupation until relatively late in prehistory. Nunn (1991:32), commenting more generally on geomorphic processes in coral-reef settings, observes that if vertical reef growth kept pace with sea-level fall, then the emergence of offshore islets would have occurred during the late Holocene regression.

The foregoing suggests that human occupations on Aitutaki's offshore islets should not date to earlier than the Holocene high stand—that is, before 2900 B.P. (based on the Mangaian evidence) to 1500 B.P. (based on French Polynesian studies). Archaeological evidence is thus far more supportive, with a survey of 11 of Aitutaki's 15 offshore islets yielding no indications of cultural activities of significant antiquity. Even in the case of the lengthy Moturakau sequence, the basalt cultural layer dates to only 700 B.P. (Beta-25767, Table 7.1). Notably, the antiquity and buildup of cultural sediments on Moturakau is atypical for Aitutaki's offshore islets and reflects the protected inner-lagoon location, the good preservation context provided by the rockshelters, and the leeward location of those shelters. Although the Aitutaki archaeological evidence is supportive of a post–1500 B.P. islet emergence, the vulnerability of these islets to storms must also be considered. Dates on noncultural sediments from Muritapua Islet ranged from modern to 470 ± 80 B.P., with the oldest date coming from a former beach ridge in the islet's interior (Stoddart 1975b). These dates suggest that in some cases the bulk of islet sediments may be relatively recent accumulations and that the lack of older cultural deposits could reflect erosional processes. Nunn (1991:32), commenting on the region at large, suggests that many offshore islets which emerged during the late Holocene, particularly those in exposed locations, may have since disappeared.

Holocene sea-level fall also affected biotic-resource distributions and abundances. After becoming stable, these offshore islets were undoubtedly colonized by seabirds and turtles, both of which are represented in the Moturakau Islet faunal assemblages. Archaeologically recovered seabirds include several species that probably once nested here, such as the extirpated Tahiti Petrel (*Pterodroma rostrata*), as well as migrants and shorebirds (Steadman 1991b; Allen 1992a). Accumulations of sandy-to-rubbly sediments around and between the islets offered new territory for burrowing bivalves as well. Early in the Moturakau sequence the Box Sunset Shell (*Asaphis violascens*) is well represented, more so than in mainland localities of comparable antiquity (Allen 1992a). Overall, islet stabilization provided new hunting and foraging opportunities and, through temporary campsites, increased accessibility to outer-reef habitats.

On the Aitutaki mainland, sea-level fall extended the coastal flats, providing new lands for settlement and cultivation but reduced inshore marine habitats. Before sea-level fall only a narrow coastal flat may have been available for settlement, and even this was not continuous along the entire coast. Inland from the Ureia marsh the volcanic slope rises steeply, and only a narrow shelf several meters above the marsh is habitable today. In contrast, inland from the Mataki site (AIT-48) to the north, a portion of the coastal plain has been stable for some time (Allen 1992a, 1994). A shovel trench approximately 200 m inland from the present coast uncovered a series of weakly developed A-horizon paleosols. Unweathered shell from the storm deposit immediately underlying the lowest paleosol at 180 cm below the surface dated to about 3300 B.P. (Beta-41061). Even allowing for a few hundred years between the storm event that deposited the shell and soil development, the date indicates that before the late Holocene regression at least a portion of Aitutaki's western coastal plain was inhabitable. Overall, however, indications are that most of this western coastal flat is a relatively recent formation that emerged about 1500 to 3000 B.P.

COASTAL SEDIMENTATION AND MORPHOGENESIS

Sediment Sources

In spite of the importance of sea-level fall, the coastal morphology of Pacific islands has perhaps been affected to an even greater degree over the past few thousand years by local sedimentary budgets (Chappell 1982:71). On Aitutaki, the western coastal plain is largely composed of marine-derived carbonate sands. Most of these materials were deposited during high-energy storms that frequently carry sediments some distance inland. Under calmer conditions, there is little coastal sedimentation, as Aitutaki's fringing reef is 0.5 km from the shore and protects the island from normal wave action. Even with storm deposits, the bulk of the sedimentary particles are medium to coarse-grained sands. Despite the small size of the sedimentary particles, fluvial deposition is strongly suggested by graded beds within the primary depositional units.

Terrigenous sedimentation on the western coast has been relatively minor, as evidenced in both the archaeological excavations and cores from the Ureia Marsh (inland of site AIT-10). In the latter, 50 to 100 cm of organic-enriched clays are underlain by coralline sands near the base of the inland slope. At the seaward edge of the marsh, less than 90 m away, these overlying volcanic sediments are only 35 cm thick. Erosion and slumping of the inland slopes appears to have accelerated in the recent past. In a few trenching localities, isolated European materials were found under 1 m or more of volcanic clays near the base of the slope. These trenches also record the interplay of marine and terrestrial deposition through time, with the former generally dominating in the lowest levels. Overall, the evidence suggests relative stability of Aitutaki's western volcanic hillslopes and cliffs in the geologically recent past.

In striking contrast to the western coast, Aitutaki's eastern shores are typified by muddy fine sands (Fig. 7.3). In particular, at Vaitupa on the northeast coast there are extensive tidally exposed mudflats. Augering near the coast at Vaipeka (near site AIT-51) encountered thick volcanic clay deposits of at least 80 cm in depth. Bellwood (1978b:113) reports a stone-lined pool (site AIT-38) in the Vaipae District (central eastern coast), identified by informants but obscured from view by sediments at the time of his survey, suggesting significant shoreline deposition in the recent past. Although both Summerhayes (1971:353) and Stoddart (1975b:38) characterize the majority of the la-

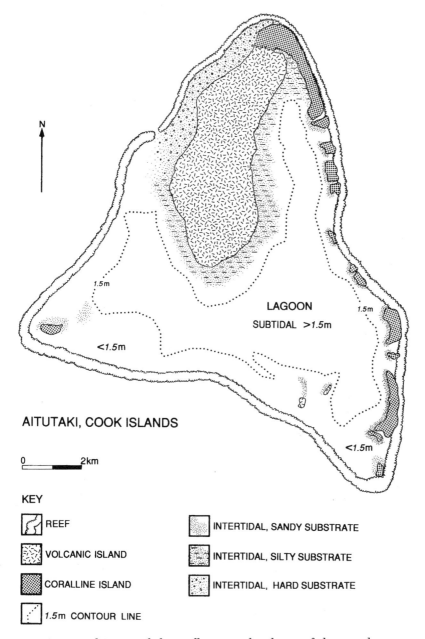

Figure 7.3 Aitutaki's marine habitats, illustrating distribution of silty near-shore areas

goon sediments as calcareous, Summerhayes notes that the deepest part of the lagoon, specifically an elongate trench with a maximum depth of 10.5 m off the east coast of the mainland, is dominated by muddy fine sands. Overall, Aitutaki's lagoon is shallow compared to other atoll environments, with more than 75% of it being less than 4.5 m deep (Summerhayes 1971). This shallowness could in part be related to limited exchange with the open sea and a restricted export of sediments. Nunn (1991:33) suggests that once Holocene sea-levels stabilized, lateral erosion once again became an important process. In areas where small lagoons were effectively enclosed by reefs, this time may have marked the beginning of lagoon infilling, largely from the introduction and reworking of reef-derived sediments. In the case of Aitutaki, the presence of a large volcanic mass may also have affected sedimentation rates. In this respect the effect of prehistoric agricultural activities on local sedimentary regimes warrants more detailed study. The issue of lagoon sedimentation is also important in terms of critical lagoon faunal resources, particularly pearl shell (see below).

Processes of Deposition

Two sedimentary processes have shaped Aitutaki's western coastal plain: aggradation and progradation. In both cases, the primary mode of deposition was through wave action associated with tropical storms. Excavation evidence from four localities indicates that sediment accumulation varied from 11 to 26 cm per century (Table 7.2). The rate and volume of sediment accumulation has been greatest at the Aretai locality (site AIT-49), followed by Hosea (site AIT-50), and Ureia (site AIT-10) (see Fig. 7.1). The data suggest a general trend of increasing sedimentation as one moves north. Based on the Ureia evidence, it also appears that the rate of sediment accumulation has varied over the past 1,000 years, being more rapid early and slowing down in the past few centuries (Fig. 7.4).

Table 7.2 *Rate of Aggradation: Central Western Coast, Aitutaki Island*

Locality (N. to S.)	Rate of sediment accumulation (cm/100 yr)	Distance from high-tide mark (m)
Aretai (AIT-49)	26	110
Hosea (AIT-50)	14	120
Poana (AIT-47)	21	60
Ureia (AIT-10)	11	90

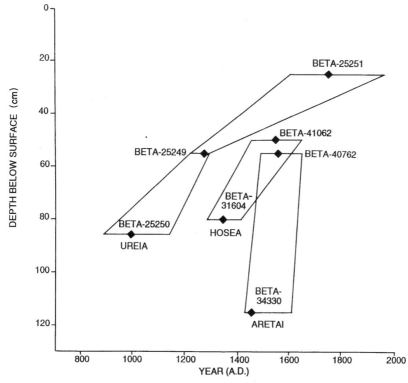

Figure 7.4 Rates of sediment accumulation, central western coast of Aitutaki. Radiocarbon dates plotted at one sigma (boxes), with intercepts (diamonds) used to calculate sedimentation rates. For dates with multiple intercepts, the midpoint of the most probable calibrated age range were used.

Progradation was also a significant process in the past. Most important for human settlement was development of a beach barrier (after Reinecke and Singh 1980:343) along the central western coast (Fig. 7.5). Initiated through progradation, this barrier separated low-lying inland areas from the reef flat, creating an environment favorable to wetland cultivation in the process. The small coastal marsh at Ureia, for example, is still used for traditional wet taro cultivation today. Based on current evidence, the direction of barrier growth is uncertain and may have been discontinuous.

The Ureia (AIT-10) data indicate that the central western barrier began to develop sometime before 1200 B.P. Basal sands at a depth of 2 m are very fine-grained and suggestive of a protected lagoonal reef flat much like that fronting the site today. These fine sands are overlain by a series of storm layers com-

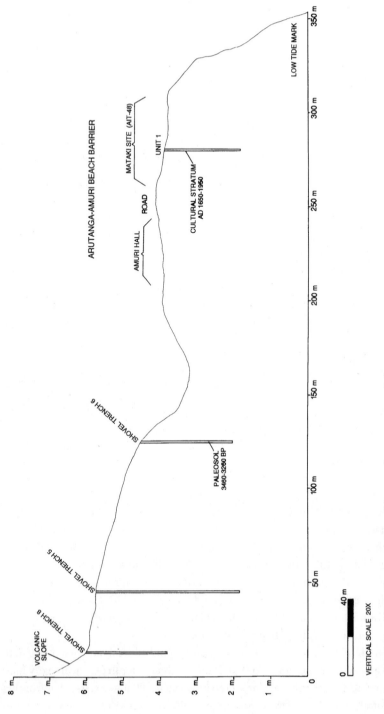

Figure 7.5 Typical profile across Aitutaki's central western coastal plain

posed of coarser sands, with the lowermost containing redeposited cultural materials; it is unclear whether these cultural sediments represent a disturbed in situ occupation or materials transported from some other locality, but the latter is considered more likely. Radiocarbon dating of these secondary cultural materials suggests that they are only slightly older than 1000 B.P., with two dates (one shell, Beta-40759; one charcoal, Beta-27439; Table 7.1) overlapping at two sigma with those from the overlying in situ 1000 B.P. occupation (see also Allen 1994).

By about 1000 B.P. the central western barrier had stabilized, as indicated by the lowermost in situ occupation at Ureia (site AIT-10). The development of this barrier was most likely triggered by the Holocene sea-level fall (cf. Yonekura et al. 1988:180 for Mangaia). Although the hypothesized timing of barrier initiation (roughly 2000 to 1200 B.P.) is not consistent with the Mangaian evidence for sea-level fall (3400 to 2900 B.P.; Yonekura et al. 1988), it does fit comfortably with the evidence from French Polynesia (post-1500 B.P.; Pirazzoli et al. 1988) and with the dated emerged microatoll from Aitutaki (Yonekura et al. 1988). Once the barrier had stabilized, growth was primarily upward, building episodically with successive storm events.

Development of the central western beach barrier had several important implications for human populations. As I noted earlier, marshy areas suitable for taro cultivation were formed and continue to be used today. The barrier itself represented new dry land that became a focus of human settlement over the next 1,000 years. As this dry land replaced reef flats, however, it probably affected the abundance of local faunal resources. These effects were probably most dramatic during the initial period of growth (e.g., ca. 2000 to 1200 B.P.), a period for which there is at present no archaeological record.

Following initiation of the barrier, progradation decreased in importance. Although the earliest occupation layer at the Ureia site (zone G) is not represented in the most seaward excavation unit (Allen 1992a), it is difficult to determine whether this lateral variation reflects shoreline growth or small-scale changes in preferences for settlement location based on other factors. Elsewhere on this western coast the excavations revealed no clear-cut age progression in the lateral location of cultural occupations, as would be expected in a prograding shoreline.

IMPLICATIONS FOR ARCHAEOLOGICAL RECOVERY

The combined effects of sea-level fall, progradation, and aggradation have also affected recovery of Aitutaki's archaeological record (Fig. 7.6). Most notably, the geological history of the area has created a bias against recovery of

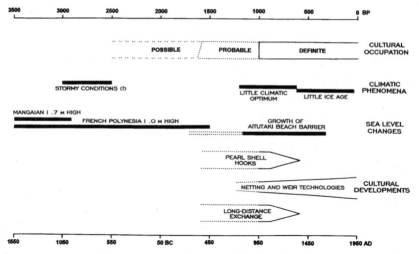

Figure 7.6 Summary of geomorphic, climatic, and cultural processes

early cultural materials from the surface. Field studies in 1989 specifically took these factors into account and used a combination of inland-to-coast auger transects and test trenches to identify the location of subsurface deposits across Aitutaki's western coastal flat. The geomorphic model developed from regional evidence, in concert with the field evidence, suggested that the most likely area of older occupations would be inland of the near-shore barrier and its associated trough. Even if subsidence was a factor, cultural deposits were expected to be at least somewhat inland from the present coast, given the evidence for progradation. Nevertheless, no cultural occupations earlier than ca. 1000 B.P. were found in this zone or anywhere elsewhere along this western coast.

Why have no pre-1000 B.P. occupations been found on Aitutaki? The palynological work on Mangaia strongly suggests a human presence in the southern Cook Islands by 2500 B.P. (Kirch et al. 1991; Ellison 1994; Kirch and Ellison 1994), whereas the 3300 B.P. paleosol identified on Aitutaki indicates that portions of the coastal plain had stabilized and could be colonized by this early date. Nevertheless, only remnants of what was probably a narrow coastal flat before the Holocene sea-level fall may remain, making discovery of early human occupations difficult. Throughout the early Holocene, when sea levels were higher than at present, growth of offshore reefs may not have been able to keep pace with the rise of sea level (cf. Nunn 1991). In consequence, coastal areas may have been more susceptible to erosion and sediment reworking. This situation would have been exacerbated during periods of greater cyclonic activity (e.g., possibly 3000 to 2500 B.P.), a period that notably coincides with the earliest expected settlement of Eastern Polynesia. Thus, the early record of human settlement on Aitutaki could be largely eroded, and isolated coastal remnants, now buried by recent sediments, may be particularly difficult to isolate and recover archaeologically.

Following the Holocene sea-level fall, sedimentary conditions changed dramatically. Sea-level fall drained shallow reef flats, and offshore reefs immediately provided a protective front. With sea-level stabilization, the central western beach barrier began to develop, and a low marshy area formed along the inland side of this barrier. As inland areas were increasingly protected from stormy high surf, depositional rather than erosional processes came to dominate at the shore. These storm deposits not only buried older cultural deposits but also reduced the possibilities of subsequent postdepositional disturbances.

EFFECTS ON BIOTIC RESOURCES

Although these large-scale natural processes undoubtedly had an impact on Aitutaki's marine and terrestrial fauna, archaeological deposits from the period of greatest change, about 3000 to 1200 B.P., are lacking. One of the most dramatic changes seen in the post–1000 B.P. archaeological record from Aitutaki is in fishing technology and associated fauna (Allen 1992b). Between 1000 and 550 B.P., an abundance of pearl-shell fishhooks, in a variety of sizes and forms, indicates the importance of angling technologies. As might be expected, predatory fishes, such as snappers (Lutjanidae) and groupers (Serranidae), are well represented in the corresponding faunal assemblages. Beginning about 550 to 450 B.P., *Turbo* hooks become more common, while fishhooks decline in abundance. From the Moturakau shelters there is also limited evidence for experimentation with other shell species, suggesting that favored raw materials are increasingly unavailable. Shortly before contact, shell hooks are apparently abandoned, presumably replaced by an expanding repertoire of nets and weirs that dominate ethnographically described fishing technologies (Buck 1927, 1944).

At the fourteenth- to fifteenth-century site of Anai'o on Ma'uke, pearl shell is used not only for fishhooks but also for other tools (Walter 1990). Late prehistoric sites on Ma'uke, however, lack pearl-shell artifacts and manufacturing debris. Given that pearl shell is exotic to Ma'uke, Walter (1990) suggests that its disappearance relates to a cessation of importation and the breakdown of exchange networks across the southern Cook archipelago. On Mangaia, a similar pattern of decline is seen, with pearl-shell fishhooks being replaced by hooks made of *Turbo* (Kirch et al. 1991).

Although declines in pearl shell elsewhere in the southern Cook Islands could relate to the breakdown of exchange relations and to patterns of increasing isolation, this explanation is insufficient for Aitutaki, where the species is locally available. On Aitutaki, the demise of pearl shell may be related to changing environmental conditions, particularly the character of the lagoon. Intes (1982) notes that pearl-shell productivity is closely tied to lagoon circulation, water quality, and depth. Deep-water habitats in particular are cited as important for protecting pearl-shell stocks against over-exploitation by predators (Intes 1982:38). Several other conditions that might affect pearl-shell abundance are substrate quality and water salinity, turbidity, and temperature, all of which could be altered by sedimentation. As I noted earlier, Nunn (1991) argues that lagoon infilling may be a natural process that accelerated after sea-level stabilization in the Holocene. Although natural lagoon infilling could be relevant here, the lag between sea-level fall and the declines in pearl shell recorded archaeologically is at least 1,000 years and possibly longer. Another factor that warrants further consideration is the role of human populations in increasing terrigenous sedimentation in the lagoon. The extensive mudflats along Aitutaki's eastern shores indicate that these processes are operative today, and that terrigenous sediments are affecting what would have been critical pearl-shell habitat, the deeper northern end of the lagoon. The particularly shallow lagoon (relative to atolls generally) may relate to Aitutaki's remnant volcanic mainland and the susceptibility of this terrain to erosion arising from agricultural activities. The onset of the Little Ice Age at 650 B.P. would have exacerbated any erosional conditions initiated by agricultural activities, and the coinciding of this climatic phenomenon with the demise of pearl-shell resources is notable. These trends, together with intensive patterns of exploitation, may ultimately have led to declines in a once abundant resource, with ramifications for patterns of technology, subsistence, and exchange throughout the southern Cook Islands.

CONCLUSION

Over the past few decades, archaeologists have increasingly recognized how different the modern environments of Pacific islands are from those of the prehistoric past. The morphogenesis of coastal regions within the time frame of human occupations has been particularly dynamic. Over the past 3,000 years, three processes have significantly affected the island of Aitutaki: sea-level fall, variable sedimentation regimes, and possibly second-order climatic trends. The period of greatest change probably dates to between 3000 and 1200 B.P., following a sea-level fall of 1 to 2 m. At this time offshore islets stabilized, coastal flats emerged, and marshy areas suitable for taro production developed. The lack of pre–1000 B.P. occupations on Aitutaki is argued to be a product of sedimentary conditions rather than a reflection of the antiquity of human settlement. Prior to sea-level fall, the Aitutaki coast was more prone to erosional processes, and only in the past 1,500 years or so have depositional processes come to dominate.

Since 1000 B.P. Aitutaki's coastal regions have been affected primarily by episodic storm accumulations of sediment on the western coast and, on the eastern shores, by gradual sedimentation from terrigenous sources. Although not widely accepted at this time, the possible impacts of two second-order climatic trends, the Little Climatic Optimum and the Little Ice Age, are nonetheless discussed. Increased storminess associated with the latter is not directly registered in the sedimentary history of Aitutaki's western coast, but it may be reflected in other archipelago-wide cultural patterns, such as the cessation of interisland voyaging and the decline of pearl shell between 550 and 200 B.P.

The decline of pearl-shell resources is argued to be tied to changing lagoon conditions, most notably sedimentation. Although the depositional history of the lagoon is poorly understood at present, increased sedimentation would have affected an array of ecological conditions to which pearl shell is sensitive. It would be simplistic, however, to argue that pearl-shell declines alone are responsible for the demise of angling technologies, especially given that angling continues elsewhere, as on Mangaia. The replacement of angling by other techniques is only one of many possible sequences of change. The character of Aitutaki's marine environment probably acted as an important selective condition in facilitating the rise of nets and weirs, technologies that were present but less important in earlier periods. The loss of timber species necessary for larger fishing vessels and a restriction of faunal catchment areas stemming from intensified agricultural activities may also have been contributing conditions (Allen 1992a, 1992b).

The model of environmental change developed here for Aitutaki also has implications for Pacific islands generally. Although a human presence is apparently registered in the southern Cook Islands by 2500 B.P. (Kirch et al. 1991; Ellison 1994; Kirch and Ellison 1994), direct archaeological evidence is lacking. The geomorphic evidence reviewed here raises the possibility that pre–1500 B.P. coastlines may have been quite active; as a corollary, earlier occupations will be difficult to recover. On Eastern Polynesia's smaller islands, particularly the Cooks, Tuamotus, Australs, and Gambiers, locating colonizing settlements may require not only geomorphically informed survey strategies (Kirch 1986) but also an element of luck.

Note

The gracious hospitality of the Government of the Cook Islands and the people of Aitutaki made this research possible. I am especially grateful to Tony Utanga, Makiuti Tongia, the Teaukura Family, Tuakura Tuakura, Bobby Bishop, Mikaela Tumu, and Maki Toko for their support and friendship. Funding for the 1987 fieldwork was provided by the National Science Foundation (Grant BSR-8607535) to David Steadman of the New York State Museum. The 1989 studies were funded by National Science Foundation Grant BNS-8822768 and Wenner-Gren Foundation for Anthropological Research Grant 5079 to M. S. Allen. The Cook Island National Museum also aided me with internal transportation costs. I thank Lisa Nagaoka, Sue Schubel, and Algernon Allen for able and enthusiastic field and laboratory assistance and Brad Evans for the illustrations.

8
Changing Landscapes and Sociopolitical Evolution in Mangaia, Central Polynesia

Patrick V. Kirch

The traditional island societies of Polynesia, though all classifiable as "chiefdoms," exhibited a most remarkable range of sociopolitical variation. Simple Polynesian chiefdoms, such as that of the anthropologically renowned Tikopia (Firth 1936), encompassed populations of 1,000 or 2,000 thousand people, for whom the bonds of kinship were stronger than the political aspirations of chiefs. At the opposite end of the spectrum was Hawaii, whose 300,000 or more people were organized by a political apparatus that Marshall Sahlins (1968, 1972), among others, has compared to an "archaic state." Between these extremes lies a fascinating range of sociopolitical structures, making Polynesia a kind of "laboratory" for studies of the transformation of human societies.

There have been several attempts to explain this range of sociopolitical variability, perhaps the most famous being Sahlin's *Social Stratification in Polynesia* (1958) and Irving Goldman's *Ancient Polynesian Society* (1970). Both of these works offer important insights, but from contemporary theoretical perspectives they are flawed by their adherence to an outdated unilinear evolutionary model. Although Sahlins and Goldman isolated quite different evolutionary dynamics, both saw the region's spectrum of societies as evolutionary stages along a single continuum from simple to complex.

These and similar efforts at synthetic explanation of the Polynesian situation were all undertaken before the development of modern archaeological work in the region, without the advantages of truly diachronic information. Among the significant advances that some three decades of archaeological investigations in various Polynesian islands have now produced is the ability to track sociopolitical transformations in consort with other changes—in population, environment and resources, agricultural production, and external exchange linkages, to name a few (Kirch 1984; Kirch and Green 1987). Rather than viewing all Polynesian societies in the period of European contact as points on a single evolutionary path, prehistorians have come to realize that the region's societies were not all on the same trajectory toward increasing hierarchy and stratification.

Considerable attention has recently been directed toward several societies that Sahlins and Goldman placed in the middle range of their classifications; these include the Marquesas, Easter Island, Mangareva, and Mangaia. Goldman (1970) called these "open" societies and pointed to their more "fluid" sociopolitical structures, in which achieved-status groups (warriors and priests, in particular) vied openly with the hereditary chiefs for control of land, resources, and power. Thomas has recently suggested that these "open" societies—far from being on a path toward increasing stratification—were "the products of a process of devolution, of the erosion of the unity of chiefship" (1990:182). This is a view seconded by Friedman (1981, 1982:191), who has furthermore pointed to some common characteristics of these societies: relatively high population density, poor ecological conditions, and intense warfare.

The prehistoric cultural sequences of the Marquesas and of Easter Island have been investigated by a number of archaeologists. The archaeological record for several other "open" societies, however, remains virtually nil. It was with the goal of expanding our knowledge of these fascinating mid-range, or "open," Polynesian societies that in 1989 David Steadman and I began an interdisciplinary project focused on the island of Mangaia in the southern Cook Islands of central Polynesia (Steadman and Kirch 1990; Kirch, Flenley, and Steadman 1991; Kirch et al. 1992 and 1995; Ellison 1994; Kirch and Ellison 1994). A rich corpus of nineteenth-century missionary documents (e.g., Williams 1838; Gill 1856, 1894) and an ethnographic study by Buck (1934) provided much material on the contact-era society, but virtually nothing was known of Mangaia's archaeology or prehistory. It was therefore necessary to establish a basic cultural history for the island, from initial colonization to

European contact. This was only preliminary to our larger objectives, however, which are to track the major impacts that colonizing humans had on this isolated island ecosystem and, reciprocally, to understand the dynamic ecological contexts within which sociopolitical transformation occurred.

Two field seasons have now been completed, and preliminary results are emerging, although much analytical work is continuing. The project is a collaborative effort involving 14 scholars at five institutions and encompassing the fields of geomorphology, palynology, and paleontology, as well as such specialties as zooarchaeology, paleoethnobotany, and geoarchaeology within archaeology.

ENVIRONMENTAL AND CULTURAL BACKGROUND

Mangaia (157° 55' E, 21° 55' S.) is the most southerly of the main Cook Islands group, which includes Rarotonga, Mauke, Atiu, and Aitutaki. With a land area of 52 km², Mangaia would normally be classified as a "high" volcanic island. Because of its geological history, however, the island displays a rather unusual, highly zoned, environment, with major consequences both for the prehistoric human population and for archaeological investigation. I refer to the ring of upraised or elevated reef limestone, called by the Mangaians *te makatea*, which encircles the island like a fortress wall. The concentric zonation of Mangaia is a function of its geological history (Marshall 1927; Stoddart, Spencer, and Scoffin 1985). A volcanic cone emerged from the sea floor about 20 million years ago and has become highly weathered, with a radial stream-drainage pattern. In the late Pleistocene, regional tectonics caused the emergence of a series of reef platforms encircling the volcanic cone, creating the makatea. Providing refuge caves and burial places for the prehistoric Mangaians, the makatea also serves as a sediment trap for the encircled volcanic drainages, providing almost ideal circumstances for the preservation of a Holocene sedimentary record.

Although some of the makatea is vegetated and has a shallow, cultivable soil interspersed between outcrops, much of it is barren and of little use. Likewise, the central volcanic cone consists of deeply weathered lateritic saprolite, which supports little more than a pyrophytic fernland vegetation. Thus the most productive zones from the viewpoint of horticultural Polynesians were the valley bottoms and slopes with their rich colluvial and alluvial soils. These valley areas provided the bulk of Mangaian subsistence, particularly through the intensive irrigated cultivation of taro (*Colocasia esculenta*).

Like other Polynesians, the Mangaians divided their island into a series of radial territories (Fig. 8.1), each cutting across the ecological grain of the is-

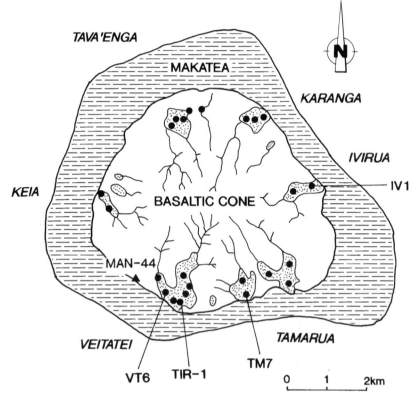

Figure 8.1 Mangaia Island, showing the traditional districts and locations of archaeological sites

land. Each pie-shaped slice included volcanic slopes, stream drainages with productive taro swamps, and the karst makatea. Each of the six districts—Keia, Tava'enga, Karanga, Ivirua, Tamarua, and Veitatei—was under the control of a chief (*pava*), while the island as a whole was dominated by a paramount (called *Te Mangaia*) whose power had to be legitimated by success in war and by the offering of a human sacrifice.

As in other parts of Polynesia, the Mangaians developed a highly sophisticated system of irrigated taro cultivation (B. Allen 1971), transforming alluvial valley bottoms into reticulate grids of permanently cultivated pondfields, watered by streams radiating from the central hill. These systems have been described as a form of "landesque capital intensification" (the term is from Blaikie and Brookfield 1987a), with permanent modification of the landscape into a topography of dams, canals, and fields. Ethnographic and agronomic studies of Polynesian taro irrigation reveal the very high productivity of these agro-ecosystems: yields of up to 50 tons/ha/yr supporting surplus levels of as much as 50% over labor input (Kirch and Sahlins 1992). Although current production levels are down considerably over those of the contact era, Mangaians still depend to a large extent on their taro fields for basic subsistence needs, and about 36 ha were being intensively cultivated in 1989. In late prehistory, these valley taro systems were the principal spoils of intertribal war and territorial conquest, documented in a wealth of oral traditions. Buck wrote of the battles over control of the irrigation systems as follows: "When the peace drum sounded and the fugitives emerged from their refuges, the victors gave them a feast (*taperu kai*) in public recognition of their safety. The conquered, however, had lost their holdings in the rich taro lands of the *puna* districts. They were awarded shares in the narrow upper ends of the valleys where they could still grow a certain amount of taro. They were also given land in the *makatea*, where sweet potatoes and paper mulberry could be produced. . . . The conquerors, in annexing food lands from the conquered, took subdistricts which were remote from their original lands. The food lands and the rule over districts and subdistricts were the important spoils of victory" (1934:107). The importance of the prime agricultural lands was heightened by the fact that Mangaia is surrounded by only a narrow fringing reef platform and lacks a lagoon. The island's marine resources are thus depauperate in comparison with most other tropical Pacific islands.

Examination of the island's major ecological zones (Table 8.1) reveals that although Mangaia has a total land area of 52 km^2, no less than 44% of it consists of either barren karst makatea or degraded fernlands. Of the arable lands, about 36% consists of those portions of the makatea on which some cultivation is possible, while 18% consists of valley slopes and interiors. The prime alluvial basins that could be converted into the most productive taro irrigation systems make up a mere 1.3 km^2, or only 2%, of the total land area, leaving little doubt why the Mangaian warriors vied so bitterly for control of these resources.

The island's current population of 1,235 is well below that reported by the first missionary census, estimated to be about 3,000 "souls" (Williams 1838). A population of 3,000 would have been equivalent to a density of about 150 persons per km of arable land. Yet, it appears likely that the population had already declined significantly before missionization, due to the effects of introduced disease. The maximum population of Mangaia in prehistory is a problem for archaeological determination, but whatever the ultimate conclusion, it is clear that the sheer pressure of human numbers on limited resources is one key factor that must be built into any model of sociopolitical change on the island.

Table 8.1 Mangaian Land-Use Zones

ZONE	AREA (km^2)	%
1. Makatea coastal strip	29.0	56
1A. Barren makatea	(10.2)	(20)
1B. Vegetated makatea	(18.8)	(36)
2. Lower volcanic slopes and valley interiors	9.2	18
3. Irrigated alluvial basins	1.3	2
4. Degraded fernlands	12.5	24
TOTAL LAND AREA	52.0	100

THE MANGAIA PROJECT: RECONSTRUCTING ENVIRONMENTAL CHANGE

The Mangaia Project began in the summer of 1989 with a pilot phase, in which our basic objectives were to establish the outlines of a cultural sequence and to track the course of mid-to-late Holocene environmental changes using archaeological, palynological, and paleontological approaches in a coordinated effort. We focused on the district of Veitatei as a representative study area, partly because of the presence of Lake Tiriara, which presented the best opportunities for stratigraphic coring and pollen analysis. In 1991, we expanded our research into the adjacent districts of Keia and Tamarua. At the close of the 1991 season, we had discovered and surveyed approximately 100 archaeological sites (Fig. 8.1) and had tested or excavated seven rockshelters and one large open habitation site complex (Kirch et al. 1992 and 1995).

The Veitatei district incorporates a typical mosaic of topographic and environmental features, including the small lake, the inner makatea cliff with its rockshelters, the irrigation systems, and the adjacent ridge spurs where open habitation sites are situated. The valley bottoms—where the alluvium was developed into intensive taro irrigation—terminate against the steep inner makatea cliff. These valleys thus form basins that trap sediments eroded from the interior volcanic hills, providing an excellent depositional record. In contrast, the interior slopes are highly eroded and degraded, covered with pyrophytic *Dicranopteris linearis* fernlands. Surveys revealed that these slopes are entirely devoid of archaeological features.

A significant component of our interdisciplinary research strategy was the recovery and analysis of stratigraphic cores that could unlock the Holocene environmental record contained within the valley-bottom sediments. We had hypothesized that the degraded condition of the island's interior was not a natural ecological climax but the result of human actions. This hypothesis had to be tested against the stratigraphic record, so in 1989 we began coring Lake Tiriara. John Flenley of Massey University in New Zealand, assisted by his students Francis Lamont and Stewart Dawson, carried out the extraction of three cores for geochemical and pollen analysis. The sedimentary record was deeper and better preserved than we had imagined. Core TIR-1 penetrated more than 15 m, and coring had to be terminated because at that depth we were in danger of losing the coring apparatus. The radiocarbon age-depth curve indicates a relatively constant rate of deposition over the past 7,000 years, providing an excellent stratigraphic record for the mid-to-late Holocene (Kirch et al. 1992), thus spanning the period before and after human colonization of the island. Pollen analysis by Lamont (1990) revealed a strikingly clear signal of human disturbance and anthropogenic vegetation change in the Veitatei catchment after about 1600 B.P. This is marked by a major decline in trees, a concomitant increase in ferns (especially *Dicranopteris*), and the apparent extinction of some taxa (e.g., *Weinmannia*). The geochemistry of the TIR-1 core documents indicated changes in the depositional basin, such as a decline in organic matter and major increases in oxides and free iron, all signals of the exposure of the central volcanic cone (Dawson 1990). Figure 8.2 summarizes the major changes revealed by the Lake Tiriara core.

It was obviously critical to test the Lake Tiriara sequence in other drainage basins around the island, to ascertain whether the TIR-1 core displayed a widespread pattern of change. In 1991 Berkeley geographer Joanna Ellison expanded the coring work around the island (Ellison 1994). Twenty-one cores were taken, with samples from each major depositional basin. The stratigraphic records reveal a consistent pattern of an earlier mid-Holocene phase, when the valley bottoms were covered by shallow lakes and peat deposition. Later, the depositional regime changed to one of rapid filling in with clay sediments. Radiocarbon dates on key cores reveal that the change in depositional regimes consistently occurred within the past 2,400 years—that is, following the Polynesian colonization of the island (Ellison 1994; Kirch and Ellison 1994). Ellison also counted the frequency of carbonized particles in her core samples, revealing that there was no evidence of burning on the island before the period of Polynesian occupation, commencing at about 2400 B.P. Thereafter, carbonized particles are present with high frequency, suggesting the use of fire as a means of forest clearance, probably in connection with shifting cultivation (Kirch and Ellison 1994).

In short, the combined palynological, geochemical, and stratigraphic analyses of the Mangaian sedimentary record leaves no doubt that a dramatic

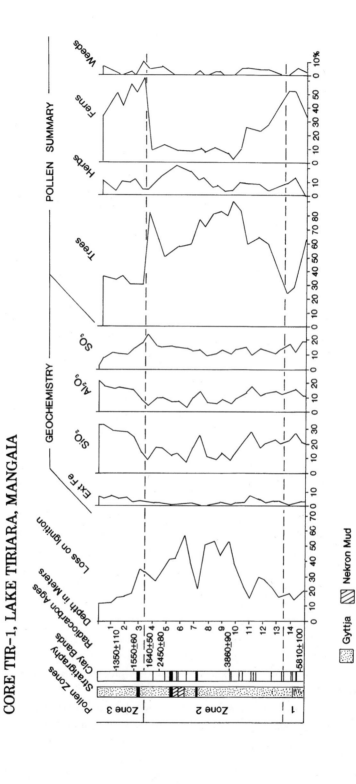

Figure 8.2 Stratigraphic, radiometric, geochemical, and palynological summary of the TIR-1 core, Lake Tiriara

series of environmental changes were initiated by the Polynesian occupation of the island. These included the deforestation of the volcanic interior, exposing the thin soil cover to erosion and leading to rapid filling in of the valley bottoms. The interior volcanic cone was in time converted into a degraded *Dicranopteris* fernland with scattered *Pandanus* and *Casuarina* trees, essentially worthless for agriculture. In contrast, the valley bottoms were greatly enhanced as potential agricultural environments through the deposition of fertile alluvium. Thus, to a very large degree, the prehistoric Mangaians were the creators of their island landscape.

THE MANGAIAN ARCHAEOLOGICAL RECORD

I turn now to some key aspects of our archaeological investigations on the island. Early in the 1989 season, we were fortunate to discover a large rockshelter site, named Tangatatau, situated within the Veitatei study area. The shelter lies at the base of the high inner makatea escarpment and offers more than 200 m^2 of floor area within the drip line (Fig. 8.3). The site lies within a half-hour walk to Lake Tiriara and the alluvial valley bottom, as well as to the lower volcanic slopes. Test excavations were carried out to determine whether a well-stratified sequence would be contained within the shelter. The 1989 trial trench revealed a complex stratigraphic sequence spanning 600 to 700 years and containing a rich faunal, botanical, and artifactual record (Figure 8.4). In 1991, we resumed excavations at Tangatatau, ultimately opening up a total area of 29 m^2 (Kirch et al., 1995). Thirty radiocarbon dates indicate that the shelter was used continuously from about the year 1000 until about 1650, just before European contact (Kirch, Flenley, and Steadman 1991; Kirch et al. 1992, in press).

The rich artifact assemblage from Tangatatau reveals a sequence of stylistic change, the first documented for the Cook Islands. Basalt adzes, for example, include nontanged, early Eastern Polynesian forms in the early deposits and classic, tanged forms in the later strata. Of particular interest is the sequence in shell fishhooks, of which more than 200 specimens have now been recovered. The early hooks were nearly all manufactured from the shell of the pearl oyster, *Pinctada margaritifera*. Significantly, pearl shell does not occur naturally on Mangaia, as the island lacks a lagoon. Thus the raw material for these hooks had to be imported from other islands several hundred kilometers away, implying long-distance exchange. After about 1300, this importation of pearl shell ceased, and hooks were manufactured of the locally occurring *Turbo setosus* shell, an inferior material. This evidence for the cessation of long-distance exchange is corroborated by work at other sites in the

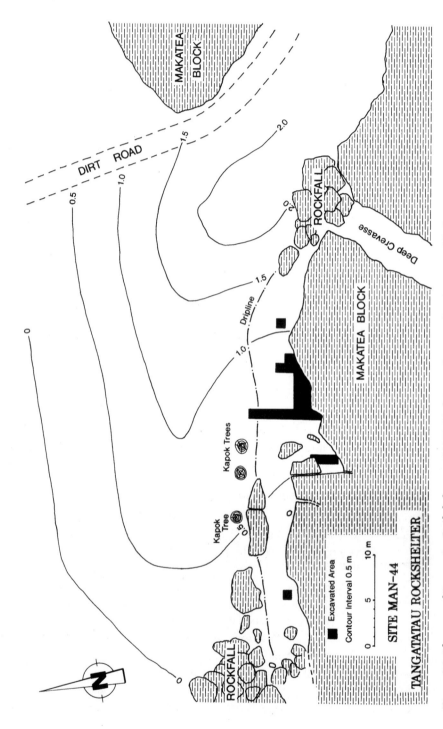

Figure 8.3 Plan map of Tangatatau Rockshelter, site MAN-44, showing the area of excavations in black

Cook Islands and elsewhere in Eastern Polynesia (Walter 1990). It appears to mark a major shift in social networks away from interisland connections to an inwardly focused perspective.

The dry, ashy sediments of Tangatatau Rockshelter preserved both faunal and floral materials extremely well, and the site provides an excellent record of changing patterns of resource exploitation and subsistence. Approximately 40,000 vertebrate faunal specimens were recovered from the excavations. Although analysis of this large assemblage is still in process, several significant trends are already evident (Fig. 8.4). First is the introduction of three exotic species, including the Pacific Rat (*Rattus exulans*), chicken (*Gallus gallus*), and pig (*Sus scrofa*). Pig appears to have increased in importance during the middle part of the sequence but then disappears from the record. This matches with ethnohistoric evidence that pig was not present at contact and was (re)introduced by the missionary Williams in the 1830s. The prehistoric elimination of pigs on Mangaia thus parallels the case of the Polynesian outlier Tikopia, where pigs also disappear from the faunal record in late prehistory (Kirch and Yen 1982:353, 358). On both Mangaia and Tikopia, it is likely that

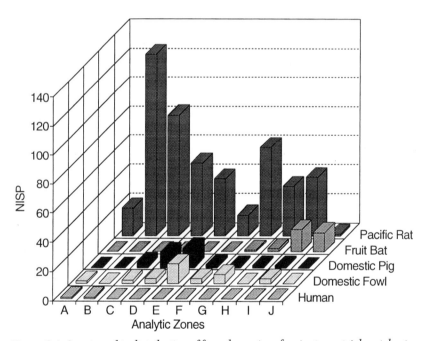

Figure 8.4 Stratigraphic distribution of faunal remains of major terrestrial vertebrate species in the MAN-44 sequence. Analytic zone J is at the base of the sequence. NISP = *number of identified specimens.*

the decision to eliminate a major domestic animal from the production system was not made lightly (especially considering the widespread *social* value of pigs as mediums of exchange and of prestation in Oceanic societies). On both islands, this decision was probably necessitated by high population density and the inability of the production systems to accommodate the trophic loss of converting potential human food (vegetable produce fed to pigs) into pig flesh.

Of particular interest is the evidence for a major decline in native food resources, especially fruit bats and birds. Fruit-bat bones are present in significant numbers only at the base of the Tangatatau Rockshelter sequence (Fig. 8.4). Today fruit bats are a highly endangered species on the island, although still hunted intensively. The bird bones reveal an even more striking pattern of heavy exploitation leading to extinction or extirpation of native species (Fig. 8.5). At least 16 species formerly present on the island became either extinct or extirpated during the period of Polynesian occupation (Steadman 1985, 1987, 1995; Steadman and Kirch 1990); these include four seabird and 13 landbird species. The extinct or extirpated landbirds include

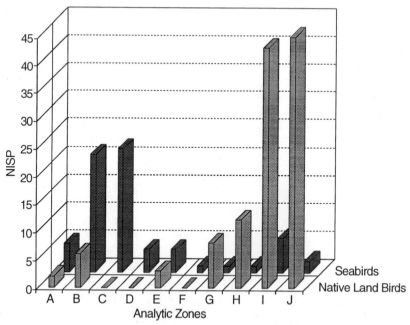

Figure 8.5 Stratigraphic distribution of faunal remains of native landbirds and seabirds in the MAN-44 sequence. Analytic zone J is at the base of the sequence. NISP = *number of identified specimens.*

Ripley's Rail (*Gallirallus ripleyi*), Mangaian Crake (*Porzana rua*), Small Mangaian Crake (*Porzana* new sp.), Mangaian Gallinule (*Porphyrio/Gallinula* new sp.), Mangaian Sandpiper (*Prosobonia* new sp.), Society Island Ground-Dove (*Gallicolumba erythroptera*), Giant Ground-Dove (*G. nui*), Mangaian Ground-Dove (*Gallicolumba* new sp.), Cook Islands Fruit-Dove (*Ptilinopus rarotongensis*), Nuku Hiva Pigeon (*Ducula galeata*), Society Islands Pigeon (*D. aurorae*), Rimatara Lorikeet (*Vini kuhlii*), and Conquered Lorikeet (*V. vidicivi*). At least four species of seabird that formerly frequented the island were also extirpated: the Black-winged Petrel (*Pterodroma nigripennis*), Polynesian Storm-Petrel (*Nesofregetta fuliginosa*), Red-footed Booby (*Sula sula*), and Lesser Fairy Tern (*Gygis microrhyncha*). Direct predation was doubtless one cause of such extinction, although the destruction and modification of bird habitats were perhaps even more important.

The Pacific Rat is represented by a large number of bones in the rockshelter deposits (Fig. 8.4). Ethnohistoric sources refer to the regular consumption of rats (Williams 1838:64; Gill 1856:140), and modern Mangaians state that their ancestors were fond of eating rats. This appears to be confirmed by the archaeological record, for a large percentage of the rat bones from the Tangatatau site are charred and fractured, clear signs of cooking.

Evidence for severe impacts on natural food resources is not confined to the terrestrial environment. The marine-shell midden, for example, shows dramatic reductions in the distribution sizes of certain food species, such as *Turbo setosus* (Fig. 8.6).

Mangaian traditions also speak extensively of cannibalism (Buck 1934:36 and passim; Gill 1894). These traditions might be discounted as "myth," and their prominence in the ethnohistoric record put down to over-zealous missionaries promoting the notion of heathen darkness. However, the faunal record from Tangatatau and other sites includes a small but significant quantity of human remains in nonmortuary contexts. We are now in the process of studying this material in light of the ethnohistoric record for cannibalism, and the results are still preliminary. At Tangatatau, however, more than 50% of the human bone shows evidence for burning, as well as a high frequency of spiral fracturing and of bashing and fragmentation of long bones. This pattern is repeated in the assemblages from other rockshelter sites. Most of this material derives from the upper strata, and it suggests that the butchering and processing of human flesh was a pattern that developed relatively late in the island's sequence.

The Tangatatau excavations also produced a large assemblage of carbonized plant remains, including: wood charcoal; *Pandanus* drupes; candlenut (*Aleurites moluccana*) endocarp and endosperm; coconut (*Cocos nucifera*)

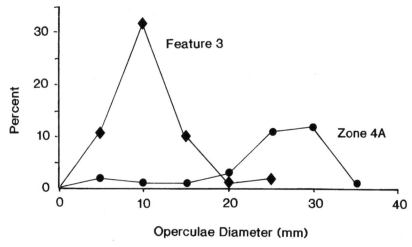

Figure 8.6 Size reduction in the reef mollusc Turbo setosus *from lower to upper levels of the* MAN-*44 sequence, as indicated by operculae diameters*

husk, leaf, bracts, and endocarp; *Cordyline terminalis* root and leaf; banana (*Musa*) leaf; *Caesalpinia major* seeds; *Hernandia* seeds; and bamboo (*Schizostachyum glaucophyllum*). Of particular interest is the presence of several Polynesian crop plants, including carbonized tubers of the sweet potato (*Ipomoea batatas*, Hather and Kirch 1991) and carbonized corms of taro (*Colocasia esculenta*) and giant swamp taro (*Cyrtosperma chamissonis*). The sweet potato remains are the oldest archaeological specimens of this South American crop plant yet recorded for Polynesia, and they appear to support Yen's hypothesis of the prehistoric transfer of the plant into central Eastern Polynesia at an early date (Yen 1974).

Although the Tangatatau Rockshelter has received the most attention, we have also worked at six other rockshelters to assess the extent to which the Tangatatau sequence is representative of the island as a whole. Three shelters were tested in Keia district, including a site with an associated refuge cave containing midden deposits. These sites span a time range similar to that of Tangatatau. Two sites were also excavated in Tamarua district. One is a rockshelter with a relatively large overhang, which unfortunately yielded little material, although it spans the interesting period of initial European contact. The second is a small overhang with evidence of intensive use and an artifact and faunal sequence that will be of considerable use in comparison with Tangatatau.

Mangaian society at the time of initial European contact was periodically wracked by wars of territorial conquest. Mangaian oral traditions, analyzed by

Buck (1934), recount a detailed record of the succession of paramount war chiefs, known as Te Mangaia. Archaeological reflections of such armed aggression are found, among other indices, in a series of refuge caves situated within the makatea limestone. One of the most prominent of these is Tautua, in the Tamarua district, recorded in tradition as the refuge of the Tongaiti tribe during times of warfare. The refuge cave is situated within a large solution cavern, and it can only be entered by scaling an 8 m high wall, thus making the cave easily defensible. The tribe's warriors were said to keep a lookout from the cave mouth over their productive taro lands in the adjacent valley bottom. The large limestone cavern contains six habitation platforms, a stone cyst grave and other burials, a *tupe* court for disc pitching, cooking areas, and numerous midden deposits. Although we were not able to excavate in Tautua, we mapped the site in detail and obtained shell and charcoal samples from various features for radiocarbon dating. These have allowed us to determine that the site was utilized in the very late prehistoric and early historic periods, matching closely the evidence from oral traditions. Indeed, the use of caves for refuge is a phenomenon that appears to have developed relatively late in the Mangaian sequence.

Early European visitors to Mangaia before missionization described a dispersed settlement pattern of habitations concentrated in the fertile valley bottoms around the margins of the taro irrigation systems. Our site surveys in the Keia and Veitatei Valleys have revealed extensive sets of artificial earthen terraces on these ridge spurs, which appear to be the vestiges of this contact-period settlement pattern. To gain some idea of the archaeological record contained within these sites and to establish the antiquity of this settlement pattern, we investigated one complex of terraces along the western margin of the principal Veitatei irrigation complex. The complex consisted of a number of artificially constructed terraces cut into the hillside just above the irrigated fields. We excavated test trenches in several of these, revealing considerable variation in internal stratigraphy. Some were virtually devoid of artifactual material. One large terrace had been dissected by a recent road cut, allowing us to clean a 30 m long profile with evidence of hearths, earth ovens, and gravel pavements. Two discrete occupation horizons were evident in the section. The later occupation included a stone-outlined and gravel-paved house floor. In the middle of the terrace complex was a smaller terrace with a coral-slab facing and gravel pavement, from which we recovered the complete stone blade of a Mangaian ceremonial adz. Trenching this terrace revealed two separate phases of gravel paving. The initial radiocarbon dates from the Veitatei terrace complex indicate an age span beginning around 1200 to 1400, with a later habitation phase dating to 1600 to 1850. Another isolated terrace in the interior of Veitatei Valley has been dated to the thirteenth to fourteenth centuries.

While our sampling of the Veitatei open sites is admittedly limited, it suggests that the pattern of dispersed settlements concentrated around the fertile valley bottoms had begun to be developed by the thirteenth century. This correlates well with the stratigraphic evidence from the coring operations, indicating that the rapid filling in of alluvium had reached a stage by this time that allowed the intensive taro systems to be developed. Thus the classic Mangaian socioeconomic pattern of intensive taro cultivation—with its attendant competition for control of prime lands—must have developed sometime during the last 400 to 500 years of the Mangaian sequence. Further excavations in the important open terrace sites of Veitatei and Keia, to expand the picture just sketched, have been the focus of continuing dissertation research by Julie Endicott.

One final class of Mangaian archaeological site deserves brief mention. These are the foundations of marae, or tribal temples, which are distributed around the island near the prime taro lands. The marae were initially recorded by Buck (1934:173–77, Table 17), who obtained considerable ethnographic information from the Mangaian elders on tribal affiliations and gods. The temples themselves consist today of earthen terraces, usually paved with coral gravel and frequently marked by upright limestone slabs and facings. These sites provide an architectonic link to the political and religious world of the Mangaians. The temples are closely associated with the irrigation systems, as in Keia district, where a string of marae follows the course of the main irrigation canal. Many of these sites have now been mapped, both by our project and by Bellwood (1978b) and Chikamori (1990). A program of systematic excavation and dating of these temples has yet to be carried out, but it could provide much information on the chronological development of the political system evidenced in the contact-period ethnohistory.

CONCLUSION

I have briefly reviewed here the major classes of geomorphological, palynological, faunal, and archaeological data that we have focused on in our interdisciplinary reconstruction of Mangaian prehistory. A major thrust of the Mangaia Project is the elucidation of the island's mid-to-late Holocene environmental history, beginning with the period before Polynesian colonization and continuing through to the arrival of Europeans and the advent of the World System. The evidence is now overwhelming for a dramatic series of environmental changes stemming from direct or indirect human impacts that radically transformed the Mangaian ecosystem (Fig. 8.7). These include the clearance of forest on the volcanic cone, initiation of erosion, and eventual development of a degraded *Dicranopteris* fernland climax on laterites. Erosion of the hillsides, however, also resulted in alluvial filling in of the valley bottoms, creating terrain highly suited to the development of a landesque capital-intensive system of taro irrigation (see Spriggs [1986] for a similar case in

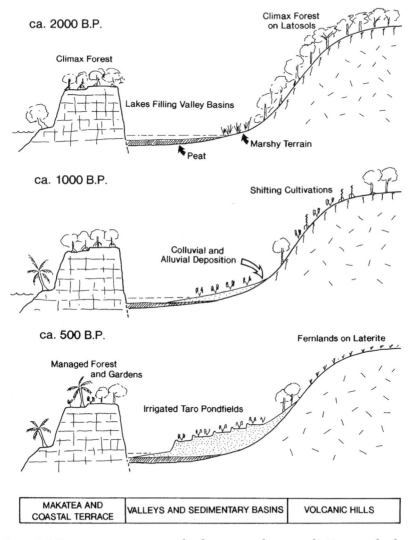

Figure 8.7 Diagrammatic summary of anthropogenic changes in the Mangaian landscape over the past two millennia

Eastern Melanesia). The effect for the human population was the concentration of the key productive resources into a set of restricted zones over which there then arose marked competition. Environmental change was not limited to the landscape and its vegetation, for the natural faunal resources of the island—fruit bats, birds, and marine foods—also suffered serious impacts.

We can now begin to situate the mytho-praxis of nineteenth-century Mangaian society (Fig. 8.8) in terms of the *longue durée,* the long run of time (Braudel 1980). The particular cultural patterns of competition for surplus production from the taro lands, the transformation of Rongo—the widespread Polynesian god of agriculture—into a god of war and taro, the transfor-

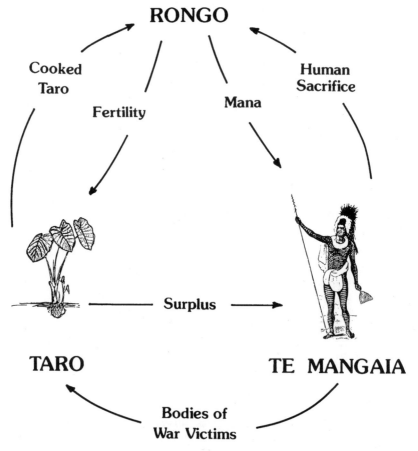

Figure 8.8 Some key relationships among chiefship, ideology, and production as expressed in the "mytho-praxis" of late prehistoric Mangaian society

mation of hereditary chieftainship into a more fluid polity favoring achieved power, human sacrifice and cannibalism—all of these take on a new significance once they are situated within a dynamic historical context. Rather than classifying Mangaian society—as Sahlins (1958) and Goldman (1970) were wont to do—along some hypothetical trajectory toward an archaic state, we may posit a separate kind of evolutionary pathway. The eighteenth-century Mangaians had their backs up against an ecological wall marked by scarcity of resources and intense competition. It was a landscape of their own manufacture, and they responded through the radical transformation of many of their cultural patterns.

This scenario should not be misconstrued as espousing some new version of environmental determinism. At European contact, the Mangaian ecosystem was a profoundly artificial, cultural world—a world actively created by preceding generations of Mangaian ancestors. Any understanding of Mangaian society as recounted in nineteenth-century ethnography requires the historical perspective of the longue durée, the reconstruction of the dynamic contexts within which Mangaians of history acted out the transformation of their social world.

Note

The research reported in part in this chapter has been supported by grants from the National Geographic Society (4001–89) and the National Science Foundation (BNS-9020750). Permission to carry out research on Mangaia was kindly granted by the Government of the Cook Islands and by the Mangaia Island Council. Tony Utanga of the Ministry of Internal Affairs in Rarotonga was especially helpful in making official arrangements. I am particularly indebted to my colleagues and collaborators in the Mangaia project whose data and analyses I draw on here: Dave W. Steadman, John Flenley, Francis Lamont, Stewart Dawson, Pia Anderson, Julie Endicott, Joanna Ellison, Virginia Butler, Jon Hather, and Peter Mills. For their assistance on Mangaia, I wish to thank Ma'ara Ngu, Diane Ngu, George Tuara, Tuara George, Tua Uria, Peter Ngatokorua, Micky Tuara, Papamama Pokino, and Nga Teaio.

9
Environmental Change and the Impact of Polynesian Colonization: Sedimentary Records from Central Polynesia

Annette Parkes

Recent archaeological and biological investigations strongly indicate that Polynesian settlers modified their island environments. The exact nature and sequence of change, however, has been less well documented. Forest clearance by fire, accelerated slope erosion (intentional or accidental), and the introduction of nonnative plants are regarded as the most important anthropogenic activities to have shaped post-settlement vegetation. Today, relict vegetation surviving in undisturbed areas, such as the *makatea* limestone of the Ngaputoru group and Mangaia in the southern Cooks (Merlin 1991), or on the steep inaccessible slopes of Mo'orea (Parkes and Flenley 1990), give some insight into the nature of pre-settlement vegetation. The analysis of fossil pollen flora, preserved in the anaerobic environments of lake sediments, provides an even more detailed chronological picture of vegetation change over pre- and post-settlement periods. Lake Te Roto on Atiu Island and Lake Temae on Mo'orea Island provide ideal natural catchment areas for the deposition and preservation of airborne pollen and spores. The accumulation of authogenic and allogenic sediments in these basins also gives some estimate of slope disturbance within the catchment, while the occurrence of known aboriginal plants within the pollen sequence can be used as markers of Polynesian influence.

METHODS

Transects of sediment cores were taken across Lake Te Roto on Atiu Island and Lake Temae on Mo'orea Island, using a hand-operated stationary piston sampler (modified Vallentyne Sampler, Walker [1964]), worked from a floating platform. A 16 m core was extracted from Lake Te Roto at a water depth of 6.8 m, while a 14.5 m sediment core was collected from Lake Temae at a water depth of 11.3 m.

Eight conventional radiocarbon dates, six from the Te Roto core and two from the Temae core, were carried out by the NERC Radiocarbon Laboratory (East Kilbride). One AMS date from the Te Roto sequence, at a depth of 3.53 to 3.52 m, was obtained with the kind support of Matt McGlone and Atholl Anderson (DSIR Land Resources, New Zealand). Dates were corrected for secular variation using the CALIB calibration program of Stuiver and Pearson (1986). Magnetic-susceptibility measurements (Thompson and Oldfield 1986) were recorded to trace changing erosion rates within the catchments, and to link these with variations in sediment chemistry and vegetation. Readings were expected to show variations in the concentration of particulate allogenic mineral matter, derived from the drainage basin and incorporated into the lake sediment. Susceptibility was measured along each core, at 2 cm intervals, using a low-frequency core-scanning "loop" sensor (Bartington Instruments Code M.S.1C).

Before the extraction of samples for pollen and chemical analysis, the sediment cores were X-rayed along their lengths in order to show any major variations in sediment density that may have resulted from washed-in clays, algal laminations, coral-sand bands, and so forth. The preservation of fine laminations, detected in the X-ray photographs of both cores, indicated that the sediments had been relatively undisturbed after deposition, a prerequisite for a clear chronological sequence. This technique also proved useful in detecting the presence of fossil mollusks, barnacles, and tube-worm casts in the Temae sequence. Major-element oxides were determined according to the fusion method of Norrish and Hutton (1969), using a Philips PW 1212 X-ray fluorescence spectrum. During the preparation for XRF analysis the percentage loss on ignition (at 600°C) was measured to establish the amount of organic material present in the samples. Independent charcoal analyses were undertaken using nitric-acid digestion of organic material, followed by heat extraction of remaining charcoal.

The standard palynological preparation techniques of Faegri and Iversen (1975) were used on both cores, along with 5 micron ultrasonic sieving (Tomlinson 1984) to remove fine particles and thus facilitate the counting of pollen grains and spores. Pollen and spore taxa were determined by comparison with modern reference material from field collections and from herbarium specimens at the Royal Botanic Herbarium, Kew, and the British Museum of Natural History, London. Palynological results were converted to percentages of the total count of pollen and spores but were also calculated in absolute terms by adding a known quantity of an exotic pollen to the sample (Benninghoff 1963). Diatom samples were prepared following Battarbee (1979). Diatom identifications were made with reference to standard floras and with help from S. E. Metcalfe (University of Hull). Scanning electron microscope techniques were used to aid identification of several diatom taxa and to confirm the presence of *Cocos nucifera* in the Te Roto core sequence. Macro-remains including mollusk shells and tube-worm cases were identified with the help of John Taylor (British Museum of Natural History).

LAKE TE ROTO, ATIU, SOUTHERN COOK ISLANDS: SITE DESCRIPTION

Atiu, located at latitude 20°S and longitude 158° 10′W (Fig. 9.1), is the third-largest island in the Cook Islands, with an area of 26.9 km^2 and a circumference of 20 km around the coast. The island is roughly quadrilateral in shape and may be divided into three distinct physiographic regions: the highly weathered volcanic interior, the raised coral limestone rim, or makatea, and the swampy lowland depression that separates the two former regions in most areas. The deeply weathered central volcanic plateau, the site of much agricultural activity both recently and in pre-settlement times, rises from moderately steep and badly eroded clay slopes to a maximum altitude of 72 m. Streams draining from the central uplands to the surrounding lowlands cut deep gullies into the slopes and then disappear at the base of the makatea cliff through sinkholes. The clays on the slopes, weathered from coarse augite-olivine-labradorite basalts (Marshall 1930), support little but the False Staghorn Fern *Dicranopteris (Gleichenia) linearis* and guava.

The makatea is an old fringing-reef surface that has been exposed through tectonic uplift (Wood and Hay 1970; Stoddart, Spencer, and Scoffin 1985) to form 3 to 6 m cliffs on the seaward margin, rising to 30 m on the inland margin. The makatea surface is rough and uneven with sinkholes, caves, underground drainage, and craggy limestone pinnacles 1 m or more high (Wood and Hay 1970). Pockets of lateritic clay soil support a largely intact native forest cover in areas regarded unsuitable for cultivation (Franklin and Merlin

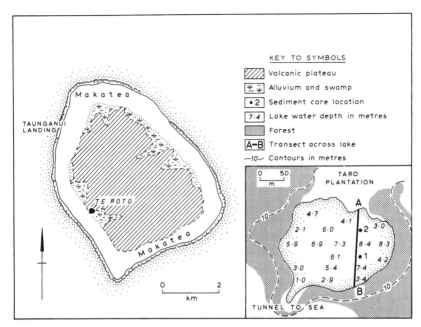

Figure 9.1 The island of Atiu and the location of Lake Te Roto; inset, the bathymetry of Lake Te Roto and the position of the core site

1992). As described by Kauta'i et al. (1984), this encircling band of makatea was once a rich source of resources for the Polynesians. Between the makatea and the central plateau is a well-defined solutional depression (Stoddart et al. 1985), parts of which are occupied by swamps (taro lands), marshes, and one lake, Te Roto (Fig. 9.2). Deposits in these swamplands mainly consist of fine sand, silt, and clay derived from the weathered volcanic hills (Grange and Fox 1953). Originally most of the natives of Atiu lived on the lower ground just inside these swamp areas, and terraces were excavated for houses from the sides of the radiating spurs of volcanic rock (Marshall 1930). The population of Atiu today (1,225 in 1981) lives on the central plateau of the island in five villages, representing the organization into land-holding districts that predates European contact.

The present-day climate of Atiu is characterized by persistent northeast, east, and southeast tradewinds and a mean annual temperature around 25°C, with a diurnal range greater than the annual range. The long-term average rainfall for the southern Cook Islands is between 1,900 mm and 2,050 mm (Thompson 1986), with the period from January to May being the wettest on Atiu. The southern Cook Islands lie within the South Pacific hurricane belt,

Figure 9.2 View of Lake Te Roto looking inland toward the central volcanic plateau, with taro swamps and coconut groves in the marshy depression below

and although little is known about the frequency of hurricanes, it is generally clear from the records of Visher (1925) that storms of exceptional severity have occurred.

Te Roto (also known as Tiroto or Tiriroto) is a small lake situated in the lowlands of the southwest quadrant of Atiu, which abuts the inner cliff of the makatea limestone. It is generally oblong in shape (ca. 183 m by 228 m) and has a maximum depth of 8.4 m as measured in 1986. The lake is about 3 to 4 m above sea level and is connected to the sea by a sub-makatea tunnel through which water is exchanged. The lake level is reported to fluctuate slightly with the tides, and the lake water becomes slightly saline after heavy storms (Kauta'i et al. 1984). The vegetation immediately surrounding the lake is today dominated by *Pandanus tectorius, Hibiscus tiliaceus,* and groves of coconut palms, with reeds and aquatic sedges growing in abundance around the lake margins. Fern scrub and grassland occur on the plateau slopes to the north.

SEDIMENTARY ANALYSES OF TE ROTO, ATIU: RESULTS

Stratigraphy and Geochemistry

Radiocarbon dates show that the Te Roto sediment sequence extends from before 7820 ± 70 years B.P. to the present (Table 9.1). This sequence, therefore, spans the range of dates estimated for Polynesian arrival in the southern Cook Islands (Bellwood 1978b; Lamont 1990; Steadman and Kirch 1990; Ellison 1993; Kirch and Ellison 1994). The age-depth sequence (Fig. 9.3, calibrated age) shows sedimentation rates between 13.98 m and 10.30 m (ca. 0.35 cm/yr), and in the upper 3.50 m of the core (averaging 0.27 cm/yr), to be generally higher than average rates in the central core section—that is, from 10.30 m to 3.50 m (0.11 cm/yr). These two zones of accelerated accumulation appear to coincide with material high in magnetic minerals (Fig. 9.3), washed-in clay bands, and sediment rich in basalt-derived oxides, such as iron, aluminum, titanium, and silica, indicated by geochemical analysis.

The stratigraphy of the Te Roto sediment core generally indicates a changing sequence from a terrestrial to a lacustrine environment. Basalt-derived material rich in iron oxide occurs at the base of the sequence (16 to 14 m) and may have resulted from a lateritic soil formed in situ. Between 14 m and 10.23 m, this highly mineral deposit gives way to a peat deposit of moss and sedge containing distinct bands of washed-in mineral clay. These clay horizons have been attributed to episodic, high-energy storm events, which resulted in in-

Table 9.1 Radiocarbon Dates from the Lake Te Roto Sediment Sequence

Code	Depth (m)	^{14}C age	Calibrated age B.P.	Calibrated age B.C./A.D.	^{13}C (‰)
SRR-3713	1.41–1.58	385 ± 45	479	A.D. 1471	−27.0
SRR-3282	2.85–2.90	540 ± 110	543	A.D. 1407	−23.7
AMS date	3.52–3.53	1420 ± 45	1310	A.D. 640	
SRR-3283	6.50–6.55	3100 ± 150	3355	1406 B.C.	−28.8
SRR-3715	7.86–7.91	4970 ± 55	5680	3731 B.C.	−29.1
SRR-3284	10.30–10.35	6730 ± 90	7579	5630 B.C.	−30.0
SRR-3084	13.88–13.98	7820 ± 70	8611	6662 B.C.	−29.2

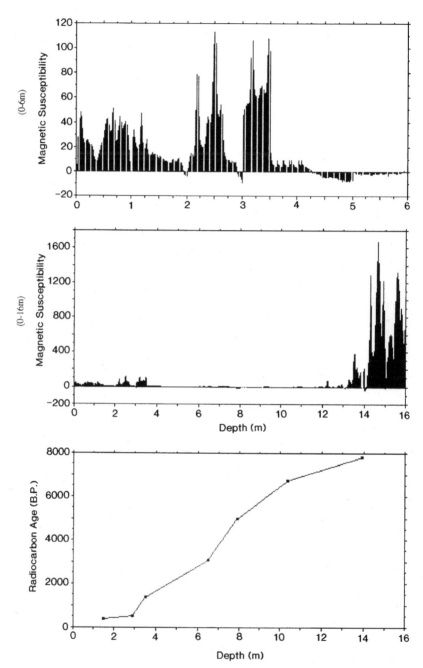

Figure 9.3 Magnetic susceptibility profile and age-depth sequence from the Te Roto sediment core

creased catchment erosion. After a transition zone between 10.23 m and 9 m of peat and organic lake mud, a predominantly organic-based gyttja with some fine washed-in clays occurs up to a depth of 3.52 m. Above 3.52 m, major washed-in, basalt-rich clays again occur in the sequence, indicating another phase of catchment disturbance, either natural or anthropogenic. The upper 2 m of the core are characterized by a return to an organic-rich lake mud, with erosion perhaps being less significant.

Geochemical analysis verified the presence of basalt-derived oxides in the clay bands characteristic of the lower 5.7 m (that is, before 7579 B.P.) and upper 3.50 m (that is, post-1310 B.P.) of the core. Also detected in the sequence were percentage peaks of sodium oxide around 10.45 m (ca. 7600 B.P.) and 7 to 4 m (ca. 3800 to 1500 B.P.). Although sodium is renowned for its solubility in water and also occurs in small amounts in basalt rock, it is believed that increased percentages of Na_2O at these depths may have resulted from natural increases in lake salinity, a view supported by the fossil diatom flora. During the period represented in the core between 7.20 m and 5.20 m, the geochemical profile for SiO_2 indicates that silica occurs in excess of the other basaltic-oxide ratios (Marshall 1930). This excess has also been attributed to an influx of saline-tolerant diatoms and marine-resting spores detected in the absolute diatom concentrations. Diatoms are able to incorporate free soluble silica into their frustules, and therefore any SiO_2 that would otherwise be lost to the system is retained in the sediment. The increased salinity inferred by the geochemistry and marine diatoms is believed to have been influenced by seawater reaching the lake through the makatea passage (Fig. 9.4). This may have resulted from increased storm frequency or higher relative sea levels. A higher relative sea level would certainly account for the transition of the Te Roto basin from a terrestrial to a lacustrine system, and it would have had a direct effect on the lake diatom flora as well as an indirect effect on island vegetation with relation to freshwater availability.

The analysis of charcoal in the Te Roto sequence proved rather inconclusive, as only low percentages were recorded, and the charcoal profile strongly resembled the profiles of basalt oxides (that is, SiO_2, Al_2O_3, TiO_2). This suggested that the levels of charcoal reaching the lake were dictated by the amount of washed-in slope material and were not necessarily indicative of the amount of burning in the catchment.

Pollen and Diatom Analysis

The percentage-summary diagram of fossil pollen and spores (Fig. 9.5) clearly shows that the vegetation on the island has changed significantly throughout the past 8,600 years at least. Before 7579 B.P. the vegetation was

Figure 9.4 Solutional tunnel through the raised makatea limestone that serves as an outlet for Lake Te Roto

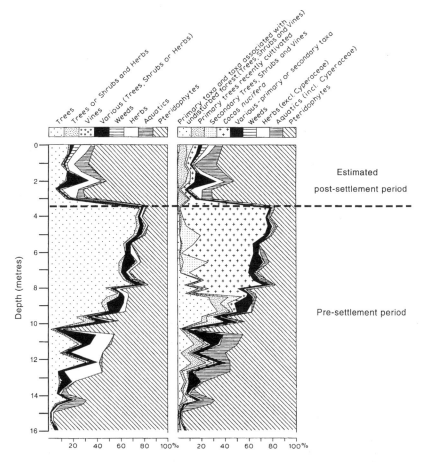

Figure 9.5 Percentage-summary diagram of vegetation types from the fossil pollen and spore record of Lake Te Roto

dominated by fernlands, which provided levels of up to 98% of the total pollen and spore count, and never below 45%. The ferns dominating this period included species typically found today in cool, high areas, such as upland Rarotonga. Arboreal, shrub, herb (mainly sedge), and aquatic taxa gradually increased during this period, possibly in response to increasing availability of moisture and/or warmer temperatures. The arboreal component of this zone was initially dominated by *Pandanus tectorius*, a species of *Calophyllum* (most probably *C. inophyllum*), and a species of *Trema* (possibly *T. cannabina*). The palm *Pritchardia vuylstekeana* and the coconut palm (*Cocos nucifera*) appeared significantly in the record above 13.15 m and 12.05 m, respectively, whereas other arboreal taxa, including *Elaeocarpus tonganus*, *Allophylus vitiensis*, and species of *Rapanea*, occurred toward the end of this period. Taxa associated with undisturbed forest conditions reached maximum percentages around 6500 to 6700 B.P., and they may represent a period of climate amelioration and stability, following what appeared to have been a relatively unstable, cool climate.

The period from 7579 to 1310 B.P. saw dramatic increases in coconut trees (up to 76%) and the swamp fern *Acrostichum aureum* (up to 42%), which dominated the vegetation along with species of fig (*Ficus*) and *Celtis*. A notable decline in several primary tree taxa corresponded to this period of the island's history and included *Pritchardia vuylstekeana*, *Trema* sp., *Calophyllum* sp., *Elaeocarpus tonganus*, *Allophylus vitiensis*, and species of *Rapanea*. *Pritchardia vuylstekeana* is no longer found on Atiu but was recently recorded on the neighboring island of Miti'aro, where it was growing in scattered clusters on the rugged makatea (Whistler 1990; Merlin, pers. comm.). Today it is listed as an endangered palm of the Cook Islands (Whistler 1990), but it apparently had a wider distribution before 5680 B.P. *Trema cannabina* is also rarely found in the Cook Islands today, whereas the other primary taxa are generally restricted to the undisturbed and coastal regions of the makatea. Absolute counts for coconut and *Acrostichum* confirmed that there was a real density increase in these plants, and that percentage increases were not just a consequence of the declining primary taxa.

The presence of *Acrostichum* and *Cocos* in such high densities, and apparently at the expense of other taxa, suggests that they were capable of out-competing the other plants, which may have been under stress from some extreme environmental factor. Because both of these taxa are capable of withstanding high salt concentrations, and in the light of the geochemistry results, it is suspected that high salinity levels were the main cause of this stress. The presence above 9.55 m of high absolute concentrations and percentages of the marine-resting spore *Chaetoceros* and other salt-tolerant freshwater diatoms, such as *Amphora coffeaeformis*, *Fragilaria brevistriata*, and *Achnanthes exigua* (Fig. 9.6), strongly supports this hypothesis. Maximum diatom concentrations for the whole core were found at a level of 5.55 m, correlating with the zone of excess silica oxide noted in the geochemical analysis.

The vegetation composition appeared to have changed abruptly from about 1310 B.P. (A.D. 640), characterized this time by a dramatic decline in coconut pollen (from 76.5% at 3.55 m to 5.4% at 3.05 m) and corresponding increases in grasses, sedges, and the fern *Dicranopteris linearis*. *Dicranopteris linearis* increases from 1% at 3.55 m to 41% at 3.05 m and is today abundantly found on all the dry hills of Rarotonga (Wilder 1931), as well as in the fern-

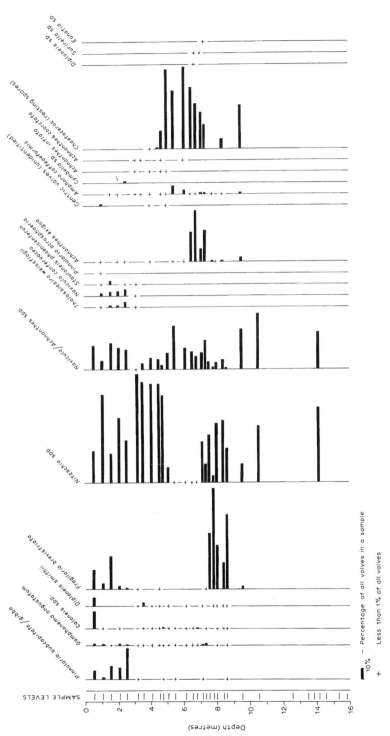

Figure 9.6 Percentage fossil diatom record from the Lake Te Roto sediment sequence

lands of Atiu, where it is strongly regarded as an indicator of disturbed and sterile soils. The summary diagram (Fig. 9.5) shows that *Dicranopteris* and other pteridophytes dominate the vegetation of this upper period, with *Acrostichum aureum* still accounting for up to 3% of the vegetation cover. The increases in the sedge *Cyperus pennatus* may have been encouraged by the formation of new alluvial areas in the moist lowlands, resulting from the increased slope erosion.

The beginning of this period also saw the first occurrence of cultivated grasses (that Gramineae in excess of 40μ m in size), plus the appearance of the ironwood (*Casuarina equisetifolia*) and the sweet potato (*Ipomoea batatas*). *Casuarina equisetifolia*, a favored tree of the Polynesians (Wilder 1931), appears to have been introduced or actively cultivated during this period, and it accounts for up to 9.5% of the pollen and spore rain reaching the lake. Whistler (1980, 1990) suggests that this tree is an aboriginal introduction to Polynesia, while Franklin and Merlin (1992) suggest that *Casuarina* may be indigenous to the Cook Islands. Ironwood was favored for making war clubs and was also used for house posts, tools, and outrigger booms for canoes. The bark is also reported to have been used for preparing a medicine (Whistler 1980). The aboriginally introduced sweet potato was and is extensively cultivated for its edible root. It is regarded as one of the most important food crops in the Cook Islands and is the only plant definitely known to have been aboriginally introduced into Polynesia from the New World. (Hather and Kirch [1992] report prehistoric, carbonized tubers of *Ipomoea batatas* from the MAN-44 site on Mangaia Island.) *Hibiscus tiliaceus*, regarded by Franklin and Merlin (1992) as a probable aboriginal introduction into the Cooks, also appears for the first time in this zone. Apart from these taxa, the only other trees significant in the pollen record of this period are *Pandanus tectorius* (reaching 2%) and *Ficus tinctoria* (recorded at 3% at a depth of 3.05 m). Throughout Polynesia *Pandanus* is regarded as second only to the coconut palm in importance and is regularly collected (Fig. 9.7) for thatching and the making of fans, hats, mats, baskets, and, formerly, sails (Whistler 1990).

The vegetation implied in this zone by the pollen and spore records strongly suggests that large-scale disturbance had occurred, resulting in the opening up of the vegetation cover to accommodate a grassland-fernland type of vegetation, along with new cultivated taxa. It seems quite feasible that the decline of the coconut was in fact due to clearance by a newly arrived or rapidly expanding Polynesian population. Clearing areas for settlement and cultivation may have resulted in an increased rate of soil erosion, accounting for the high density of basaltic-clay laminations at the onset of this zone. A sizable

Figure 9.7 Pandanus *leaves are still used for thatching and remain the second most important source of material after the coconut*

fragment of coarse basalt, containing large crystals of quartz, was found in the sediment core at a depth of 3.47 m, again supporting the idea that erosion on the central volcanic slopes was a significant factor. Newly disturbed slopes provided ideal sites for the establishment of pioneer fern, grass, and sedge species, all well established during this period.

The fossil diatom content of the core sequence above 1310 B.P. also changes significantly (Fig. 9.6). Taxa like *Pinnularia gibba, Gomphonema angustatum, Stauroneis phoenicenteron, Thalassiosira weissflogii,* and *Navicula confervacea* all point to a lake environment influenced by high nutrient levels, suspended sediments, and perhaps poor oxygen content. *Pinnularia gibba* can occur in eutrophic or oligotrophic water conditions and is capable of living in oxygen-poor water (Lowe 1974), while *Stauroneis phoenicenteron* var. *phoenicenteron* has a wide range of ecological tolerances and usually occurs in water with a temperature greater than 30°C. This diatom is commonly found living in polluted and eutrophic water conditions (Lowe 1974). The diatom *Thalassiosira weissflogii* is frequently found in rivers or at river mouths, where water-level changes or tides have an influence, and where concentrations of suspended sediments are high (H. Håkansson pers. comm.). Its occurrence in this zone may have resulted from increases in the levels of suspended sediment within the lake.

The presence at 1.05 m of *Bidens pilosa* and *Ageratum conyzoides*, both weeds regarded as early European introductions, probably marks the first European contact with the island. Both of these "weedy" composites are natives of tropical America and are common in waste places and croplands. Other introduced weeds, occurring for the first time in the top 1 m of the sequence, include *Passiflora quadrangularis, Mikania micrantha, Psidium guajava, Sida rhombifolia, Acalypha,* and *Elephantopus mollis.* European contact with Atiu occurred in 1777, which does not conflict with the age range of this upper zone. The lack of clay laminations in the sediment stratigraphy of the top 1 m may indicate a decline in agricultural activity around the lake basin, resulting perhaps from a declining population. Alternatively, movement of main settlement sites away from the lowlands and onto the basalt plateau may have released pressure from the land around Lake Te Roto.

Perhaps one of the most surprising discoveries made from the pollen analyses was the presence of *Cocos nucifera* at a depth of 13.80 m, a level that closely corresponds to the radiocarbon date of 7820 ± 70 (SRR-3084; 8611 B.P. corrected date). This record is crucial in considering the much-debated question of human or natural dispersal of the coconut in the Pacific. Bearing in mind present archaeological and biological data on Pacific migrations (e.g., Bellwood 1978a; Kirch 1982a, 1986; Steadman and Kirch 1990), it seems inconceivable that people could have reached these islands and introduced the coconut at this early date. This pollen record, therefore, controversially suggests that natural dispersal was the mechanism for the establishment of the coconut on Atiu. The presence of coconut remains dated at 5420 ± 90 B.P. from Aneityum Island in Vanuatu (Spriggs 1984) also supports the natural dispersal theory of coconut to the Western Pacific.

LAKE TEMAE, MO'OREA, SOCIETY ISLANDS: SITE DESCRIPTION

The island of Mo'orea, considered to be the geographical center of the Pacific Ocean, is situated at latitude 17° 30' S, longitude 149° 50' W (Fig. 9.8). The island has an area of 132 km^2 and is roughly triangular in shape, with 61 km of coastline (Galzin and Pointier 1985). Mo'orea is the surviving south rim of an ancient and considerably eroded basaltic caldera, which forms a jagged semicircular ridge interrupted by a series of peaks, the highest of which is Mount Tohivea, at 1,207 m. This old crater rim encircles a central plain that opens to the sea along the northern coast at Cook's Bay and Opunohu Bay (Fig. 9.9). On the seaward side of the main ridge, secondary ridges radiate toward the coast, formed by deep valleys dissecting the former volcanic cone. The island is encircled by a barrier reef that encloses a generally shallow lagoon and by a coastal plain composed of both coral-beach material and alluvial deposits. The plain supports a restricted native flora, including *Hibiscus tiliaceus*, *Thespesia populnea,* and *Barringtonia asiatica,* mixed with introduced species like *Terminalia catappa* and *Casuarina equisetifolia.* Other taxa characteristically growing on the coastal flats include coconuts, banyan trees, and *Calophyllum inophyllum* (Florence 1982). Indigenous mangroves are absent from Mo'orea, but *Rhizophora mucronata* has been introduced on the Vaiare coast in an attempt to stabilize the sediments. On the lower hillslopes in the zone of intense human influence, where fire, deforestation, and soil erosion are common, a very degraded mesotropical vegetation exists, dominated by a *Psidium-Dicranopteris* scrub. The entrances of the lower valleys are dominated by *Hibiscus tiliaceus, Ficus* species, and *Inocarpus fagiferus (mape),* while the less disturbed upland valley slopes support a hygrotropical vegetation, with *Metrosideros-Dodonaea* forests occupying the sheltered interfluves (Florence 1982). Unfortunately, many of the hygrophile forests of Mo'orea have been progressively replaced by the highly aggressive weed *Miconia calvescens* (Fig. 9.10), which was introduced into the Society Islands in 1930.

The climate of Mo'orea is characterized by high humidity, average temperatures between 25 and 30°C, and an annual thermal variability that is less than the daily variability. However, a cool and dry season from June to Sep-

182 Annette Parkes

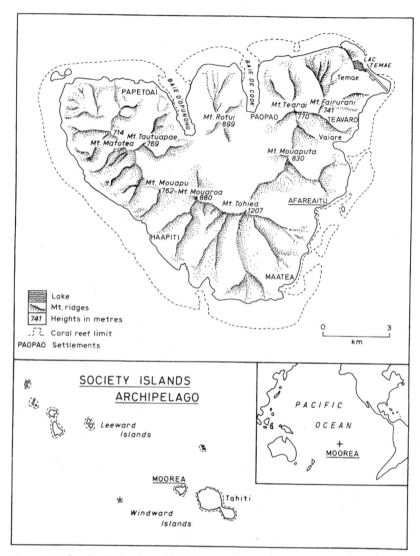

Figure 9.8 The island of Mo'orea and location of Lake Temae

Figure 9.9 View toward the interior of Mo'orea as seen from Vaiare Ridge. Remnants of the old crater rim encircle the central plain.

Figure 9.10 The broad-leaved Miconia calvescens *progressively replaces the hygrophile vegetation on even the highest ridges on Mo'orea*

tember (mean temperature 24.5°C) can be distinguished from a hot (26.5°C) and humid rainy season from October to April. Mean annual rainfall recorded at the meteorological station at Poapoa is 2,733 mm, but rainfall varies considerably according to local factors, such as altitude and aspect. The dominant winds reaching the island are the northeast to southeast tradewinds. The relatively dry southeast trades (the *mara'amu*) blow consistently from May to August, varying to easterlies from September to December, but they reach only the southeast and southwest regions of the islands. The moist northeast trades (the *to'erau*) blow from January to April, coinciding with the hurricane season. Tropical cyclones accompanied by heavy rainfall are generally rare in Mo'orea, with the high number of cyclones recorded in 1982 and 1983 being exceptional.

From archaeological studies throughout the Society Islands (Emory and Sinoto 1964, 1965; Sinoto and McCoy 1975; Semah et al. 1978; Sinoto 1979, 1983a, 1983b), Kirch (1986) concluded that by at least the ninth century A.D. the Society Islands were occupied by several communities sharing a material culture similar to that found in the Hane-Ha'atuatua site in the Marquesas. Captain Cook was one of the first European explorers to land in Mo'orea, in 1777, but the first known census on Mo'orea was taken by the missionaries in 1804, at which time the population was 2,533. Thirty years previously the population had been estimated at 20,250. In 1988, the census recorded 8,801 people, with the majority of them now living on the coastal plains.

Lake Temae, the only lake on Mo'orea, is situated in the northeast corner of the island and is separated from the sea by a coral ridge (Figs. 9.11 and 9.12). Today this brackish lake is approximately 1,000 m long and 400 m wide, with a maximum depth of 11.30 m measured in 1985. Early maps of the island, however, show that Lake Temae was formerly larger in area and lay between two smaller lakes. Lake Motuiti, to the northwest (seen in Fig. 9.12), is now totally dry for most of the year and supports a vegetation dominated by *Pandanus*, while the sediments of Lake Varea, to the southeast, now lie beneath Temae airport. Judging from its coastal position, it seems feasible that Lake Temae was formerly part of the lagoon system, and that the coral ridge was once an island on the barrier reef. Now part of the mainland, the ridge supports a littoral vegetation along with a plantation of coconut palms and *tiare* plants (*Gardenia taitensis*). Coconut groves and *Pandanus* trees have also been planted around the shores of the lake, while the aquatic reed *Typha angustifolia* dominates the marshy lake margins and inland delta area. On the landward side of the lake the Temae Valley rises abruptly up to Mount Fairurani at 741 m. The surrounding two ridges, reaching the coastal plain on either side of the lake, maintain only a sparse cover of vegetation, dominated

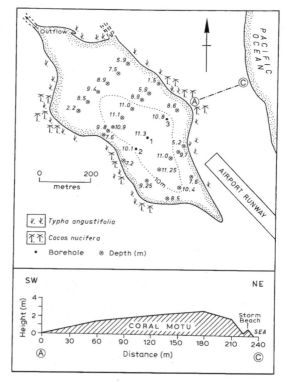

Figure 9.11 Bathymetry and location of the core site on Lake Temae. Below, profile over the coral ridge that separates the lake from the sea.

by grasses and ferns. The village of Temae, which today consists of a few scattered houses situated at the base of the valley, was one of the main sites of ancient settlement on the island.

SEDIMENTARY ANALYSES OF LAKE TEMAE, MO'OREA: RESULTS

Stratigraphy and Geochemistry

The two radiocarbon dates from the Lake Temae sediment core (Table 9.2) show that rates of sediment deposition in the basin were fairly rapid, particularly in the section below 7 m (1.86 cm/yr). No radiocarbon date was obtained from below 11.50 m, due to the low availability of datable carbon in the calcareous sediments. By taking the average rate of sedimentation recorded between 11.50 m and 6.85 m, however, an extrapolated date of approximately

Figure 9.12 View of Lake Temae from the top of Temae Valley. The coral motu supports cultivated tiare plants, while Pandanus and Typha grow around the lake periphery. Lake Motuiti, now filled in, indicates the former extent of the lake.

Table 9.2 Radiocarbon Dates from the Lake Temae Sediment Sequence

Code	Depth (m)	^{14}C age	Calibrated age B.P.	Calibrated age B.C./A.D.	^{13}C (‰)
SRR-3088	6.8–6.9	1210 ± 90	1160	A.D. 790	−27.7
SRR-3288	11.45–11.56	1540 ± 100	1411	A.D. 539	−15.9

1572 years B.P. was estimated for the base of the core at 14.50 m. This predates the date of archaeologically established Polynesian presence in the Society Islands (ca. 1100 B.P.). The sediment record should, therefore, give insight into pre-settlement environments as well as changes related to human settlement. It must be noted that the sediment samples used in the two radiocarbon dates were obtained from levels rich in allogenic material, which may have slightly increased the average age of the dated horizons.

The general stratigraphy (see Fig. 9.15) and geochemistry (Fig. 9.13) of the Lake Temae core show that the lower 7 m of the sequence is dominated by a fine, buff-colored, calcareous mud interrupted at intervals by coarse coral-sand horizons. These horizons contain high concentrations of the marine mollusk *Nassarius fraudulentus* (Marrat), barnacles, and tube-worm casts (probably of *Neopomatus uschakovi* Pillai), and they are believed to represent periodic storm events that transported and deposited outer-reef material into the Temae basin. Two darker zones, containing mineral-derived clays and detrital material, occur at 11.60 to 11.18 m (dated around 1411 B.P.) and at 9.91 to 9.65 m (ca. 1330 B.P.). These two subzones, which also correspond to peaks in magnetic susceptibility (Fig. 9.14), probably indicate periods of slope disturbance, resulting in an influx of mineral material from the Temae Valley. Interestingly, these horizons still have a significant percentage of calcareous material, along with a coarse-sand fraction, marine mollusks, and barnacles, which indicates that material had been brought in from the marine environment as well as from inland. This perhaps suggests natural climatic rather than anthropogenic disturbances, and the horizons are believed to represent severe storm events.

The lower section of calcareous mud abruptly ends with the influx of basaltic-based material and detrital fragments at 7.10 m (ca. 1160 B.P.) to 6.40 m. Significant peaks in the basalt oxides occur, with SiO_2 reaching a maximum of 34%, Al_2O_3 18%, Fe_2O_3 16%, and TiO_2 4%. The highest susceptibility readings in the core sequence also correspond to this band (Fig. 9.14). Unlike the previous disturbance horizons, this band appears to contain very little calcare-

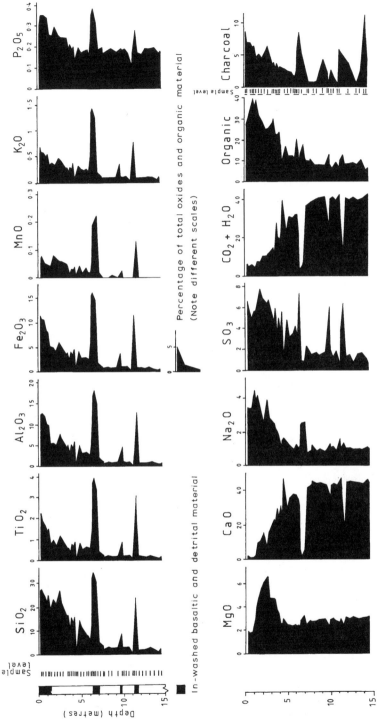

Figure 9.13 Analysis of major-element oxides, organic content, and charcoal in the sediments from Lake Temae

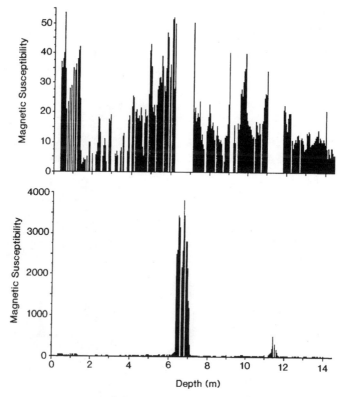

Figure 9.14 Magnetic susceptibility readings from the sediment of Lake Temae

ous material (2%) and no marine organisms, which might have been expected if a strong storm or cyclone was responsible for the slope disruption. On the other hand, slope erosion may have been initiated by deforestation, perhaps by burning, as charcoal percentages increase significantly at this time, to 8.6% (Fig. 9.13).

A transition back to a fine calcareous-based matrix interrupted by coral sands occurs between 6.40 m and 3.93 m, after which a gradual increase in algal laminations, magnesium oxide, organic content, and mineral oxides is seen in the next 2 m. The isolation of Lake Temae from the sea, which is believed to have taken place during this period, would have resulted in an increase in water temperature similar to that observed by Galzin and Pointier (1985). An increase in temperature, along with an increase in the nutrient status of the lake, may have encouraged the growth of algae. The presence of similar red-and-yellow nekron mud has been noted from other coastal Pacific lakes.

Environmental Change and Polynesian Colonization 191

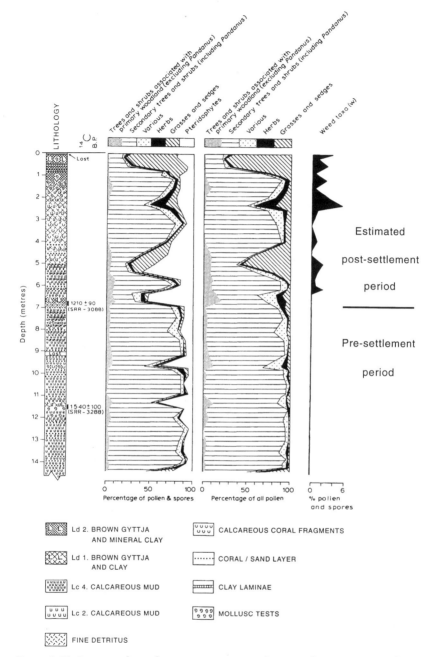

Figure 9.15 Stratigraphy and percentage summary diagram of vegetation types from the fossil pollen and spore record of Lake Temae

These include Lakes Rotonui and Rotoiti on Miti'aro in the Cook Islands and Lake Pala and Red Lake on Aunu'u Island in American Samoa (Parkes et al. n.d.). This algal production may well indicate the onset of isolation of a lake from marine influence and would also account for the high organic and MgO content in this zone. The excess of SiO_2 relative to the other basaltic minerals that occurs at a depth of between 3 and 1 m is believed to have resulted from an increase in diatoms, again in response to higher temperature and nutrient levels within the lake.

Around 1.78 m to the top of the core there is a further change, to a dark, organic- and mineral-based mud with only a small calcareous component. The calcareous component in this case is made up of abundant freshwater mollusks (*Melania incisa* Reeve), with the highest concentrations between 1.70 and 1.26 m, 0.77 and 0.72 m, and 0.09 and 0.05 m. No freshwater mollusks were found below 1.80 m. The peaks of basaltic mineral oxides toward the top of the sequence, which correspond to relatively high magnetic-susceptibility readings, may again indicate a period of slope disturbance and soil influx. This is supported by the appearance of fine clay laminations shown in both the sediment stratigraphy (Fig. 9.15) and in X-ray photographs of this section.

Pollen Analyses

The percentage-summary diagrams of fossil pollen and spores (Fig. 9.15) indicate the proportion of trees, shrubs, herbs, and pteridophytes, as well as the distribution of primary, secondary, and "weedy" taxa throughout the record. The pollen record from 14.48 to 7.18 m is dominated by *Pandanus tectorius*, which constitutes 39% to 86% of all pollen and spores. This hardy tree occurs throughout the island today and has successfully been cultivated on the coastal plain. Other taxa significant in this zone are *Cocos nucifera* (reaching 13%), *Trema orientalis* (up to 13%), *Ficus tinctoria* (up to 11%), *Macaranga tahitensis* (up to 7%), *Neonauclea forsteri* (up to 6%), and members of the Urticaceae/Moraceae family. Most of these trees are today associated with secondary forest formations on Mo'orea, but *Neonauclea forsteri* and *Ficus tinctoria* are found in isolated stands in the mid-altitude hygrophile forests of Tahiti, where human disturbance is reduced. The presence of *Cocos nucifera* pollen in these lower levels suggests that this palm was introduced into Mo'orea before the arrival of the Polynesians.

Pollen samples associated with the two lower zones of slope disturbance indicated in the geochemistry (between 11.51 m and 9.87 m), contain a significant percentage of degraded pollen grains and spores, mainly of taxa characteristic of primary upland hygrophile forest environments. These include *Ascarina polystachya*, *Weinmannia* sp., *Eugenia rariflora*, *Metro-*

sideros collina, and the shrub *Polyscias guilfoylei*. The species of *Weinmannia* is most probably *Weinmannia parviflora*, which along with *Metrosideros collina* forms the dominant species of the mid- and high-altitude hygrophile forests on Mo'orea and Tahiti today (Florence 1982). Degraded upland ferns also present in the pollen record include the tree ferns *Cyathea affinis* and *Cyathea horrida*, both associated with the undisturbed cloud forests of the Society Islands and generally found in deeply shaded areas on stream banks or near waterfalls; *Lycopodium phlegmaria*, a fern common at altitudes above 300 m; and members of the *Cyclosorus/Hypolepis* group. *Macropiper latifolium*, an indigenous shrub present in the montane forests, represents 8% of the grains in these horizons.

The presence of degraded grains of primary upland taxa, in conjunction with much detrital material and the appearance of the disturbance indicator *Dicranopteris linearis* (up to 4%), supports the idea that two relatively severe events took place high up on the humid slopes of Mo'orea. This is believed to have resulted in the removal of the topsoil, containing quantities of upland forest pollen and leaf litter, which was then redeposited on the lower plains and in the lake/lagoon. In these "disturbance" zones there is also a noticeable decline in relative percentages and absolute counts of *Pandanus* pollen, and littoral forest species—including *Barringtonia asiatica* and *Messerschmidia*—are noticeably absent. It is interesting to note that *Myrsine affinis*, a small tree of the high-altitude hygrophile forests, is no longer detected in the pollen record above 9.96 m. However, the main primary forest component does appear to have maintained a consistent level in the periods following the disturbance events of this zone. The lowest pollen sample of the sequence, at 14.48 m, also contains some degraded upland taxa and significant counts of sedge, grass, *Dicranopteris*, and coconut pollen. The sedge *Fimbristylis cymosa* accounts for 20% of the total count, while other sedges account for another 10%. These sedges, plus grasses at 4% and *Dicranopteris linearis* at 6%, all suggest that an open and disturbed vegetation existed during this period (ca. 1570 B.P.), at least in the vicinity of the lake. This level may, therefore, represent the close of another disturbance event in the catchment, with the formation of new alluvial land followed by colonization of pioneer taxa.

The pollen record from 6.93 to 6.58 m, corresponding to the band of dark-brown sediment rich in mineral oxides, shows a peak in degraded upland forest taxa, similar to those seen in the previous disturbance horizons. Additional primary taxa that briefly occur in this zone include the endemic tree composite *Fitchia*, *Rapanea tahitensis*, *Eugenia* sp., *Alstonia costata*, *Astronia fraterna*, and *Freycinetia impavida*, all found today in the high-altitude hygrophile and cloud forests of Tahiti but less frequently on Mo'orea (where

Fitchia is now totally absent). In this zone the tree ferns *Cyathea affinis, Cyathea horrida,* and other Cyatheaceae increase to their highest levels (17%, 3%, and 8%, respectively). After the initial increase in upland forest taxa in this zone, and unlike the record in the periods following previous disturbance events, several primary woody species decline to undetectable amounts, including *Alstonia costata, Astronia fraterna, Fitchia* sp., *Premna tahitensis,* a species of palm, and *Metrosideros collina.* Other species, such as *Ascarina polystachya, Rapanea tahitensis, Weinmannia* sp., and *Weinmannia vescoi,* decline to some extent but appear to recover again toward the top of the core.

The record in this zone also shows a fall in the percentage and absolute counts of *Pandanus tectorius* and *Cocos nucifera* as well as several secondary taxa, such as *Macaranga tahitensis, Neonauclea forsteri,* and *Trema orientalis.* There is a corresponding appearance of the swamp reed *Typha latifolia* (2%) and members of the Cyperaceae family (2%). *Colocasia esculenta* (taro), believed to have been an aboriginal introduction and widely cultivated throughout the alluvial swamplands of Polynesia, occurs for the first time in this zone (2%), along with *Cordyline terminalis* (3%), an erect shrub with an enlarged edible root. Although probably native to the Western Pacific, *Cordyline* is believed to have been carried by Polynesians to the eastern island groups, where it was subsequently cultivated. The presence of these two taxa, as well as the appearance of the introduced weed *Acalypha* (at 6.33 m), strongly indicates the first presence of people on the island. The increase in alluvial swamp vegetation in this and the following zone perhaps indicates further colonization of freshly deposited alluvial material. The redistribution of soil material from the steep, relatively inaccessible valley slopes to the flatter coastal areas would have provided newly arrived Polynesian settlers with a fertile and elevated coastal plain, ideal for the cultivation of crops like coconut, *Pandanus,* and taro. The presence again of *Dicranopteris linearis* from around 7.18 m to the top of the core may be a direct response to vegetation and land disturbance.

Between 6.33 and 4.90 m the reed *Typha angustifolia,* the sedge *Fimbristylis cymosa* (up to 23%), and grass species all increase. Two other swamp taxa, *Acrostichum aureum* and *Stenochlaena palustris,* occur at low levels and perhaps represent the initial stage of a natural transition from the pioneer species of sedge to the slower-colonizing swamp ferns. *Casuarina equisetifolia* occurs for the first time in the pollen record of this zone, perhaps supporting the view that this tree was introduced by Polynesian travelers rather than being indigenous to the islands. *Calophyllum inophyllum* also appears in the record at this time. Other significant changes in the pollen composition of this period include the fluctuations in coconut and *Pandanus* counts. These fluctuations may have resulted from changes in cultivation techniques by the Polynesians and their preference for one of the two trees over the other. A peak in *Cocos nucifera* (24%) occurs around 4.90 m, while *Pandanus* falls to a low of 6%. Coconut then falls dramatically from 4.56 to 3.37 m, replaced by *Pandanus,* which dominates the vegetation (64 to 71%) up to 3.37 m. Other useful tree species that occur include *Hibiscus tiliaceus* and the introduced shade tree *Terminalia catappa.* At present, *Hibiscus* is common everywhere from sea level to highlands, and it grows particularly well at the mouths of large river valleys. This versatile tree was cultivated for its large leaves, which were, and still are, used as a covering for native ovens. The light wood is suitable for making canoe outriggers and paddles, while the fiber from its bark is used to make cordage, mats, and *kava* strainers (Whistler 1990).

Between depths of 2.76 and 1.42 m, and closely corresponding to a stratigraphy dominated by algal material, the pollen record indicates the reintroduction or recovery of several primary plant taxa. These include *Weinmannia* sp., species of Elaeocarpaceae, and *Eugenia rariflora.* Apart from *Pandanus* (between 34 and 63%), however, it is the secondary trees and shrubs, along with pteridophytes like *Dicranopteris linearis, Angiopteris evecta,* and *Acrostichum aureum,* that dominate the vegetation of this zone. The robust king fern, *Angiopteris evecta,* today forms an important component of the valley forests and the secondary upland forests on Mo'orea. Also noteworthy is the increasing diversity and quantity of ruderal herbs, including *Mimosa pudica* and *Bidens paniculata,* both introduced cosmopolitan weeds, common today in gardens and sunny open pastures and along roadsides in Mo'orea. *Achyranthes aspera,* a weedy plant of coastal plantations, also occurs briefly at 2.54 m. There are significant increases in the percentages and absolute counts of cultivated plants as well. These include *Colocasia esculenta* (5%), the *tiare* plant *Gardenia tahitensis, Cocos nucifera,* the large shade-producing tree *Barringtonia asiatica,* and *Casuarina equisetifolia.* The slight recovery of primary upland taxa and the corresponding increase in swamp cultivars (e.g., *Colocasia* taro) could indicate a shift of importance from hillslope activities to the lowland plains. This perhaps supports the idea (Spriggs 1985) of a late intensification of agricultural activity on the now more extensive alluvial plains.

A final vegetation zone occurs between depths of 1.20 and 0.25 m and shows a further increase in introduced ruderal herbs. These herbs include *Bidens pilosa, Ludwigia octivalvis, Ageratum conyzoides, Elephantopus mollis,* and the prolific *Stachytarpheta,* all of which are found in disturbed and open areas of Mo'orea today and have been associated with European introduction. The highly competitive shrub *Miconia calvescens* also appears for the first time in this period. It is this extremely prolific species, introduced to Mo'orea only 60 years ago, that poses the greatest threat to the island's remaining natural vegetation (Fig. 9.10). Another invasive plant, originally introduced in 1850 for its fruit and now occupying vast scrub areas with *Dicranopteris linearis,* is the guava plant (*Psidium guajava*). The sudden influx of *Typha angustifolia* pollen into the lake above 1.20 m (reaching up to 59%) again suggests that there was a increase in the availability of open, flat land. Alternatively, the increase in this freshwater reed may have resulted from a decrease in the salinity of the lake, with *Acrostichum* (more tolerant of saline conditions) being naturally replaced by *T. angustifolia.* Today Lake Temae is almost surrounded by *T. angustifolia,* which has colonized nearly all the available peripheral mud. The composition of pollen and spores found in this uppermost zone closely resembles the floral composition currently found on Mo'orea.

The very existence of the lake in its present form is probably dependent on human activity. The location of the lake suggests that it could formerly have been part of the lagoon behind the coral barrier reef, and that it has been cut off from direct marine influence by a massive influx of sediment at both ends of the lake. This sediment may have been derived from the seaward reef, as coral fragments were broken away during storms and later accumulated in the lagoon, or more probably from material washed in from the steep basaltic slopes above the lake during periods of disturbance (anthropogenic and/or natural). The idea of a marine origin for the lake is strongly supported by several other lines of evidence. First, the lake is at sea level (Fig. 9.11) and is still slightly brackish. Second, the bulk of the lower part of the core consists of calcareous mud containing marine organisms, which disappear dramatically above 7 m. This calcareous mud is typical of sediments found in the sheltered lagoon environments of Mo'orea today (Galzin and Pointier 1985).

Unfortunately, the vegetation and sedimentary record from the Lake Temae core does not extend far enough back in time to see any natural trends resulting from long-term climatic changes. The record does, however, indicate that extreme natural events, such as large-scale storms, did have short-term effects on the indigenous, pre-settlement vegetation. Nevertheless, taxa diversity appears to recover fairly rapidly after these events. The record also indicates that Polynesian settlers modified the native floral composition, being responsible for both the decline of several upland forest taxa and the formation of the shrub and fern vegetation so typical of the lower regions of Mo'orea today. The *Hibiscus-Ficus* forests typical of the lower valley regions were probably also formed during this period. Recovery of the primary forests was further jeopardized by Europeans with the introduction of highly competitive pioneer taxa, such as *Miconia calvescens* and *Psidium guajava.*

CONCLUSION

The stratigraphic record from Lake Te Roto on Atiu shows that several distinct changes have occurred in the lake environment and surrounding vegetation during the past 8,611 years. Before 1310 B.P. (A.D. 640) changes have been attributed to natural variations in climate and salinity levels linked to relative sea-level change. In the lowest period of the sequence—that is, before about 8600 B.P. (ca. 6662 B.C.)—a vegetation dominated by ferns and club mosses existed in a cool, perhaps dry, climatic regime. The period from 8600 to 7600 B.P. was characterized by an increase in temperature and/or the availability of moisture, which encouraged the growth of indigenous trees like *Pritchardia vuylstekeana, Pandanus tectorius,* and *Cocos nucifera.* The geochemical record, however, implies a relatively unstable climate, with periodic storms perhaps checking the full development of the vegetation and causing severe soil erosion.

A more stable climate is suggested by the development of forest taxa from 7600 to 6500 B.P., with optimum conditions around 6500 to 6700 B.P. During this same period diatoms, vegetation, and geochemical records indicate a gradual increase in salinity within the lake and the island system. Saline-tolerant diatoms, including *Chaetoceros,* dominate the fossil records from around 6000 to 2500 B.P., while saline-tolerant plants, such as coconut and *Acrostichum,* dominate the vegetation from 6000 to 1310 B.P. Increased salinity during this period has been attributed to a relative rise in sea level, affecting the diatom population within the lake and causing stress to intolerant plants. This stress resulted in the natural decline of several primary forest trees, including *Pritchardia vuylstekeana, Calophyllum* sp., and *Elaeocarpus tonganus,* around 4500 B.P.

A dramatic decrease in coconut trees and corresponding increases in aboriginal taxa, grasses, sedges, and the fern *Dicranopteris linearis* after 1310 B.P. (A.D. 640) are believed to represent the period of initial colonization by a Polynesian population. The removal of coconut trees and the opening up of the vegetation resulted in increased soil erosion from the catchment slopes. Lake Te Roto may also have become slightly eutrophic at this time.

Based on the available dates of the Mo'orea sequence, several preliminary conclusions have been reached. During the period between approximately 1572 and 1160 B.P. there were frequent small- and medium-scale storm events that scoured material from the seaward side of the barrier reef and deposited it into the protected lagoon environment. These manifest themselves in the formation of coarse coral-sand bands sandwiched between finer calcareous mud, deposited during calmer periods. A rough estimate suggests that small-scale events occurred once every five to seven years, while medium-scale events occurred once every 80 to 90 years. These events, however, were not severe enough to have affected the inland vegetation. Two or three major storm events also occurred within this time period, evident at levels 11.50 m (dated at 1411 B.P.) and 9.87 m (ca. 1300 B.P.), and possibly also at the lowest level, 14.42 m (ca. 1570 B.P.). These storms (averaging one every 140 years) were severe enough to affect both coastal and inland areas. Upland vegetation was disrupted, and soil became exposed to strong winds and heavy rain, resulting in slope erosion. Nevertheless, this upland vegetation generally recovered after these events, and loss of taxa was minimal.

Around 1160 to 1180 B.P., the disruption of vegetation and resulting soil erosion have been attributed to either the arrival of the first Polynesians on the island or perhaps a shift to more-intensive cultivation practices on the catchment slopes if an indigenous population already existed. Plant taxa associated with Polynesian migration also appear around this period. The coconut palm, in contrast, appears to have already become naturally established before human arrival. Throughout this period of increased soil erosion, the lake was slowly being isolated from the rest of the lagoon by material washed in from the bluffs. The lack of calcareous sediment and coral sands in the lake during this period indicates that the marine environment was not disrupted.

On Mo'orea, increases in the indigenous population from 1160 B.P. (ca. A.D. 790) onward may have encouraged the practice of forest clearance, perhaps initially on the lower valley slopes, where it was easier to settle and cultivate. Interesting questions, however, arise when we consider why the upland vegetation, on steep, less accessible slopes, was also disrupted. Evidence from Tahiti and Mo'orea indicates the existence of cultivation high in the interior of the island at least up to the time of Cook's visit (Beaglehole 1967), and during times of interisland conflict populations may well have retreated to the upland areas. Alternatively, the Polynesians may have purposely induced soil erosion in these upland areas to encourage the accumulation of fertile alluvial plains down toward the coast (see Chapter 5).

In the period up to about 300 years ago, there were further changes in the vegetation composition, probably as a result of changing cultivation practices. The progressive human disturbance and opening up of the landscape, perhaps exacerbated by storms, influenced the colonization of ruderal herbs and grassland species. The formation of new, flat, and highly fertile land encouraged the cultivation of more coconut and *Pandanus* on the sandy areas, along with taro in the deltaic swamps. The lake was finally cut off from the surrounding lagoon during this period, marked by a transition from calcareous-based sediment to organic- and mineral-based sediments in the lake. Marine dinoflagellates, mollusks, and barnacles also disappear from the sediments. Algal muds then developed in the lake with continuing input of nutrients and a possible increase in temperature.

The arrival of the Europeans some two centuries ago is marked by the further introduction of aggressive weed species and by a slight recovery in the primary forest taxa. The recovery of an upland forest may have resulted from a decline in the indigenous population through disease and war. Alternatively, or in addition, the settlement of coastal, as opposed to inland, areas may have released pressure from the upland forests. Today all the settlements are located around the coast, where most of the cultivation takes place. There is a continuous increase in mineral input during this period, which peaks toward the top of the core. This perhaps reflects the latest population increase, during the twentieth century.

Note

Many thanks to Sarah Metcalfe and John Flenley for their advice and support during this project. I am also indebted to John Garner and Keith Scurr for help with photographs and diagrams. Special thanks go to James Algret and family, Claude Peyri, Margaret Hicks, Mr. Hanerie, and Mr. Koro, and indeed all of the islanders who made fieldwork on Mo'orea and Atiu such a pleasure.

10
Human Occupation and Environmental Modifications in the Papeno'o Valley, Tahiti

M. Orliac

In this chapter, I show how environmental modifications appear in the archaeological record of the Papeno'o Valley in Tahiti from its settlement in the fourteenth century A.D. until its abandonment at the beginning of the nineteenth century. Observation of these modifications allows one to set up hypotheses concerning their origin in human actions and natural causes. The main evidence comes from excavations carried out between 1978 and 1982 in a rockshelter at Putoa, 6 km inland. Other observations were made between 1976 and 1983 on natural profiles in the banks of the Papeno'o River and its tributaries (Orliac 1984a, 1984b). I shall also refer to data from recent salvage excavations in the high Papeno'o basin, on the banks of the Tahinu River (Orliac 1990).

ENVIRONMENTAL EVOLUTION: FIRST RESULTS OF THE PUTOA
ROCKSHELTER EXCAVATIONS

Putoa Rockshelter, in the Papeno'o valley 6 km inland (Fig. 10.1), is formed by the sharply overhanging wall of a 50 m high cliff that delimits the Tupa pla-

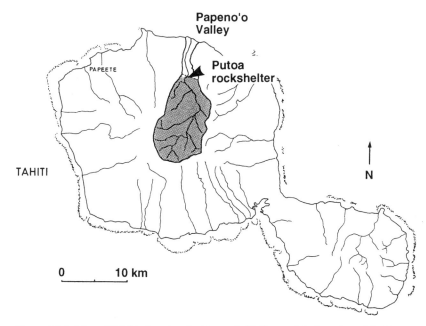

Figure 10.1 Map of Tahiti, showing the Papeno'o Valley and the location of the Putoa Rockshelter

teau. In front of the rockshelter, cemented alluvium forms a promontory covering an area of about 6,300 m². In its southeastern part, the promontory dominates the Papeno'o River by 10 m and, to the north, the little Vai Tupa stream. The promontory is closed off to the west by the overhanging cliff (Fig. 10.2). This broad, flat expanse occupies a dominant, easily defensible position. The rockshelter and the little dry stone enclosure inside were described by K. P. Emory (1933) in 1925. In 1976, J.-M. Chazine (1978) drew a plan of the numerous stone structures on the promontory: *marae,* low walls, pavements, and house borders. The topographic setting of this small religious and probably residential group of structures suggests it was assigned to a person of distinction. The structures are part of a relatively dense dwelling complex, settlement of which was favored by the existence of vast fluvial terraces, both on the left bank of Vai Tupa and on the right bank of the Papeno'o River. The Putoa tract is a part of the *mata'eina'a iti Tauaro i roto i te fa'a,* one of the most important of the Papeno'o mata'eina'a, or land-holding descent group (Orliac 1986).

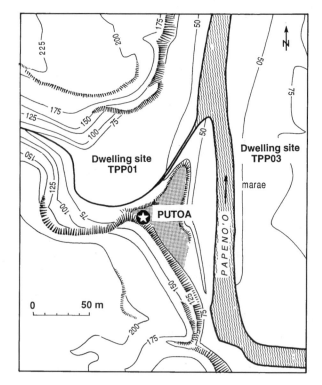

Figure 10.2 Topographic map of the environs of the Putoa Rockshelter

Excavations at Putoa Rockshelter

Two test pits were excavated in Putoa Rockshelter. The first (Units D16 to D22), 8 m² in area, was cut into the driest part of the shelter, which is never wetted by rain (Fig. 10.3). Surface vegetation is composed only of occasional grasses (Gramineae); nevertheless, a *ti* (*Cordyline fruticosa*) and a large *mape* tree (*Inocarpus edulis*) grow along the cliff. The archaeological levels, 1.70 m thick in this locality, contain numerous features (postholes, pits, hearths, and ovens) and many artifacts (debris from flaking stone; stone, bone, and shell tools; terrestrial and marine mollusks; fragments of shellfish; scales and bones of fishes; and bones of turtles, lizards, birds, rats, large mammals, and humans). In the upper layers, glass fragments, nails, a gun flint, and a clay pipe indicate a post-European contact age.

A few meters south of the first pit but still within the overhang of the rockshelter, another test excavation (F25), 4 m² in size, was cut in a place where water drips from the wall and percolates into the soil; *opui* plants (*Amomun*

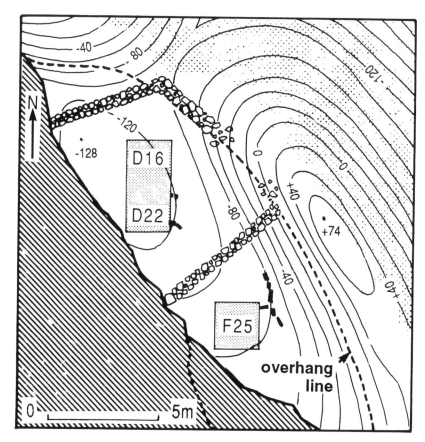

Figure 10.3 Plan of the Putoa Rockshelter, showing the enclosing rock wall and the location of excavation units. The dashed line indicates the overhang of the shelter roof (the drip line)

cevuga) grow here. The archaeological layers, 1.70 m thick, contain numerous occupation features, in particular large ovens. Flaking products abound, but organic remains have disappeared, except in the uppermost layers. In this excavation, the stratigraphy enables us to understand better than in Units D16 to D22 how the deposits accumulated.

Dating

Radiocarbon. Chronological subunit 400 (Unit IV), containing traces of the first human occupation of the shelter, has not been yet dated. Two dates were obtained in 1986 from samples from Pit D22 by the CNRS-CEA laboratory in Gif-sur-Yvette. The first date (Gif 7129), on charcoal from chronological subunit 610 (JTL), one of the earliest in Unit VI, is 580 ± 60 B.P., (A.D. intercepts 1284 and 1425 [Stuiver and Reimer 1986] or A.D. 1288 and 1429 [Pazdur and Michczinska 1988]; see Fig. 10.4). The second date (Gif 7128), 160 ± 60 B.P. (A.D. 1645 and 1950) was obtained on charcoal from chronological subunit 820 (PR.A oven), showing a new occupation of the shelter after a rock fall and a sterile washed-in layer. Subunit 820 does not contain European artifacts.

Archaeological dating of flaked stone. The presence of flaked stone provides information on activities and also on chronology, because in Tahiti stone tools were very quickly abandoned in favor of metal tools. Some 2,743 flakes and stone tools came from the D16 to D22 pit. Their distribution (Fig. 10.5) reaches a maximum in Unit VI and at the base of Unit VII (15 to 25%). There is a certain homogeneity until the 1000 to 1041 subunits (10%); this homoge-

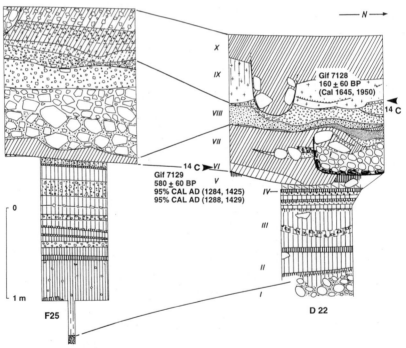

Figure 10.4 Stratigraphy of excavation units F25 and D22 in the Putoa Rockshelter, with the location of radiocarbon samples

neity is broken by the minimum of Unit VIII (5%), a layer of which (810) is sterile. In the uppermost layers (1042 to 1044), the scarcity of flaked stone (1.6%) shows the replacement of stone tools by metal tools. Nevertheless, artifacts from these layers are too numerous to have originated through disturbance of the underlying layers; they are certainly temporary tools, only occasionally used. Chronological interpretation of these data leads us to assign the layers of Unit IX and those at the base of Unit X to a period during which stone was still being used for tools. It seems, however, that the use of stone was declining during the period represented by the upper layers, which therefore probably date from the end of the eighteenth century or to the early nineteenth.

Remains of the European-contact period. The stone-tool data agree with those from intrusive remains. The earliest evidence of European arrival is the presence of introduced terrestrial mollusks in subunits 912 to 920. The first foreign artifact is a flaked flint (gun or lighter flint) from subunit 940; the foreign presence is really attested in subunits 1010 and 1020 by the bones of goats and European rats and in subunits 1030 and 1040 by nails and glass frag-

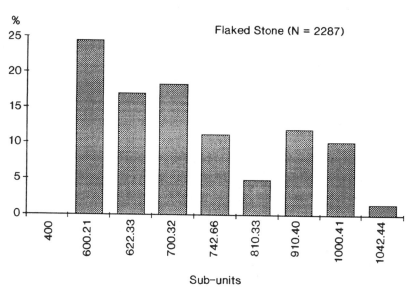

Figure 10.5 Frequency distribution of flaked stone in Putoa Rockshelter (oldest deposits on the left, youngest on the right)

ments. Three fragments of the same clay smoking pipe, scattered over 8 m² of subunits 1030 and 1040, date the final occupation of the shelter to the end of eighteenth century or the very beginning of the nineteenth.

The Depositional Sequence

After the fluvial terrace was deposited and cemented, its surface was disaggregated (Unit I). This was followed by an accumulation of fine deposits, 1 to 2 m thick, composed of silt with some small stones (<1 cm), alternating with thin beds of clayey silt (Units II to V) containing leaf prints and algae. The first occupation of the site, as yet undated, is represented by the upper part of the fine deposits. During a brief stay, the first visitors to the shelter lit a fire near which they ate crabs, birds, fluvial mollusks, and mussels. In this period, the Pacific Rat (*Rattus exulans*) was already plentiful. The nonmarine mollusks in this deposit are those of the primary forest. After this brief, initial occupation the prevailing sedimentary conditions were not affected and continued for a time not possible to estimate accurately, but sufficiently long for the deposition of 15 to 20 cm of sediment (Unit V).

During the fourteenth century, the main settlement under the shelter and probably on the promontory was marked by construction of structures (indicated by postholes), pits, hearths, and ovens (Unit VI). The site was at this time the center of various activities, including cooking, the flaking and sharpening of adz blades, and the shaping and using of various bone and shell tools. The occupants enjoyed sufficient status that they possessed a tattooing comb. The sediment that preserved these remains, 15 to 20 cm thick, derives from the disturbance of underlying sterile layers mixed with anthropogenic deposits composed mainly of oven stones. Color and texture of sediment were scarcely modified, because there was little charcoal in the cultural layers. This indicates that the intensity or duration of occupation was not great.

A black group of layers (Unit VII), 50 cm thick, follows abruptly after the former unit. Dwelling structures (pits, hearths, ovens, sometimes 1.5 m in diameter) occupy the whole area of the test pit; the deeper and middle parts of this unit are formed by a tangle of structures, the chronology of which is possible to establish; only the upper part of this unit can be considered a true occupation level. The density of structures and the complexity of ash beds in the ovens underscore the intensity and long duration of occupation. The exceptional presence of two flakes of volcanic glass—of which there are only five other known examples in Tahiti—might be linked to ritual activities (such as circumcision?).

At the time these intensive activities were occurring, a rockfall (Unit VIII) resulted in a layer of blocks (60 cm thick) in the southern part of the shelter

(Pit F25). Fine sediments were washed in around this stone fan, and they are 50 cm thick in the southern part of the shelter (F25) and 25 cm thick in its center (D22). The surface of Unit VII was preserved intact by these sediments. This deposit, which exhibits several phases, first provoked a complete break in activities; then, after a period of uncertain duration, some hearths and ovens were lit here and there. Tools were flaked in their vicinity, and human remains were deposited on the ground, where they were scattered over several square meters. This sporadic reoccupation dates to later than the middle of the seventeenth century (based on a radiocarbon age determination). The presence of dog and probably turtle remains confirms the high social status of the shelter's occupants. The deposits of human bones confer on the shelter a sacred character, which was probably well established before this period. The shelter was then gradually reoccupied. Unit IX is formed of ashy layers, 30 to 50 cm thick; activities other than cooking are of slight intensity. This unit is wholly contemporaneous with the period of European contact.

Unit X, 20 to 50 cm thick, comprises very black layers, deriving from activities as diversified as in Units VI and VII: cooking, and shaping and using various stone, bone, and shell tools. This unit dates from the end of the eighteenth century to the early nineteenth. The upper layers were weakly affected by pedogenesis after the shelter's abandonment.

The rockshelter chronology can be summarized as follows: At an undetermined time, the shelter received a first visit (Unit IV), before the primary human occupation in the fourteenth century (Unit VI). While occupation was intense, a rockfall interrupted the use of the site (Unit VIII). Reoccupation of the shelter took place after the middle of the seventeenth century. The upper 30 cm of the shelter's deposits probably span the last third of the eighteenth century and the early nineteenth centuries (Units IX and X); sedimentation stopped when the site was abandoned.

Modifications in Flora

Prehistoric floral modifications can be detected through direct (pollen, charcoal) and indirect (nonmarine mollusks) evidence, although these data are of unequal value. Pollen is not well preserved in the site's deposits, while charcoal sometimes reflects human choices and at other times the natural vegetation. Nonmarine mollusks, despite their brittleness, are in this case the most reliable witnesses of the vegetal environment.

Present-day vegetation. Two surveys made in 1982 by Jacques Florence, a botanist at the Centre ORSTOM de Pape'ete, Tahiti, in the neighborhood of the

Table 10.1 Stratigraphic distribution of identified wood charcoal from the Putoa Rockshelter

Fuel	Tahitian name	Family	Scientific name	VI 610 D22 JT.A	VI 621 D22 JT.J2	VI 622 D22 JT.B	VII 732 D22 VP.A	VII 732 D22 VP.A1	VII 761 D22 MC.A	VII 764 D22 MC.C	IX 910 D22 S.B	IX 920 D22 R.G	X 1040 F25 B.Dfond
?	?	Undetermined	?	X									
?	?	Gramineae	?			X	X		X			X	X
excellent	?	Gramineae	Schizostachyum galucophyllum					X	X				X
excellent	Purau	Malvaceae	Hibiscus tiliaceus	X							X		
excellent	Mara	Rubiaceae	Neonauclea forsteri		X	X	X	X		X	X	X	X
excellent	Toro'ea	Rubiaceae	Plectronia barbata		X	X	X	X					?X
excellent	Toi	Rhamnaceae	Alphitonia zizyphoides	?X				X	X		X	X	X
good	Tafano	Rubiaceae	Guettarda speciosa			?X			X				
good	Apape	Anacardiaceae	Rhus tahitensis	?X	X		X		?X		X	X	?X
medium	Mape	Leguminosae	Inocarpus fagifer					X					
?	Atahe	Apocynaceae	Alstonia costata				X	X	?X		X		X
?	Hutureva	Apocynaceae	Cerbera manghas		X	X		X	X	X			
?	?	Apocynaceae?	?			?X							
poor	Maiore	Moraceae	Artocarpus altilis		X		X	X	X		X		
?	?	Moraceae	?					?X					
?	?	Moraceae	Ficus sp.				X	X	X				X
?	Tuitui	Euphorbiaceae	Aleurites sp. (nut)	X	X	X	X	X	X				X

shelter enabled him to enumerate species by habitat and abundance. Based on these data, Florence concluded: "These two surveys made on a fluvial terrace point out a *Hibiscus-Neonauclea* mesophilic grouping, characteristic of river banks, enriched by more xerophilic elements, such as *Xylosma, Rhus* or *Premna*. It is still essentially a climax vegetation, despite the intrusion of *Tecoma stans* or *Spathodea campanulata*. The presence of *Barringtonia* and *Alocasia* provide evidence of an ancient human settlement." A floral survey was made four or five years after the clearing of the promontory. Second-growth and herbaceous strata had built up again between this clearing and the survey.

Pollen. Twenty samples from each of the stratigraphic units were analyzed by Anne-Marie Sémah, an archaeologist at ORSTOM, Centre de Noumea. In the best-preserved samples, 100 pollen grains and spores were counted. Spores had been well preserved, but there were only a few well-preserved pollen grains, and they did not provide any information about vegetational history.

Charcoal. Several thousand fragments of charcoal, from ten ovens of Units VI, VII, IX, and X, were analyzed by Catherine Orliac, and various species were determined (Table 10.1). The size of the fuel could be estimated through charcoal morphology and the examination of the convergence of rays, indicating that these were mostly twigs or little branches. Nevertheless, large branches, several decimeters long and about 10 cm in diameter, were burnt in large ovens (*umu ti*). We may therefore infer that wood gathering was opportunistic.

The function of the combustion structures must be taken into account in order to interpret the results of the anthracological study. The use of "specialized" ovens, such as large umu ti, required great expenditure of energy; in their case fuels offering the best calorific power (as in the MC.C oven) were chosen. In the commonly used smaller ovens, the choice of fuels was wider. In other respects, combustion effects were dependent on the nature of the fuel; some fuels burned completely into ashes, while others produced considerable charcoal. For example, Catherine Orliac has shown that *Hibiscus* gives more ash than charcoal (Orliac and Wattez 1989), so that this more common and abundant fuel will be the rarest in determinations of oven macro-remains. On the other hand, woods that do not burn easily, when used occasionally or mixed with good fuels, will produce much large, easily recognizable charcoal. But there are also excellent fuels producing solid charcoals, as in the case of *Neonauclea,* which will tend to be over-represented. In consequence, the information obtained is more useful for interpreting the function of ovens than for understanding eventual modifications of the vegetal landscape. As was predictable, *Hibiscus* is paradoxically scarce; conversely, *Neonauclea* is present in nine of the ten ovens studied. The trees that currently grow in the area of the shelter are the same as those that were burnt in the structures studied. Nevertheless, one can also add *Guettarda speciosa* (*tafano*), *Alstonia costata* (*atahe*), and *Alphitonia zizyphoides* (*toi*). Nowadays, *G. speciosa*, absent from the shelter's surroundings, is limited to the littoral zone, while *A. costata* and *A. zizyphoides* grow at a slightly higher altitude. *Cerbera manghas* (*hotureva*) and *Alphitonia*, among the largest Tahitian trees, became rare in the middle valley, no doubt because of the excessive felling that followed European arrival.

The breadfruit (*Artocarpus altilis*), not currently growing in this part of the valley, appears at the beginning of the settlement phase (oven JT.J2; subunit 621; fourteenth century); its presence reveals an environmental transformation corroborated by the nonmarine mollusk fauna. Its absence is not significant, because *Artocarpus* fire logs make poor fuel (Orliac 1991). In combustion structures, breadfruit appears only as twigs that presented no difficulties for burning. *Artocarpus* grew in the area of the shelter when subunit 910 was deposited that is, after the middle of the seventeenth century (according to ^{14}C dating) and after European arrival (according to the evidence of nonmarine mollusks).

The charcoal thus indicates that the fuels used between the fourteenth and nineteenth centuries give an incomplete picture of a tree flora that was undoubtedly more complex than the present one. The breadfruit, present during the settlement phase, indicates that humans were affecting the local environment by the fourteenth century; breadfruit is sporadically present between the middle of the seventeenth and the beginning of the nineteenth centuries. In spite of human actions affecting the environment, the traditional economy preserved large trees until the beginning of the nineteenth century, when excessive pressure, probably from whalers and European navies, led to their disappearance.

Nonmarine Mollusks. Nonmarine mollusks (land snails) from the D16 to D22 pit are well preserved and relatively numerous. Corinne Cherel-Mora and Simon Tillier of the Muséum National d'Histoire Naturelle, Paris, examined 1,065 specimens. The composition of the nonmarine-mollusk faunas reflects in a fairly precise manner the aspects of the vegetation to which they are closely linked. The presence of their shells in the rockshelter is partly independent of the cultural deposits, as these snails are inhabitants of the natural leaf litter or wall of the shelter and also live on organic refuse. Shells can be washed down with sediment or brought in on fuel. However that may be, the picture they give is that of the specific or local place where they lived, or of its near surroundings.

Regarding environmental change, the most interesting molluscan fauna is that linked to the primary forest, whose abundance reveals a landscape without transformation. Next come the endemic species from Tahiti and representatives of primary Polynesian fauna, less strictly linked to the primary forest; their dominance shows that eventual competition with introduced species favors the native species. The fauna indicating deep changes is that which comes with humans, introduced first by Polynesians during their transoceanic voyages and then by Europeans, who spread numerous species. Probably since the eighteenth century, but mostly in the nineteenth, these species (largely of neotropical, especially Southeast Asian, origin) have replaced many indigenous species.

These different groups are represented by the following species:

Group 1, primary forest: Endodontidae (several species [° aff.] close to species already known have not yet been described and are extinct today); °*Mauthodonta* aff. *parvidens* (Pease 1861); °*Nesodiscus* aff. *cretaceus* (Garrett 1884); *Libera microsoma* (Solem 1976); *Libera bursatella* (Gould 1846); *Libera spuria* (Ancey 1889); °*Libera* aff. *dubiosa* (Ancey 1889); *Libera garettiana* (Solem 1976); *Libera incognata* (Solem 1976); °*Libera* aff. *incognata* (Solem 1976).

Group 2, endemic to Tahiti: Helicarionidae; *Trochomorpha cressida* (Gould 1846); Truncatellidae; *Taheitia porrecta* (Gould 1848).

Group 3, primary fauna not linked to the primary forest: Partulidae; *Partula hyalina* (Broderip 1832); *Partula otaheitiana* (Bruguière).

Group 4, Indigenous? (subfossils only): Assimineidae; the species living today on the shelter wall (sp. "A") has not been found as a subfossil and is probably a recent introduction; the one found as a subfossil (sp. "B") is smaller; *Assiminea* sp. "B."

Group 5, species introduced by the Polynesians: Achatinellidae; *Lamellidea oblonga* (Pease 1864); *Lamellidea pusilla* (Gould 1847).

Group 6, neotropical origin (Solem 1964), or origin not well known (Christensen 1981), ancient introduction (before 900 B.C. according to Christensen 1981): Subulinidae; *Lamellaxis gracilis* (Hutton 1834).

Group 7, Neotropical Origin (The Antilles?), recent introduction: Subulinidae; *Subulina octona* (Bruguière 1792).

Group 8, Southeast Asian origin, recent introduction with coffee or sugarcane (Solem 1959): Bradybaenidae; *Bradybaena similaris* (Férussac 1822).

Group 9, recent introduction: Subulinidae; *Opeas pumilum* (Pfeiffer 1840).

The diagram (Fig. 10.6) expressing the relative importance of each of these groups in different chronological subunits shows, on both sides of the rockfall (subunits 800 to 830), a marked contrast in faunal composition. There is a high proportion of indigenous species under the rockfall, followed by a high proportion of introduced species above it.

In units situated under the rockfall, the molluscan story begins at subunit 400, with 99% of the fauna composed of indigenous species (two individuals—*Lamellidea oblonga* and *L. pusilla*—belong to Group 5, probably disseminated

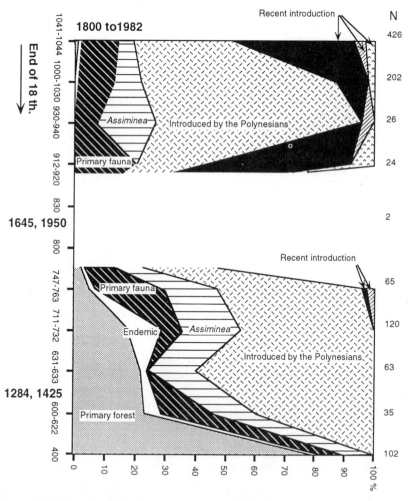

Figure 10.6 Temporal changes in nonmarine mollusks in the Putoa Rockshelter deposits. Bold numbers indicate radiocarbon dates.

by Polynesians). Moreover, 80% of this fauna is made up of primary forest species, four of which (now extinct) have never been described by malacologists. Obviously, this fauna is linked to the natural vegetation of the valley prior to human settlement. The absence of the introduced species *Lamellaxis gracilis*, which apparently followed humans and gardening during the conquest of the Pacific Ocean (it is present on Tikopia by 900 B.C. [Christensen and Kirch 1982]), argues in favor of a "pristine" environment. After the departure from the shelter of its occasional visitors, the sedimentary conditions that existed before their arrival returned for some time, indicating environmental stability.

During the fourteenth century, at the beginning of the settlement on the promontory and in the shelter (subunits 600 to 622), indigenous nonmarine mollusks fall off strongly in relation to the introduced *Lamellaxis gracilis*, which then becomes the dominant species. In spite of the environmental change this implies, primary forest species still constitute 20% of the fauna. The environment remained relatively stable until chronological subunit 732. Then, in subunits 747 to 763, when human activities were very intensive in the rockshelter, forest species suffered a new setback; nevertheless, indigenous forms remain strongly in evidence. Two species (*Subulina octona* and *Opeas pumilum*) indicative of recent or modern introductions are present in these units; as each is represented by a single individual, they probably indicate contamination rather than European arrival.

Only two *Partula* shells were found in the wash-in deposits (Unit VIII) succeeding the rockfall. The relative absence of other mollusks could be explained by the rapidity and nature of the deposit, as it seems that an excess of wetness quickened their destruction, so that only the solid *Partula* shells were preserved. Scarce hearths of the first reoccupation (Unit 820) date from A.D. 1645 and 1950; the molluscan fauna are equally absent from their vicinity.

Mollusks occur weakly again in subunits 912 to 920 (24 individuals). Despite the low number of individuals in Unit X, faunal composition expresses a real uniformity in Units IX and X, where it is completely different from the faunal composition of subunits 747 to 763. The primary forest species disappear entirely; henceforth, indigenous mollusks represent only 20 to 25% of the faunal composition. In the first subunits of Unit IX, mollusks introduced by Europeans are numerous. European contact is attested by the mollusks *Subulina octona*, *Bradybaena similaris*, and *Opeas pumilum* before the occurrence of the first introduced artifact, found in chronological subunit 940. Despite the absence of European artifacts, the abundance of introduced mollusks leads one to suppose that this episode is not earlier than the end of the eighteenth century; a more recent date would hardly be credible, because stone tools are still very numerous.

Rare shells of Endodontidae (*Libera garettiana*, one individual, and *L. spuria*, four individuals) found in the surface layer of Unit X could be interpreted both as a stronger occurrence of a very numerous population (426 individuals) and as a new dense vegetational colonization of the site that followed the abandonment of this part of the valley at the beginning of the nineteenth century.

In conclusion, the nonmarine mollusks provide significant information about environmental transformations in this part of the Papeno'o Valley. Before the fourteenth century, the evidence allows us to paint a picture of a virgin landscape, occupied by primary forest. During the fourteenth century, human settlement definitively altered the environment. The dominance of a nonmarine mollusk species introduced by Polynesians (*Lamellaxis gracilis*) reveals a strong human effect on the landscape (clearing and gardening). Nevertheless, primary forest still shows real strength, although it diminished when human activities became intensive. After a rockfall that broke off occupation of the rockshelter (provoking an abandonment whose duration is impossible to estimate), reoccupation of the shelter took place in an environment where species linked to the primary forest had completely vanished and recently introduced mollusks formed more than half of the fauna. This episode can be dated to the end of the eighteenth century, before European artifacts spread to this part of the valley. Sedimentary deposition ended at the beginning of the nineteenth century, and with it the recording of environmental changes.

These data reveal that human settlement was the direct cause of the environmental modifications in the vicinity of Putoa Rockshelter. Moreover, the modifications seem proportional to the intensity of human activities (as suggested by the diminution of the proportion of mollusks linked to the primary forest from subunits 711 to 732 and 747 to 763). On the other hand, it is difficult to interpret the changes after the rockfall and washed-in deposits. Site use became less intensive than before, but the mollusks reveal a strongly modified environment, corroborated by the presence of breadfruit tree charcoal (subunit 910). A hypothesis that emerges from this observation is that the environmental alteration could be linked to the event that provoked the rockfall. For instance, cyclonic actions on an environment weakened by human action could involve irreversible modifications. Meanwhile, the presence of *A. zizyphoides* in subunit 1040 shows that this large tree was preserved until the final occupation of the valley.

Faunal Modifications

Before interpreting environmental modifications evidenced by the faunal data, I must express some reservations. Faunal remains in an archaeological level—in particular near combustion structures—are generally interpreted as food remains. To be certain this is the case, however, it is necessary to examine them from both qualitative and quantitative points of view. One must consider whether their deposition is natural or indisputably linked to human activity. This issue mainly arises for species that may have lived and died in the area of the site, such as lizards, rats, and, to a certain extent, birds. From the quantitative point of view, the taphonomy of the various kinds of "food" remains must be considered: whether they derive from natural actions (corrosion, the actions of microorganisms), the cleaning of the floor by people or domestic animals (pigs, dogs, chickens), or quick burial in pits (voluntary or not). Finally, depending on these parameters, it is necessary to evaluate whether the quantity of faunal remains could provide a substantial part of the human diet.

The goal of the following discussion is to define the conditions under which the faunal remains were deposited in Putoa Rockshelter; they therefore concern first the quantitative aspect of the question. The excavated area of the site was suitable for faunal preservation, and brittle mollusk shells are well preserved, as are the bones and scales of fishes. Faunal remains did not suffer from chemical or biological actions, and, at most, corrosion acted in a more intensive way in the upper layers of Unit X than in others. Successive floors seem not to have been cleaned, for there were numerous remains on them. Refitting of shells, bones, and stone fragments revealed that they suffered only minor displacements. Remains are in equal quantities on floors and in the fill of structures, and they are no more numerous in voluntary fills (obviously the case for large umu ti) than in other contexts. Bones bearing gnawing marks are scarce, while mollusk shells show no evidence of having been pecked by chickens.

These observations argue that the Putoa faunal remains are representative of the animals that were eaten at the site, although they doubtless form a tiny part of what was cooked in the shelter and served elsewhere. From the quantitative point of view, the number of bones and shells preserved in the shelter is very low when divided by the five centuries of occupation. But quantity is balanced by diversity, for 18 marine-mollusk species are present, not to mention urchin and crayfish, three fluvial molluscan species, freshwater shrimp, and two edible river fishes, several sea fishes, at least five bird species, turtle, dog, and pig, to which goat is added in the upper layers.

Hence the quantitative data argue for a low amount of meat eating in the shelter, whereas qualitative data show a great diversity of rather small prey, from various terrestrial and marine environments. This diversity expresses a concentration on very varied resources, the food value of which was low. According to a sociological hypothesis, the remains indicate people of low rank eating little food. But this does not fit with the presence of dog and turtle, symbolic animals that were reserved for the higher social classes. The most satisfying hypothesis is that the food, parsimoniously selected in all its diversity, was used as propitiatory offerings "for the oven." According to an alternative hypothesis, the rockshelter was a sacred place in which bones of ancestors were deposited, in connection with the *marae* (temple) situated at the opposite end of the promontory; an offering altar (*fata ai'ai*) would have been set up in the shelter, where various foods were presented to ancestors and gods.

The edible mollusks. A total of 9,981 mollusk fragments were found (7,419 marine mollusks, 2,562 fluvial mollusks). Twenty-one species are represented, from the river and its tributaries, the reef, the sand of the river mouth, and the sandy beach of Papeno'o. Species living on the reef include *Spondylus* sp., *Turbo setosus, Hipponyx* sp., *Modiolus auriculatus, Conus* sp., *Cymatium* sp., *Pyrene scripta,* and *Morula* sp. Species dwelling on the reef or on pebbles and rocks in the upper part of the shore beaten by waves include *Nerita plicata* and *Patella* sp. One species occupies low water with small madreporic formations: *Cypraeacea* sp. Sand-dwelling mollusks include: *Periglypta reticulata, Gafrarium pectinatum, Asaphis violascens, Scutarcopagia scobinata, Terebra crenulata, Strombus mutabilis,* and *Pinctada margaritifera.* The *Pinctada* pearl shell in Putoa, present in the form of hooks and ornaments, was imported, as its environment (a closed lagoon) is not present in Papeno'o.

Three fluvial-mollusk species were brought into the shelter: *Clithon spinosus, Neritina canalis,* and *Septaria porcellana.* These species nowadays are occasionally eaten by some Papeno'o inhabitants. The ratio of the three fluvial-mollusk species fluctuated notably during the shelter's occupation (Fig. 10.7b).

The molluscan faunal composition shows important variations during the shelter's occupation (Fig. 10.7a). The fluvial mollusks predominate in Units IV (86%) and X (64 to 78%). Between these units, mussels represent 64 to 85% of mollusks. After the rockfall (Unit VIII), *Turbo* became the most abundant marine mollusk after the mussel.

The vertical distribution of the different molluscan species allows us the following conclusions: (1) The shelter's occupants have always kept links with the seaside; even the first occupants during their short stay (Unit IV) brought mussels, to which they added their gatherings in the river. (2) During the maximum occupation of the site, the numerous genera are represented by a small number of individuals, often of no food value. Were they perhaps used in some rit-

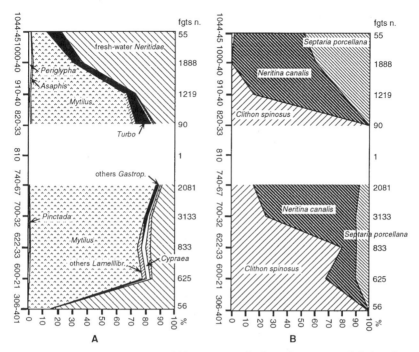

Figure 10.7 Temporal changes in edible marine mollusks in the Putoa Rockshelter deposits

ual or as ornaments? (3) Mussels were preferred to all the other species during the pre-European period. Their replacement by fluvial mollusks at the base of Unit X corresponds to a drastic change in the diet of the shelter's occupants; as a matter of fact, in this unit, remains of the other animals are the more abundant. (4) *Turbo* became relatively important after the rockfall (Unit VIII); this change appears before the replacement of mussels by fluvial mollusks. The increase in *Turbo* does not seem to have only an alimentary cause, for the individual numbers remain low. (5) *Clithon spinosus,* the endemic fluvial species, strongly diminished when the occupation intensity of the shelter was at its maximum, which possibly represents the consequence of strong collecting pressure. This mollusk—decorated with black stripes and little spines—is very pretty and could be used for ornaments, as with *Neritina canali.*

Fishes. The presence of bones and scales of fishes in Putoa Rockshelter can only be attributed to human activities. One thousand well-preserved fish remains (of which 125 are scales) were found in Units VII to X. Depending on the area dug, the majority (55%) come from the base of chronological Unit X, 28% from the upper part of Unit VII, 10% from Unit IX, and 7% from the upper part of Unit VIII. They were very rare at the base of Unit VIII and at the upper part of Unit X (Fig. 10.8). The high concentration of fish bones at the base of Unit X corresponds to a concentration of mammal bones, whereas their absence from the base of Unit VIII accords well with the rapid deposition of this layer and the subsequent abandonment of the rockshelter. The low proportion of fish bones in the upper part of this unit coincides with the infrequent use of the site, shown by the scarcity of other remains indisputably linked to human activities. Their great rarity in the upper part of Unit X equally confirms a diminishing frequentation, if not a change in activities.

A comparison between the bones of modern edible river fish, *nato* and *pui* (eels), and remains found in archaeological layers allowed me to recognize these two genera, in the form of vertebrae, rays of the dorsal fin, and scales for the nato and vertebrae and cranial pieces for the pui. Both genera form at least 10% of the fishes of Units VII and IX and at least half of those from Units VIII and X. Certain of the other remains come from small marine fishes, mostly Scaridae (parrot fish).

The quantity of fish bones does not represent a great number of individuals. For instance, an eel skeleton contains 80 vertebrae, whereas only 10 vertebrae were found in Units VII to X. In other respects, bones and scales

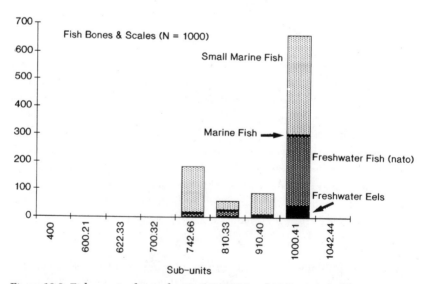

Figure 10.8 Fish remains from subunits 742 to 766 and 1042 to 1044 of the Putoa Rockshelter

attributed to the nato come from individuals of medium or rather small size. The eels are not of a kind to make a fisherman proud: most are 50 cm long or less; four individuals exceeding this size are not more than 1 m long. This is rather unimpressive in Tahiti where river eels are currently 1.30 m and more in length.

How can one interpret the low number and small size of individuals? Are they a result of (1) high fishing pressure or of (2) the social status of the shelter's occupiers? Or (3) were fishes offered to gods or ancestors in this sacred place? It is difficult to choose among these hypotheses, even if observations tend to indicate the third. Nevertheless, an obvious point asserts itself: without fishing rules, the probably high human population living in the valley from the fourteenth century would have caused the extinction of the poor fauna in its narrow river. But there were taboos protecting flora and fauna. One of the main functions of the leaders consisted of setting out rules to be respected; that the nato and pui survived five centuries of human occupation of the valley proves that these rules were respected.

Birds. A total of 570 bird remains were studied by Joelle Pichon and Jean-Christophe Balouet. These were in fragments rarely more than 1 cm long, and in consequence their determination was difficult, especially as comparative series are poor in France. However, their chronological distribution and the few determinations that were made enable us to assess both the extent to which birds were preyed upon and the evidence for environmental modifications (Fig. 10.9).

Most of the bones are those of small birds; chicken is absent. Both seabirds and landbirds are represented. From the lower part of Unit VI are a Tahitian Sandpiper and a Red-billed Rail; from the lower part of Unit VII, a parrot, a Red-billed Rail, and a shearwater; from the top, a petrel; from unit VIII, a parrot, a Tahitian Sandpiper, and a petrel; from Unit IX, a shearwater; and from Unit X, three petrels.

The origin of these bird remains poses a question: Are they the result of natural deposition (carcasses thrown out of nests, pellets of birds of prey) or of human activities (alimentary or technical refuse)? The very nature of the site argues in favor of the first proposal. Currently, swallows fix their nests to the shelter wall, and in the upper part of the cliff there are numerous hollows that could be used by petrels for nesting. Some bird remains could be those of individuals that died during brooding time, but the cliff's morphology would not allow large bird carcasses to be deposited in the sheltered part of the site (see limit of overhang, Fig. 10.3), and the scarcity of bird bones in upper layers does not accord with this hypothesis.

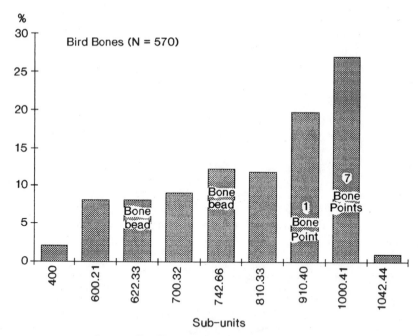

Figure 10.9 Distribution of bird bones in the Putoa Rockshelter deposits

The cliff could also have sheltered a member of the Strigidae (cf. *Strix delicatula*), currently known in the Austral Islands (Bruner 1972), the spread of which may formerly have been wider: Teuira Henry mentions the *o'oea*, or *pareva*, a "night bird flying along the coasts . . . whose species is now extinct in Tahiti" (1951); the extreme fragmentation of the bones is the same as in pellets. The lack of bird bones in superficial layers could correspond to the disappearance of this little bird.

Yet fowling was certainly practiced. It is difficult to assume from remains that birds have been eaten, even if it is likely in some cases. Heating marks visible on many bones are not relevant; they are often too strong or not well located. This reveals their casual origin in an area where combustion structures are contiguous. It is reasonable to ask questions about the food value of a little parrot, when the symbolic importance of its feathers is well known. Bird bones were used for making beads and points; some bear cut marks or wear-polish. A tubular bead was found in the upper part of Unit VI, another in the upper part of Unit VII. In Unit VIII there are a diaphysis bearing polished areas and another bearing cut marks; from Unit IX come two diaphyses with wear-polish and one point (from a composite hook?); seven other points come from the lower and middle parts of Unit X.

The presence of the Tahitian Sandpiper in Units VI and VII, 6 km inland, is more suggestive of fowling by humans than as prey for a small member of the Strigidae. To conclude, it seems that the presence of bird remains is due mainly to human activities.

Rats. Almost 2,000 rat bones were found. The origin of their presence in the rockshelter is as problematical as that of the bird bones: Are the rat remains of natural or of anthropogenic deposition? When Europeans arrived in Tahiti, rats were not eaten there, contrary to the case in other Polynesian archipelagoes (New Zealand, Tonga, the Cook Islands, Easter Island, Tuamotus). However, it is not impossible that they were eaten in the Society Islands in the distant past (Henry 1951:462-63). Nothing allows us to think it was the case during any phase of the rockshelter's occupation since the fourteenth century. There is no appreciable difference in the nature of the bones, in their frequency, or in the location of traces of their burning between the upper and the deepest layers.

It is not unusual to find rat skeletons in dry rocky crevices. In Putoa Rockshelter, one could add to the natural deposition of rat bones the pellets of the hypothetical member of the Strigidae I mentioned earlier. The fragmentation of rat bones argues for this hypothesis; as a matter of fact, from Units VII to X, intact specimens represent only 10 to 30% of the measured bones. In the top chronological subunits (1042 to 1044), this ratio exceeds 50% (Orliac 1981). As for bird bones, this decreasing ratio of fragmented bones could be linked to birds of prey vanishing at the beginning of the nineteenth century. Be that as it may, the vertical distribution of the 2,000 rat bones argues strongly for a natural origin. Furthermore, this distribution shows similarities with that of the bird bones. Obviously, some of the bird bones come from human activities, hence it seems normal that the proportion of rat bones is higher than that of bird bones in layers of Unit VIII, where human activity is weakest (Fig. 10.10).

A new study of rat bones was carried out by Judith Dubois (1991). In a population mainly related to *Rattus exulans*, this study shows higher values for humerus lengths in Units IV to VII than in Units IX and X. Tibias reveal an opposite trend: their dimensions increase from the bottom to the top of the stratigraphy. These changes are as yet difficult to interpret. Do they result from an internal evolution of the rat population or are they to be linked to environmental modifications? The final two units correspond to an important change in the terrestrial molluscan fauna and the vegetational environment.

The largest rat teeth (upper M1, upper M2, lower M1) and bones are all concentrated in the upper layers of Unit X (subunits 1020 to 1044, Orliac 1981); their dimensions, often greatly exceeding those of bones found in lower

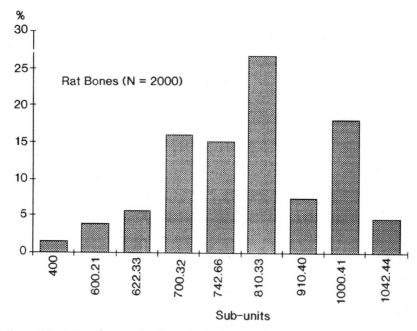

Figure 10.10 Distribution of rat bones in the Putoa Rockshelter deposits

layers, indicate the arrival of *Rattus rattus*. Remains of another large rat, probably *Rattus norvegicus*, were found only on the present ground surface.

Large vertebrates. The bones of large vertebrates (dog, pig, turtle, goat, human) were studied by François Poplin (Laboratoire d'Anatomie Comparée du Muséum National d'Histoire Naturelle) and Catherine Orliac. They consist of several hundred fragments, most of them not exceeding 3 cm in length; 166 were identified, mostly teeth. Large-vertebrate remains, scarce in Units VI to IX, are concentrated in the lower part of Unit X, comprising 70% of identified bones and teeth (Fig. 10.11). In all units except Unit VIII, the best represented animal is the pig, but more than a third of the remains are fetal bones or bones of suckling pigs, of no or very low food value. The same applies to the dog remains. The presence of dog and turtle bones is a sign of the high social rank of the site's occupants, further indicated a tattooing comb was found in Unit VI.

The composition of the large-vertebrate fauna of the upper levels of Unit VIII, corresponding to a period of low frequentation of the rockshelter, leads us to propose a model for the deposition of at least a part of the bones. In these deposits were found the scattered bones of a human skeleton: three vertebra

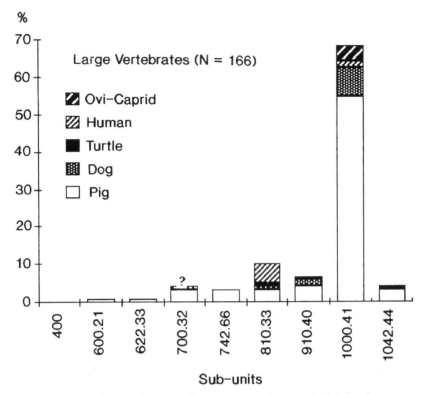

Figure 10.11 Distribution of large vertebrate remains in the Putoa Rockshelter deposits

fragments, two hand bones, one patella, and a femur fragment; in the same layer, there were also a dog rib and, from a pit dug into this level, probably a turtle bone. The pig bones consist of a fetal skull, a mandibular branch fragment, a maxillary fragment, and a phalanx. One should recall that birds are relatively numerous in this level, where parrot, sandpiper, and petrel remains were found. These data allow us to consider the fauna not only for its alimentary qualities but also for its symbolic function as a ritual offering. (Remains of suckling pigs as well as mussels and other marine mollusks were found in two of the burial sites in a small marae of the Tahinu Valley.) Turtle and sandpiper remains come from the seashore.

Sediment Modifications

It is interesting to compare the deposits predating human settlement (Units II to V) with those accumulated during the occupation of Putoa Rockshelter. Sedimentation probably began after rockshelter formation several millennia ago. Deposits vary in sediment thickness from 2 m (Pit F25) to 1 m (Pit D22), and the anthropogenic deposits attained a maximum thickness of 1.70 m during the roughly five centuries of human occupation. This simple observation indicates a faster rate of sediment deposition during the period in which the shelter was occupied.

In spite of this increased rate of deposition during the phase of human occupation, sediment texture differs only slightly between the two phases (before and after occupation of the shelter). Granulometry of the sediments was investigated by Jean Trichet (Laboratoire de Géologie de l'Université d'Orléans). Sediments from Units II to V are fine clayey silts, while the anthropogenic levels from Unit VI are silts, as are the washed-in deposits of Unit VIII. In the cultural levels, the coarse fraction of sands is a little more dominant than in the sterile levels. This can be interpreted as a relatively active washing effect on slopes lacking dense vegetal cover. Except in Unit VIII, sedimentation was relatively constant during the shelter's occupation. In other respects, it was notably increased by anthropogenic deposits (mainly oven stones). As soon as occupation ceased, however, sedimentation stopped too, allowing soil formation (upper part of Unit X). When one compares sediment composition before and after occupation of the shelter by humans, environmental modifications appear only as a higher rate of sedimentation accompanied by scarcely coarser sediment grains.

During the shelter's occupation, however, a major event marks the sedimentation story. It concerns the deposits of Unit VIII, a large rockfall, 0.60 m thick in Pit F25, the surface of which was intensively washed and covered by stratified beds of silt, 0.30 m thick in Pit D22. How can one explain the origin of this coarse slope deposit? Is it linked to human activity or to natural phenomena? The shelter is located at the foot of a cliff more than 50 m high, unfavorable to gardening; hence the rockfall is not likely to have originated from excessive forest clearance, and its depositional origin would seem to have been natural. Major rockfalls happen from time to time, as a result of slope equilibrium reaching breaking point. The fall of a tree growing high on cliffs or steep slopes sweeps away a significant quantity of rock and sediment. I observed a number of these rockfalls during 15 years of assiduous exploration of the valley. Moreover, these accidents multiply during high winds, and more so during tropical depressions and cyclones. Observations made in the Papeno'o Valley after the 1983 cyclones revealed how frequent rockfalls were and how vast their extension (Orliac 1984b).

Both sedimentary and archaeological indications argue in favor of a cataclysmic origin of the Unit VIII deposits. Indeed, the rockfall was followed by intense fluvial action, evidence of strong stream flow, as produced when diluvial rains follow cyclones. These deposits fossilized a cultural floor showing all the signs of intensive activities. This occupation, abruptly interrupted, would later recommence only weakly, as if the previous catastrophic event had severe consequential effects for settlement in this part of the valley.

BURIED ARCHAEOLOGICAL LEVELS IN THE PAPENO'O VALLEY

A few hundred meters east of Putoa Rockshelter, the right bank of the Papeno'o River displays a complex stratigraphy, 2.50 m deep, composed mainly of sand or silt flood deposits (Fig. 10.12). These deposits, in which charcoal appears from the bottom, had buried three superimposed pavements, and they contained numerous products of flaked stone. A pavement was laid at the top of the lower third of the deposits, after which a coarse colluvium was deposited from the hillside, which disturbed the cultural floor associated with the pavement, making the terrace higher. The origin of this thick, high-energy deposit is probably the same as that of the rockfall in Putoa Rockshelter (Unit VIII).

A few hundred meters downstream, a cultural layer with combustion structures was buried under very coarse, high-energy deposits, 1.80 m thick, dating from 220 ± 80 years B.P. (Gif 4405, A.D. 1435 to 1950). On top of the alluvial fan of the stream that feeds these deposits, there were house remains, in the neighborhood of which were found stone tools. This construction cannot be later than the end of the eighteenth century, as the event which provoked the formation of this coarse deposit is situated between the fifteenth century and the end of the eighteenth. One more kilometer downstream, combustion structures dated to 220 ± 60 years B.P. (Gif 4405, A.D. 1510 to 1950) were buried under coarse flood deposits 3 m thick (Fig. 10.13). Here a small shrine was built on the surface of the deposits not later than the very beginning of the nineteenth century. Similar observations and dates come from the middle valley, 7 km inland (250 ± 80 B.P., Gif 4404, A.D. 1425 to 1950, Site TPP05). After cyclone Veena in 1983, archaeological survey of the banks of the Papeno'o and its tributaries enabled us to find 11 other places where cultural layers were deeply buried.

At the mouth of the Papeno'o Valley, research on marae Tetuahitia'a has shown a cultural layer spreading beneath the present monument and a colluvium deposit 1 m thick. At the other end of the valley, in the very heart of the caldera, salvage excavations in the Tahinu basin have revealed, in a level containing structures dated to the fourteenth century, the existence of vast spreading "sheets" of thin or coarse colluvium, 0.15 to 0.50 m thick; at least three of these sheets are covered by structures dating from the end of the eighteenth century.

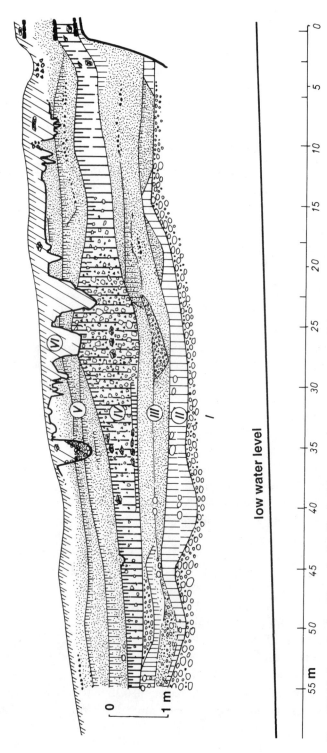

Figure 10.12 Stratigraphy of the right-hand bank of the Papeno'o Stream, in front of the Putoa Rockshelter. Note the three superimposed stone pavements at the far-right edge of the section, and the coarse slope deposit (Layer IV).

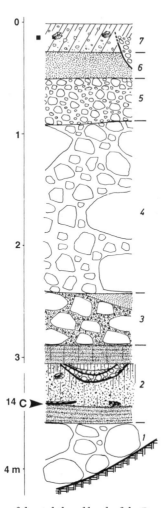

Figure 10.13 Stratigraphy of the righthand bank of the Papeno'o Stream, 1 km downstream from the Putoa Rockshelter. Note the deeply buried archaeological floor with combustion structures. The ^{14}C age from the sample indicated by the arrow is 220 ± 60 B.P.

It is pointless to attempt to synchronize all these deposits before the end of the eighteenth century with the same cataclysmic event, as there were clearly several phases of rapid deposition. It is obvious that no part of the valley was spared. It is even likely that the strongest of these events was fixed in oral tradition in the form of a deluge myth (Henry 1951); in fact, this narrative described phenomena precisely corresponding to cyclone actions. Unfortunately, the traditions do not give any clues allowing us to pinpoint the event. Notwithstanding the uncertainties of ^{14}C dating, it is possible to assign the cataclysm to the sixteenth or seventeenth century.

The evidence of catastrophic events linked to cyclones during this period leads us to reexamine interpretations of slope deposits or of the abrupt decline of trees in pollen diagrams. The origin of this environmental damage is generally attributed to human activities, mainly to clearance by fire. But this interpretation cannot be proved in the case of flood deposits or vast spreads of silty colluvium, because these phenomena can only be the consequence of a soil erosion out of proportion to any clearing done by Polynesians. Vast fire clearances are quite impossible to carry out in the upper part of the Papeno'o basin, where rainfall totals 10 m per year!

A deficit of trees in pollen diagrams often raises the same difficulties of interpretation as slope deposits do. On Easter Island (Flenley and King 1984), a severe decrease in the tree ratio appears long before any human presence. Pollen analysis from Vaihiria Lake in Tahiti (Parkes and Flenley 1990) also indicates deficits; but it would be astonishing if people cleared the abrupt slopes surrounding the lake, as these do not present any possibilities for gardening or settlement. In both these cases, it is at least as likely—if not more so—that environmental modifications originated in natural causes rather than in human activities. This is a much-debated question (McFadgen 1985; Grant 1985; McGlone 1989) to which I do not claim to bring a definitive solution, even if my affection for the Polynesians incites me to choose natural actions to explain the damages suffered by the environment.

CONCLUSION

In the Putoa Rockshelter, the impact of human settlement on the local environment first appears obviously in the fourteenth century, through increasing alteration of the primary forest. Nevertheless, this transformation becomes really important only after several catastrophic events occurred, probably cyclones, during the sixteenth or the seventeenth century, followed by more cataclysms after European arrival at the end of the eighteenth century. Large trees in the valley barely survived intensive cutting during the nineteenth century. The fauna does not reveal significant changes before European arrival, after which several bird species were to vanish. Strict regulations permitted the river fauna to survive during the five centuries of the valley's occupation. It seems evident that the environment suffered more from cyclones and the European arrival than from the Polynesians' activities.

Note

Excavations of the Putoa Rockshelter were supported by the Département d'Archéologie du Centre Polynésien des Sciences Humaines; in particular, staff members Raymond Teri'iero'o Graffe, Ruddy Tevivi, and Taverio Vaira'a. I also received help from Marie-Christine Deprund, Tiare Ebb, Catherine Orliac, Pierre Ottino, Barry Rolett, and Alain Renard. Specialized studies were carried out by Georgette Delibrias (data), Judith Dubois (rats), Jacques Florence (botany), Catherine Orliac (large vertebrates and charcoal), Joelle Pichon and Jean- Christophe Balouet (birds), François Poplin (large vertebrates), Anne-Marie Sémah (pollen), Simon Tillier and Corinne Chérel-Mora (terrestrial mollusks), and Professor Jean Trichet (sediments). I am grateful to Paul G. Bahn for kindly reviewing my translation.

11
Pre-Contact Landscape Transformation and Cultural Change in Windward O'ahu

Jane Allen

Archaeological evidence collected since the Fosberg symposium, "Man's Place in the Island Ecosystem," in 1961, strongly supports the view that the inhabitants of island areas throughout the Indo-Pacific region have traditionally transformed their environments in order to moderate nature's influences on culture (e.g., Allen 1990; Brookfield 1972; di Piazza and Frimigacci 1991; Kirch 1982b, c). Archaeological studies of both tropical and other areas tend to emphasize the effects humans had on the physical environment, probably for two main reasons. One is that anthropologists focus on humans and on cultural technologies that were designed in part to control nature. The other is empirical: human impacts are often easy to recognize archaeologically, whereas the impacts of nature on humans in any but disastrous situations are generally visible only indirectly, through the mediating screen of culture. Even significant climatic and sea-level changes may be difficult to recognize empirically (e.g., Nunn 1991), especially when the available evidence represents isolated or widely separated times.

One advantage of archaeology as a method is that it applies a diachronic perspective to studies of ecological change (see Christensen 1989). In this chapter, I present geoarchaeological evidence from windward (eastern) O'ahu Island, demonstrating both cultural influences on nature and environmental influences on culture through time. The evidence comes from four windward valley systems that dominate four traditional and important *ahupua'a*, land-tenure units that reached from mountains to sea, crosscutting varied environmental and economic zones (Fig. 11.1). Waimanalo Bay, Maunawili Stream, and Kawai Nui Marsh in Kailua ahupua'a, the Luluku Stream and Kamo'oali'i Stream valleys in Kane'ohe, and Kahana Valley in Kahana ahupua'a offered Polynesian colonists such varied resource zones as beaches, protected bays, permanent streams, coastal and inland forests, and arable streamside lands. Waimanalo, Kailua, and Kane'ohe were among the earliest areas settled on O'ahu, around A.D. 400 to 600. Colonists may have settled in Kahana Valley even earlier, by 30 B.C. (Beggerly 1990:327).

Much archaeological evidence for environmental change in Hawaii, and throughout the tropical Pacific, comes from areas that were used for extensive and/or intensive cultivation before Euro-American contact (1778 in Hawaii),

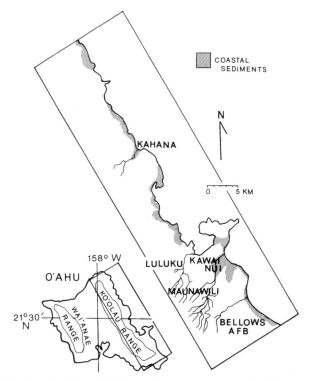

Figure 11.1 Map of windward O'ahu, with site areas mentioned in the text

often over several centuries. Agricultural sites offer several distinct advantages for geoarchaeological and other environmental studies. Agricultural sites are extensive, permitting data collection and horizontal correlation across broad areas. They often incorporate complex vertical sequences that represent land use and change over long periods. And because rural areas are often the last to be developed, large numbers of agricultural sites still exist in relatively unmodified condition.

CHANGING LANDSCAPES

Bellows Air Force Base, Waimanalo

Site o18 (Fig. 11.2), located on a relict, dune-capped, marine barrier beach 50 m inland from the current shoreline on Waimanalo Bay, was among the first Hawaiian sites to produce pre–A.D. 1000 dates (Pearson et al. 1971). Three

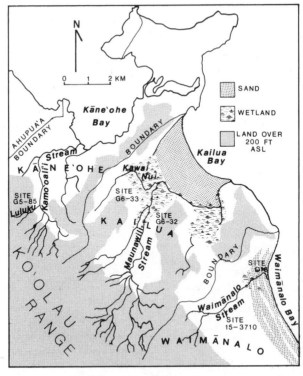

Figure 11.2 Waimanalo, Maunawili, Kawai Nui, and Luluku, Kane'ohe: sites and environmental features

main buried cultural layers excavated during the first archaeological season (1967) produced artifacts resembling early Marquesan types (e.g., reversed triangular adzes, a *Conus*-shell coconut grater; Pearson et al. 1971). The site also produced a radiocarbon sequence that, although inverted, suggested occupation by A.D. 600.

Because of the site's apparent antiquity, several additional attempts (Cordy and Tuggle 1976; Kirch 1974; Miller 1989; Tuggle et al. 1978) have been made to date the layers through both radiocarbon assay and volcanic-glass hydration rind analysis. Eight ^{14}C ages and 37 hydration dates suggest strongly that o18 was first occupied sometime between A.D. 300 to 500 and 1300.

The o18 beach ridge today forms a sand barrier at the seaward edge of an extensive floodplain built primarily by Waimanalo Stream and two main tributaries, which drain the uplands surrounding the area and merge in the backplain. The combined stream today empties into Waimanalo Bay through modern drainage beside the o18 barrier; it was formerly blocked by the barrier and forced to flow southward in a swale for some distance before breaking through to reach the bay. The backplain comprises narrow, streamside, finger wetlands separated by large tracts of marine-sand deposits for more than 1.6 km from the beach. Inland from that point, the floodplain broadens as two main stream valleys enter it from the hills inland.

The marine-sand tracts that punctuate the Bellows backplain have been modified during modern times but undoubtedly represent former barrier beaches that, like the o18 barrier, once paralleled the coast, trapping marine (calcareous) and terrigenous (basaltic) sediments in swales. Historic maps of the area (E. D. Baldwin's 1917 map of Waimanalo Plantation and Waimanalo Sugar Plantation's 1935 map, both reproduced in Athens 1990 [Figs. 6, 8]) suggest at least four barriers, with intervening swales, for areas north of Waimanalo Stream near the coast and o18. The backplain here, then, was not a broad lowland expanse or a basin but a series of concentric, prograding barriers and swales like those characteristic of many coastal areas in the Indo-Pacific region (e.g., Jane Allen 1988, 1990; Stargardt 1983).

Artifacts of the early types found at o18 have also been recovered from areas behind the o18 barrier beach (Leidemann and Cleghorn 1983). They had apparently been deposited on high ground formed by barrier sediments, not in a floodplain. Site 50-80-15-3710 (State of Hawaii site number) produced two fifteenth- to sixteenth-century ^{14}C dates (490 ± 70 B.P., 520 ± 60 B.P.) processed on charcoal and a charred log recovered from a gleyed wetland clay layer overlying calcareous sands (Athens 1988). These are the earliest dated terrigenous soils in the Bellows backplain. Athens concludes (1987:63; see also Beggerly 1990) that the site was a saltwater wetland until A.D. 1400, quickly filling with terrigenous sediments after that date to become a brackish or freshwater wetland by 1500.

On current evidence, the development of arable, terrigenous soils in most of the Bellows backplain postdates A.D. 1300, probably also postdating abandonment of site o18. But areas farther inland would have filled in with soils earlier and, as suggested by Kirch (1985:74), could have supported agriculture even during the colonization period. A *kukui* (*Aleurites moluccana*) nut recovered from the beach site (o18), which has produced a ^{14}C age of 1330 ± 230 B.P. (Spriggs, pers. comm., 1991), almost certainly represents cultural exploitation of inland hillslopes and stream valleys. Kukui typically grows in upland areas, and kukui nuts are frequently recovered from inland pondfield soils in windward O'ahu (e.g., Allen et al. 1987:172).

Although no dated sequence is yet available for the inland Waimanalo, slopes and valleys in nearby Maunawili Valley have supported dryland and irrigated agricultural complexes since A.D. 1000 or earlier (Allen, ed., 1994). The lower slopes bordering the Bellows floodplain were probably also under cultivation relatively early in the o18 sequence.

The Bellows swales and later the floodplain filled with terrigenous and marine sediments trapped behind the former barriers. The probability that humans were present in the Waimanalo watershed before 1000 introduces the possibility that the filling in of the backplain was influenced by human activities, including cultivation of the inland areas. For the Bellows Air Force Base area, this suggestion remains a model. More conclusive evidence comes from Kawai Nui Marsh, north of Waimanalo.

Kawai Nui Marsh, Kailua

Two radiocarbon dates, 1210 ± 215 B.P. and 1500 ± 145 B.P. (conventional radiocarbon ages, uncorrected), reported by Clark (1980, 1981) suggest that the burning of vegetation to clear dryland fields on slopes beside Kawai Nui Marsh began between A.D. 305 and 955. A pit feature at site 50-oa-G6-33, west of the marsh, produced a third early date, 1220 ± 90 B.P., suggesting occupation between A.D. 640 and 820. The earliest cultural dates from the marsh basin itself are much later, reflecting cultivation of taro (*Colocasia esculenta*) in the fourteenth through seventeenth centuries (Jane Allen 1990, 1992; Allen-Wheeler 1981; Cordy 1977, 1978). As suggested by geomorphologist John C. Kraft (1980a, 1980b), Kawai Nui ("big water" in Hawaiian) apparently held a bay and subsequently a lagoon until sometime after the initial occupation of the surrounding slopes by Polynesian colonists. No soil cover existed in most of the basin until relatively late in the pre-contact cultural sequence.

Geological corings in the marsh basin began with Dames and Moore's

(1961) transects, which investigated open marsh areas and the margin of the sand beach barrier that underlies Kailua town. Archaeological studies conducted in the wetland include Cordy's (1977, 1978) excavations below the southeast slope; Kraft's (1980a, 1980b) coring studies at 13 localities throughout the basin (Kirch 1985: 75); excavations below the southeast slope and 80 m into the marsh (Allen-Wheeler 1981: Fig. 5); Hammatt et al.'s (1990) corings in open marsh areas and below the southeast and western surrounding slopes; and Athens and Ward's (1991) study, which investigated a levee constructed inland behind the Kailua barrier, and two open marsh areas (Cores A and B).

The Kawai Nui evidence indicates that a bay occupied the basin between 4000 B.C. and A.D. 200. Basal marine bay sediments contain coral rubble, shells of bay mollusks, including *Tellina* (Hammatt et al. 1990), *Brachiodontes* (Athens and Ward 1991), *Trochus* and *Pinctada,* and shark teeth. The marine layer has been ^{14}C-dated to 3910 ± 90 B.P. (Allen-Wheeler 1981) on fresh marine shell. Hammatt et al. dated sediment samples from Cores 5A, 6, and 7, producing the following dates, respectively: 2075 ± 80 B.P. (390 B.C. to A.D. 200); 5370 ± 90 B.P. (4425 to 3890 B.C.); and 2430 ± 70 (780 B.C. to 400 B.C.). Athens and Ward (1991) report the following ranges (conventional radiocarbon ages not reported) for marine bay sediments in Core B, at an open marsh locality: Layer VIII, silty loam with shell, bulk sample dated 1520 to 960 B.C.; Layer VII, clay loam with shell, coralline residue dated 1710 to 1016 B.C.; and Layer V, clay loam with shell, bulk sample dated 770 to 230 B.C.

Kraft (unpublished data, cited in Hammatt et al. 1990:20–22) suggests that by A.D. 350 to 650 the Kawai Nui bay was transformed into a lagoon, progressively blocked behind the Kailua sand barrier and the reef tract that forms its substrate. The Kailua barrier formed and emerged from the sea after initial Polynesian colonization of the area, reaching approximately its current length and breadth by A.D. 1750, shortly before Euro-American contact.

Between A.D. 600 and 1300, a thick peat layer accumulated at the southeast marsh localities that I tested in 1981. This peat formed as a floating vegetation mat and provided a base for later sediments and soils, which have now buried the peat to depths of between 2 to 5 m at localities tested by Hammatt (1990), Athens and Ward (1991), and myself. By 1400 to 1500, terrigenous sediments filled large areas of the basin, supporting a grass- and sedge-dominated marsh (Hammatt et al. 1990). Available sedimentary and dating evidence indicates that most filling in of the basin with terrigenous sediments occurred after 1000. As occupation of slope areas had begun about A.D. 500 to 600, and charcoal from adjacent slopes had been deposited in the basin before 1000, it appears likely that humans played a role in bringing about the filling in of the lagoon.[1]

Athens and Ward (1991:114–16) have identified, from the peat layer dated to about A.D. 650 to 1300, pollen and spores from *Cibotium, Metrosideros, Ilex, Pelea,* and possibly *Pritchardia;* underlying lagoonal (coralline) sediments produced all but *Ilex* and *Pelea.* Forest communities therefore characterized the nearby slopes until around A.D. 1300. After 1300, sedges and then grasses increased in dominance, suggesting deforestation. One pre-contact tree surge suggests forest regeneration during the period of marsh-floor pondfield taro cultivation and hillslope use. Ward's (1990) and Athens and Ward's (1991, 1993; Athens et al. 1992) results support these findings. Grass pollens largely replaced *Pritchardia* and other tree pollen by A.D. 1410 to 1650, again suggesting deforestation during the pre-contact sequence.

Athens and Ward (1991:99) recovered no charcoal fragments from their core samples, with the exception of Core A, 192 to 196 cm below surface. The absence of charcoal, the researchers feel, argues against extensive use of fire to clear vegetation on surrounding slopes during the pre-contact period. Other explanations may exist, however. First, I have generally found that although charcoal is common in agricultural soils, it is rarely found in the hillslope sediments and soils that manage to invade terrace complexes during periods of abandonment or fallow. I have suggested recently (Allen 1992) that the reason for this is that most charcoal in agricultural fields, even pondfields, is primary, not secondary. Another possible explanation for the lack of charcoal in Cores A and B is their location in areas where stream waters flow freely, presumably carrying light, fine particles farther seaward. A third is that core sample size may be too small to catch all charcoal. During trench excavations in 1981, charcoal was recovered from all layers down to the peat layer (VI: A.D. 600 to 1300) in Trench C, below the southeast slope, and through Layer IV (probably fourteenth century A.D.) in Trench B, at an open marsh location (Allen-Wheeler 1981).

In sum, the filling in of the Kawai Nui basin with terrigenous sediments after A.D. 1200 to 1300 occurred during a period of expanded agricultural use of hillslopes inland, including Maunawili Valley (Allen, ed., 1994). Sediments eroded from Maunawili's slopes would have been carried by Maunawili Stream and its large network of tributaries into Kawai Nui Marsh on their way to Kailua Bay. The arable soil that formed on sediments deposited in the Kawai Nui basin eventually supported taro pondfield agriculture in the marsh floor until around A.D. 1850, and irrigated rice for 75 years after that.

The Luluku and Kamo'oali'i Watersheds, Kane'ohe

Cultural landscape transformation reached near-monumental proportions along Luluku Stream, Kane'ohe, where extensive complexes of cut-and-fill pondfield terraces constructed during the pre-contact era created a stepped landscape on a slope that exceeds 20 degrees in many areas. Stream waters diverted from Luluku Stream through irrigation ditches were dispersed by gravity across the gently sloped fields, and they flowed from field to field through gap and mound spillways. Large terrace sets, many now destroyed, were cultivated along the entire Kamo'oali'i Stream network (McCoy 1976; Shun et al. 1987).

The earliest radiocarbon dates from site 50-OA-G5-85 in Luluku indicate that irrigated agriculture began in downslope areas: the earliest pondfield (Feature 35, Layer VIII) was cultivated at a floodplain locality about A.D. 500 to 700 (two samples on charcoal, 1560 ± 100 B.P., 1330 ± 150 B.P., conventional ages). This pondfield soil, which now lies buried beneath 120 to 150 cm of later agricultural soils and colluvial sediments, occupied a floodplain at the time in question (Allen 1991b; Allen et al. 1987). Most buried pondfield terraces farther upslope postdate A.D. 1200. Complex buried sequences excavated at the site represent up to five periods of pondfield cultivation, separated by thin colluvial deposits that invaded the terraces during fallow or abandonment. In one area, the lowermost cultural layer is a cemented, oxidized terrace agricultural soil that represents dryland cultivation.

The construction of terraces dramatically changed edaphic and hydrological regimes in this inland hillslope zone, trapping soils, sediments, organic matter, and water behind large, rock facings. Where the forest had not previously been cleared, the terraces encouraged renewed vegetational stability but supported a much-altered floristic community of primarily cultigens.

Prehistoric landscape change in the Luluku and Kamo'oali'i drainages is also reflected by nonagricultural archaeological finds. Williams (1989, 1992), conducting research in areas farther upslope, found that even ridge tops here are not simply erosional environments. Since A.D. 1200, hillslope soils and sediments have buried abandoned fire pits and other cultural features to depths of 50 cm or more in many cases. Erosional and depositional regimes even on ridge tops may have been enhanced in Luluku by the clearing of fields short distances downslope. When hillslope forests are burned or otherwise cleared, the integrity of the soil cover is damaged. Soil structure and colloidal strength are weakened in both topsoils and upper subsoils, with the result that they erode more easily during subsequent rains. In a chain reaction, damage sustained in cleared areas downslope may also affect areas upslope, where soils increasingly slide into newly created erosional voids.

Farther downslope in Luluku, below the main agricultural terrace zone, landscape modification has been dramatic during the cultural period. The floodplain locality that supported a taro pondfield about A.D. 500 to 700 now lies buried beneath 1.2 to 1.5 m of later field soils and hillslope sediments.

Luluku is the only upland site area of the four cases discussed in this chapter. The main period of terrace use in the Luluku watershed began after A.D. 1200 and was approximately contemporaneous with the sedimentary filling-in of the Kawai Nui Marsh basin and other lowland areas, including Kahana Valley. I shall discuss possible relationships between this increasing sedimentation and the extensive terracing of inland slopes in some areas after I have presented the evidence from Kahana Valley.

Kahana Valley

The floor of Kahana Valley (Fig. 11.3) held a bay between 1700 and 320 B.C. (Beggerly 1990:253). As at Kawai Nui, this bay became a lagoon and then eventually, after peat formation, filled with arable sediments and soils. Evidence from the overlying layer, Stratum VI, (dated to 2680 ± 110 B.P., on shell), suggests that the lagoonal transition took place between 430 and 30 B.C. (Beggerly 1990:251–52, 267–68). The terminal date for lagoon formation re-

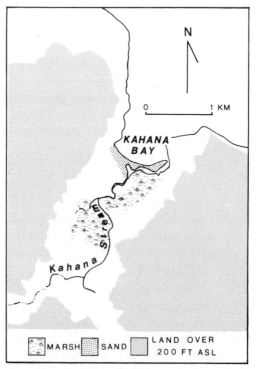

Figure 11.3 Kahana Valley: environmental features (based on figures and data in Beggerly 1990, Hommon and Barrera 1971, and Hommon and Bevacqua 1973).

mains uncertain, as Stratum V flood deposits created an abrupt discontinuity (Beggerly 1990:269) that appears erosional. The formation and emergence of the barrier that blocked the lagoon is not yet fully dated, although large, prograding sandbars located up to 800 m inland from the current shoreline have formed since 1300 B.P. (Beggerly 1990:271–72).

Stratum VI, the lagoonal layer, contains *Tellina macoma dispar* and *Theodoxis neglectus* shells in calcareous marine and terrigenous sediments; very finely fragmented charcoal; kukui (*Aleurites*) nuts; and pollen grains representing, in order from the base up, kukui, ti (*Cordyline*), and taro (*Colocasia*). The pollen samples also contain spores representing *Cibotium* (Hawaiian tree fern, *hapu'u*) and two ferns that colonize disturbed areas: *Nephrolepis* (sword fern), and *Dicranopteris* (false staghorn, *uluhe*). Uluhe in particular is nearly ubiquitous in recently burned or eroded areas throughout the tropical Indo-Pacific region (Carlquist 1970:304; Kirch 1984: 139–41; Allen et al. 1987: 26; see also Chapters 8, 9, this volume).

Beggerly interprets the charcoal in Stratum VI as contemporaneous with the dated shell and suggests that cultivation began during the late first millennium B.C. The charcoal, which has not yet been directly dated, is very finely fragmented, suggesting breakage either during cultivation or transport. Since no evidence exists for cultivation at this locality at the time in question, I suspect that the charcoal, like the terrigenous sediments and probably most of the pollens mentioned above, had been transported to the basin by water and probably wind from areas inland. If the charcoal is in secondary deposition, it may not be precisely contemporaneous with the shell but may have been deposited anytime before Stratum VI was buried by Stratum V sediments. Stratum V could not be directly dated, but it predates A.D. 895, the earliest date for Stratum IV. The charcoal in Stratum VI suggests cultural use of fire in areas nearby before 895.

Stratum V consists of fluvial or overland floodborne deposits of subangular and angular rocks up to cobble size, suggesting environmental instability at this locality. The area continued to receive fine marine sands in small quantity, as well as terrigenous silts and clays (Beggerly 1990:250). No charcoal, marine shells, or other organic materials were recovered. Beggerly (1990:317) suggests that the absence of charcoal and cultigen (taro, ti) pollens in Stratum V reflects abandonment of the valley by the area's agriculturalists, in order to allow inland forests to regenerate. Uluhe spores decrease, and *koa* (an endemic forest tree, *Acacia koa*) reappears in the pollen sequence. Interestingly, pollen grains representing *Ludwigia octivalvis*, a weed closely associated with taro, are present in low frequencies.

If the charcoal found in Strata VI and VII was deposited in situ at the basin locality, its absence from Stratum V might indeed suggest abandonment. Although the transportation of charcoal from slopes into lowland fields seems to have been a relatively rare occurrence in Hawaii, it remains possible that the Stratum VI and VII charcoal has been transported by floods from slope fields a short distance away. In that case, the absence of charcoal from the Stratum V flood layer may reflect other processes: for example, lightweight materials including charcoal may have been carried farther downstream. Another possible explanation for the lack of charcoal might be that the hillslope fields were by now terraced, preventing erosion of soil and vegetation except during the most violent floods.

Overlying Stratum V, Stratum IV is a peat layer that formed between A.D. 895 and 1255 (^{14}C, corrected, on organic matter, at two standard deviations), approximately contemporaneous with the formation of the Kawai Nui peat layer (A.D. 640 to 1310). Strata I to III, overlying the peat, are soils formed on silts and clays transported into the wetland by low-energy floods. All contain organic matter including charcoal fragments. Stratum III formed after A.D. 1250 (AMS date on organic matter). Strata I and II postdate A.D. 1720 (Beggerly 1990:249). Soil and pollen data suggest that Strata I to III were actively cultivated at the basin locality tested. Both taro and *Ludwigia* pollen are present in all three layers, although the uppermost 50 to 60 cm (Stratum I) produced neither. *Cordyline* pollen continues to the modern surface.

The Kahana sequence includes three peak periods of charcoal deposition in the basin: the lagoonal period (Stratum VI), ca. 430 to 30 B.C.; the earliest soil-forming period (basal Stratum III), ca. A.D. 1250 to 1600; and the post-contact period (Strata I, II), post–A.D. 1778. If the earliest charcoal deposits were transported in overland floods from hillslope fields, they represent significant erosion and redeposition during the first millennium A.D. Between the deposition of Stratum V and A.D. 1200, virtually no charcoal was deposited on the valley floor; terrigenous sedimentation decreased in the valley. Terrigenous sedimentation then resumed and accelerated, filling the basin with sediments that supported arable soils and cultivation sometime before A.D. 1600, as I interpret the evidence from Stratum III.

The history of environmental change in Kahana resembles in several particulars the sequence discussed for Kawai Nui. Secondary charcoal transported from areas inland reached soils in both basins during the first millennium A.D., before cultivation was possible in the basins themselves. A peat layer formed on lagoonal sediments in Kahana between A.D. 895 and 1255, in Kawai Nui between A.D. 600 and 1200 to 1300. And, in both basins, arable soils eventually formed on terrigenous sediments overlying the peat base. Soil and pollen evidence suggests cultivation within the Kahana basin after A.D. 1250 and within the Kawai Nui basin and the Bellows backplain by A.D. 1500.

SOIL WEATHERING, CULTURAL PROCESSES, AND LANDSCAPE CHANGE

The lateritic soils, primarily oxisols and ultisols, that characterize extensive hillslope areas in Waimanalo, Kailua, upland Kane'ohe, and Kahana (Foote et al. 1972) are intensely weathered, leached, and fragile. These soils, even when covered with lush vegetation, are not inherently fertile: they depend on vegetation to recycle critical nutrients. If the vegetation cover is reduced, as it is when areas are clear-cut or burned to clear fields for cultivation, the nutrients that had been provided to new plants by existing plants quickly leach downward, out of the reach of low growth, shrubs, and even small trees (Birkeland 1974; Dudal and Moormann 1964; Young 1976).

Clear-cutting an area strips away most nutrients. In contrast, burning to clear fields releases nutrients that had been locked within the former plant cover: the resultant burnt vegetation (charcoal), if it is conserved and not allowed to erode away, supplies ample nourishment to new growth. Underlying soils, however, remain infertile and weakened. The former root systems had not only recycled nutrients but had bound inorganic particles, increased soil moisture, and enhanced colloidal strength and structure. If newly burnt vegetation is allowed to wash away or become depleted before new vegetation takes hold, erosion of soils becomes an ever-increasing threat.

The stripping of vegetation from hillslope fields, and pre- or post-abandonment failure to provide adequate soil and moisture retention, must therefore be considered among the most significant potential cultural contributors to landscape change in windward O'ahu. In the absence of terracing or some other stabilizing technique, the fragile, degraded soils of the hillslopes have eroded downward into the lowlands in a continuous process that, although generally slow, occasionally attains catastrophic proportions (e.g., Wentworth 1943). The windward lowlands under discussion filled with these terrigenous sediments and soils until former bays and estuaries became lagoons and swales, then swamps or marshes, and eventually arable, dry or drainable lowlands.

GEOLOGICAL PROCESSES AND LANDSCAPE CHANGE

Although geoarchaeological evidence from many areas in the Hawaiian Islands indicates that humans have accelerated colluviation and the alluvial transport of sediments into coastal lowlands, natural processes would eventually have filled the windward O'ahu lowlands even without human assistance. Indeed, noncultural processes associated with late Holocene glacial events in the high latitudes and with sea-level and climatic changes around the world have recently been advanced by Nunn (1991) as the primary sculptor of landscape change in the tropical Pacific during recent millennia. According to this thesis, two second-order sea-level changes may have produced significant climatic perturbations and landscape changes since the initial colonization of the Hawaiian archipelago. The so-called Little Climatic Optimum (LCO) took place between A.D. 750 and 1300; the Little Ice Age (LIA) followed, between A.D. 1300 and 1900 (Nunn 1991:17–22).

Climatic changes accompanying first-order late Quaternary glacial and interglacial fluctuations in high latitudes are well documented: glacials are accompanied by cooling, interglacials by warming. Changes in air moisture and precipitation appear to vary from place to place (e.g., Bloom 1978). In the tropical Pacific, for which relatively few data exist, the Last Glacial period (about 70,000 to 10,000 B.P.) apparently produced a drier climate on islands, including Easter Island (Flenley and King 1984) and Aneityum (Spriggs 1981; Nunn 1991). Wetter conditions are believed to have accompanied first-order interglacial periods on at least some Pacific islands.

The second-order changes described by Nunn for the LCO and the LIA were accompanied in the tropical Pacific by climatic fluctuations that varied from the general patterns documented for first-order fluctuations. Nunn (1991:18) suggests that the LCO, rather than bringing wet conditions, was actually drier in the tropical Pacific than at present. As the LIA began, rainfall increased, declining only later in the glacial period (Nunn 1991: Fig. 4). The sea-level and climatic changes produced in the tropics by the LCO and LIA are relatively minor and therefore difficult to document. Few conclusive data yet exist for either the LCO or LIA in the Pacific; nonetheless, we cannot rule out the possibility that these fluctuations may have influenced human settlement and land use, especially in coastal areas. Both the LCO and LIA took place after the initial settlement of the four windward O'ahu sites discussed in this chapter. I shall summarize the changes outlined by Nunn (1991) for the LCO and LIA in terms of the possible influences they may have had on landscape transformation in these areas.

The Little Climatic Optimum (A.D. 750 to 1300)

During interglacial periods, marine transgressions effect sea-level rises and lateral inundations, with erosion of new marine shelves; air temperatures rise, while cyclones and other storms decrease. The climate became drier during the LCO in at least some portions of the humid tropics (Nunn 1991:17–19).

Polynesian long-distance voyaging may have taken advantage of the LCO, as storms decreased and westerly winds became more frequent.

Inundation by a rising sea would have drowned river mouths during the LCO, creating enlarged estuaries. The elevation possibly reached by the sea in Hawaii during the LCO remains uncertain, although Nunn (1991: Fig. 3) suggests a +1.75 m higher sea level in Fiji during this period. Kraft (1989) found no evidence for an elevated sea level in Kawai Nui during the past 2,000 to 3,000 years. While sea-level rise during the LCO could have increased salinity in estuaries blocked behind barrier beaches (as at Kawai Nui and Kahana), no reversion for salinity is reported for such areas on O'ahu. As sea level restabilized later in the LCO, lateral erosion and the cutting of marine terraces would have commenced. Such terraces would have become exposed when sea level dropped during the LIA, and then served as platforms that collected colluvial and alluvial sediments, creating coastal plains. The marine sandbars that formed in Kahana Valley during the past 1,300 years could have accumulated on a marine terrace dating to the LCO.

As the climate became drier during the LCO, surface soils and buried regoliths would have lost soil moisture. The drying of moist windward slopes would have produced the greatest instability, subjecting these to increased sheet erosion and landslides, as suggested by Nunn (1991:18–19; see also Wentworth 1943). Some evidence exists for sheet erosion on hillslopes in Luluku during the LCO period. The downslope pondfield that had occupied a floodplain in A.D. 500 to 700 at site 50-Oa-G5-85 was by A.D. 1300 buried beneath more than 50 cm of hillslope sediments. Larger colluvial clastics were deposited on upper portions of the same site until terracing activities began around A.D. 1200 (Allen et al. 1987: Figs. 26, 30). However, no evidence exists for a marked increase in the deposition of terrigenous sediments between A.D. 750 and 1300 in the other three O'ahu study areas. In Waimanalo, Kawai Nui, and Kahana, the major periods of lowland infilling all postdate A.D. 1300. At most, the LCO may have served primarily as a catalyst for culturally induced environmental changes (such as the impact of feral pigs on vegetation, as well as human-ignited fires). If, for example, windward slopes did undergo drying out during the LCO, any soil damage effected by subsequent cultural activities, such as forest clearing, would likely have been exaggerated.

The Little Ice Age (A.D. 1300 to 1900)

During glacial episodes, sea levels fall, causing streams to incise their drainages as they seek a lower base level. Slope gradients increase in channels and on valley walls, which become unstable; sediment wasting and an increase in stream sediment load result. The cooler, wetter, and stormier conditions that accompanied the first century or so of the LIA in the tropical Pacific are documented by Nunn (1991:19, Fig. 4), who suggests that most Polynesian long-distance voyaging ceased around A.D. 1300 because of increasingly stormy conditions.

Nunn (1991:22) argues that at the onset of the LIA, slopes of the drier, leeward sides of islands in the trade-wind belt experienced the greatest instability, as formerly dry sediments and soils became saturated with rainwater. In windward areas like those under consideration here, hillslope erosion continued, increasing in recent centuries unless checked by cultural terracing. During the LIA, fluvial sediments would presumably have reached stream mouths in increased volumes, to be trapped on coastal features eroded during the preceding LCO. On windward O'ahu, the magnitude of the infilling after A.D. 1300 suggests that the LIA and cultural transformations together share responsibility for landscape change in the lowlands.

Natural or Cultural? Environmental Change in Windward O'ahu

Nunn (1991) argues for the primacy of geological influences in effecting landscape change around the tropical Pacific during the time span of human occupancy. However, the evidence I have just presented from windward O'ahu suggests that natural and anthropogenic processes share responsibility for landscape change in pre-contact O'ahu. Moreover, anthropogenic change dominated—both in areal extent and rate of advance—after A.D. 1000. Evidence from the four windward study areas also suggests that certain processes often occurred on a localized scale, and that landscape change was not simultaneous or synchronous in the four areas. For example, while portions of the current barrier beach at Kahana have formed since A.D. 750, the barrier at Waimanalo (on which site o18 is situated) had stabilized before A.D. 600. These regional variations along the same windward coastline argue against Nunn's hypothesis that uniform geological processes throughout the tropical Pacific were the main agents of landscape change during recent millennia. Future investigations will need to test these conclusions and attempt to determine whether geological, pedological, and other natural processes may have predisposed certain areas to cultural transformations. Disentangling the effects of second-order natural changes from those of extensive and intensive cultural land use will be difficult.

INTERACTION AND FEEDBACK IN WINDWARD O'AHU
LANDSCAPE CHANGE

Considerable evidence suggests that the extensive terrace complexes that blanket large slopes in Maunawili, Kawai Nui, upland Kane'ohe, and Kahana were constructed by the Hawaiians in part to reverse the process of mass wasting of terrigenous soils. Stratigraphic profiles and radiocarbon date sequences from all four areas suggest that erosion followed the clearing of hillslopes for cultivation, a process that reached significant proportions sometime after A.D. 1000. Archaeological evidence from agricultural sites in pre-contact Hawaii now includes several pre–A.D. 1000 radiocarbon dates, on samples collected directly from agricultural soils (Allen 1992). These dates represent both pondfield cultivation on low-lying, level land where terracing was unnecessary and unterraced dryland cultivation on slopes (Allen et al. 1987; Clark 1980, 1981; Dicks et al. 1987; Schilt 1980).

During the first few centuries after initial colonization of the archipelago, hillslope cultivation probably resulted in minimal erosion and redeposition in lowland areas. Nonetheless, the burning of forest vegetation may have weakened soils over areas larger than those under cultivation alone. Once the indigenous Hawaiian population increased to the level where fields needed to be markedly expanded, erosion would have increased, perhaps dramatically. Although the filling in of basins like Kawai Nui and Kahana eventually provided large tracts of fertile land for intensive cultivation, in the short term the traditionally cultivated inland hillslopes were losing their ability to support crops.

Demands for agricultural produce certainly increased until at least A.D. 1500, first for demographic and then for sociopolitical reasons. Many facets of Hawaiian culture and society underwent dramatic transformations between A.D. 600 and 1500; land use and tenure are among the cultural subsystems that changed most dramatically (Kirch 1985), with the result that landscape change increased in response. Burning continued as a common method for clearing fields, but after 1200 or so, the cultivation of unterraced slopes in windward O'ahu seems to have declined. Rather, extensive terracing of hillslopes became standard practice in Luluku and Maunawili, and probably also in Kahana, by 1300 to 1400.

Terracing—and the effective retention of slope soils—brought about remarkable transformations in both the uplands and the lowlands of the windward O'ahu landscape. The effectiveness of traditional Hawaiian terraces at retaining moisture and sediments, and at enhancing soil fertility, are illustrated by the fact that these terraced areas were among the most frequently claimed parcels in the mid-nineteenth century (during the land distributions known collectively as the Great Mahele; see Kirch and Sahlins 1992). These prehistorically constructed terrace complexes were converted to rice cultivation later in the nineteenth century, and some continue to support thriving stands of cultivated banana and feral coffee. Terracing and the diversion of stream waters through terraced agricultural fields required sophisticated construction and hydraulic skills, familiarity with sedimentary and pedogenic processes, and significant labor inputs. Terracing became a standard technique in the Hawaiian agronomic repertoire—despite high labor requirements—because it is effective, conferring benefits from virtually the first day of use, restoring soil moisture and structural integrity, enhancing fertility, and halting soil erosion and water loss.

CONCLUSION

Geoarchaeological investigation in four major windward valley systems on O'ahu (Waimanalo, Maunawili–Kawai Nui, Luluku-Kamo'oali'i, and Kahana) has provided evidence that these areas were colonized by Polynesians by A.D. 600 or earlier. Cultivation of hillslopes near each of the early settlement areas was under way soon after. Significant slope erosion soon began, brought about by processes including the burning of forested hillslope lands. Additional cultural influences may have included disturbance by human-introduced pigs, dogs, and rats. Landscape change may also have been increased by drier conditions resulting from the LCO sometime after A.D. 750.

In the coastal lowlands, both marine and terrigenous sediments gradually began to fill the Waimanalo, Kawai Nui, and Kahana bays. These bays were transformed into lagoons that became blocked behind barrier beaches after A.D. 600 at Waimanalo, after A.D. 200 in the Kawai Nui basin, and after 30 B.C. in Kahana. The Waimanalo and Kailua barriers apparently formed prior to the advent of the LCO, whereas the Kahana sandbars dated thus far probably formed during the subsequent LIA. Most infilling of lowland areas with exclusively terrigenous sediments postdates A.D. 1200; infilling in the Waimanalo backplain postdates A.D. 1400.

In unterraced slope areas in Luluku, the deep burial of cultural features beneath later sediments suggests that erosion and redeposition continued unabated throughout the premodern era. In the intensively cultivated Luluku fields, however, erosion virtually ceased between A.D. 1200 and 1600, as extensive agricultural terrace complexes were constructed, transforming steep, unstable slopes into staircases of level, fertile fields that could be irrigated with dammed and channeled stream waters. Terracing provided a successful solution to the problem of meshing cultural needs with environmental constraints. The archaeologically documented landscape histories of windward O'ahu suggest that the Hawaiian builders of the terrace systems effected a successful response to the needs of a changing society in a rapidly changing environment.

Note

1. Athens and Ward (1991) have suggested that the Kailua barrier had formed and even emerged above sea level by 200 B.C., during the precultural period, producing a claimed transition from a saline to a "mostly freshwater" (1991:59) environment. In my opinion, this conclusion is not supported by the cited evidence; space limitations, however, preclude a full refutation of this claim here.

12

Hawaiian Native Lowland Vegetation in Prehistory

J. Stephen Athens

This chapter focuses on the reconstruction of native lowland vegetation on O'ahu, one of the eight major Hawaiian Islands (Fig. 12.1).[1] My intention is to provide a baseline account of the pre-Polynesian record for this topographically diverse island of 1,575 km², and then to show how, when, and what kinds of changes occurred with the advent of Polynesian settlement during prehistoric times. In so doing, I shall attempt to determine aspects of native plant distributions and to answer questions relating to vegetation change on O'ahu, demonstrating the value of paleoenvironmental studies and, in particular, the microfossil record for an understanding of Hawaiian archaeology and prehistory. Finally, with the completion of a number of new paleoenvironmental coring projects, I seek to build and expand on earlier published accounts by Jerome Ward and myself (Athens et al. 1992; Athens and Ward 1993).

The Hawaiian Islands, nearly 3,000 km from the nearest continent, comprise the most isolated archipelago of comparable size (16,642 km²) and topographic diversity (sea level to 4,205 m) on the globe (see Loope and Mueller-Dombois 1989:260). Evolution in its subtropical, ocean-moderated environment (between about 19° to 22° N latitude for the eight main islands)

Hawaiian Native Lowland Vegetation in Prehistory   249

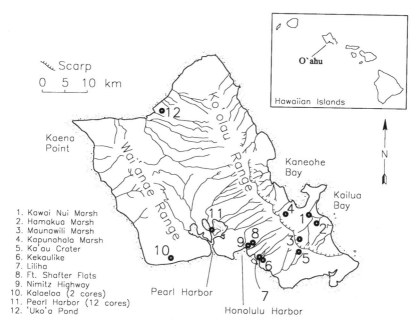

Figure 12.1 Map of O'ahu showing paleoenvironmental coring locations mentioned in the text

has proceeded in nearly total isolation, resulting in a high degree of endemism in the flora and fauna. The extreme vulnerability of island biota and ecosystems to alien invaders has long been recognized (e.g., Darwin 1859; Fosberg 1963). Additional geographical, environmental, and natural-history background information for Hawaii may be found in a number of easily obtainable publications and will not be considered further here (e.g., Carlquist 1980; Macdonald and Abbott 1983; Sanderson 1993; and the introduction to Wagner et al. 1990).

The vegetation today in the few remaining "wild" lowland areas of O'ahu is absolutely different from what would have been in these areas in prehuman times (cf. Gagné and Cuddihy 1990; Cuddihy and Stone 1990). Indeed, Kirch (1982:34–35) has estimated that by A.D. 1600, late in the period of prehistoric Polynesian occupation, "probably 80 percent of all of the lands in Hawaii below about 1,500 feet in elevation had been extensively altered by the human inhabitants."

While Kirch's figure of 80% may sound rather extraordinary, we now believe the true figure to be closer to 100%, if not quite 100%. This, of course, is amazing. The lowland vegetation of O'ahu was in fact completely altered from its original and pristine condition. And it has been this way since *before* the time Captain Cook arrived in Hawaii in A.D. 1778. Western plant introductions (Cuddihy and Stone 1990; Nagata 1985), while of great consequence to the look and makeup of the lowland vegetation landscape seen today, only added to the substantial change that had already occurred.

BACKGROUND TO PALEOENVIRONMENTAL STUDIES

To obtain information on past vegetation, our investigations depend primarily on well-dated palynological sequences obtained from wetland sediment cores. The fundamental assumption for most paleoenvironmental coring investigations is that wetland sedimentary basins in the lowlands should in many cases provide a complete and uninterrupted record of sedimentary deposition, including plant pollen, from the surrounding watershed. Under ideal circumstances such deposition should encompass a long period of time, dating from when such basins formed and continuing to the present. Preservation of pollen in the anaerobic depositional contexts usually found in wetland environments is also generally good. In the Pacific, many such basins should date from just after the end of the last glacial period, when sea level rose to its modern level, creating a depositional rather than erosional coastal sedimentary regime. This means that in theory it should be possible to locate wetland or lake basins in favorable coastal areas that have complete and continuous sedimentary sequences covering both the time before Polynesian settlement and the entire period of prehistoric Polynesian settlement.

One of the crucial goals of our work has been to obtain baseline information on the pre-Polynesian period so that the changes occurring with Polynesian settlement could be securely documented, evaluated, and understood. Indeed, virtually all previous archaeological research had overlooked this crucial element for understanding questions about the effect of Polynesian settlement on the Hawaiian environment.

After our successful investigations at both Kawai Nui Marsh on the windward side of O'ahu (two deep cores and other related investigations) and the Ft. Shafter flats on the leeward side in 1990 (profiles and samples from a deep construction trench and a core from the trench base), there could be no question about the value of sediment coring and paleoenvironmental research in Hawaii (Athens and Ward 1991; Wickler et al. 1991). Essentially, these data indicated that *Pritchardia (loulu)* palms (Fig. 12.2) were once quite common in the lowlands of O'ahu and could even have been a major component of the lowland forests. By about A.D. 1000—perhaps 250 to 300 years after Polynesian colonists arrived on O'ahu—the endemic Hawaiian palms went into a

Hawaiian Native Lowland Vegetation in Prehistory 251

Figure 12.2 A lone Pritchardia *sp. palm in the upper Palolo Valley, Honolulu. Photo by author and J. Ward*

steep decline. Naturally occurring *Pritchardia* palms, of course, are at present relatively rare on O'ahu, and have been since before the time of Captain Cook. There were also a number of other significant findings from the Kawai Nui and Ft. Shafter flats cores, and I shall summarize these with our more recent coring data.

Nearly six years have gone by since the investigations at Kawai Nui Marsh and the Ft. Shafter flats. At present we have completed investigations or have in progress work at a total of 12 coring locations on O'ahu (Fig. 12.1). Other minor projects have also been conducted at various additional localities and on other islands, and several colleagues have also made contributions. The cumulative data are most interesting. Each coring location offers different opportunities for research in terms of the time period covered by the core, the geographical setting, preservation and diversity of microfossils, and other factors.

Although analysis has been completed for a number of our cores, it continues for many others—though there is enough information already available for us to know the general character of the data and perhaps a few details of special interest. Examples in this category include the Nimitz Highway core on the leeward side of O'ahu, which has an especially dense and well-preserved palynomorph assemblage covering both the pre-Polynesian and the prehistoric Polynesian periods. The Kapunahala Marsh core from the interior lowlands of the windward side is another very important core that to date has only been partially analyzed (the remaining analyses are now under way). In addition, continuing projects at Kalaeloa (Barbers Point) on the 'Ewa Plain of O'ahu (two cores) and Pearl Harbor (12 cores from fishponds throughout the harbor area, several of which are up to 20 m in length) are expected to provide major sources of new paleoenvironmental information.

Figure 12.3 shows the time period covered by the different cores. Ka'au Crater, the only higher-elevation or montane core (ca. 518 m elevation), covers the period from about 36,000 to about 7000 B.P., though the other cores are all Holocene in age. Several of the cores have gaps in their records, containing only the pre-Polynesian and late prehistoric parts of the sequence. These lacunae, usually beginning in about 2200 B.P. and lasting until about 500 B.P., are likely the result of lowered sea level following the mid-Holocene rise and the consequent establishment of a new depositional equilibrium (see Jones 1992; Athens and Ward 1993). Nevertheless, several complete depositional records from the time prior to Polynesian settlement through at least the late period of prehistory have been obtained from the O'ahu coastal lowlands. Besides the previously mentioned Kawai Nui cores, we have cores from 'Uko'a Pond, Kapunahala Marsh, and Nimitz Highway, though analyses are not yet com-

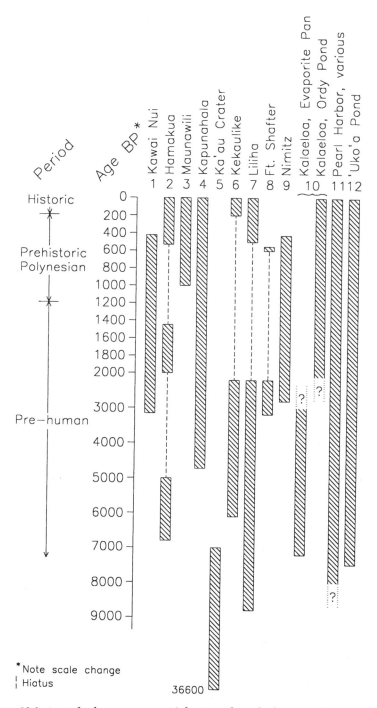

Figure 12.3 Ages of sediment cores on O'ahu. Note the scale change on the graph

plete on the latter two. This is also the case for the cores at Kalaeloa and is likely to be the case for at least some of the Pearl Harbor cores. The Kalaeloa cores are particularly interesting both because the area represents what is today an extremely dry environmental zone and because of the rich fossil record of avian and nonmarine-snail remains documented for this area during the prehistoric Polynesian period (Olson and James 1982:24–31; Christensen and Kirch 1986).

I now discuss the findings from several major areas of our continuing research. Detailed documentation may be found in our various technical reports, all of which are on file at the State Historic Preservation Division and the Hamilton Library of the University of Hawaii (both in Honolulu). As should be evident, the information presented here is from a program of research that is still in progress, and our ideas and interpretations may change as new data are obtained.

PRE-POLYNESIAN VEGETATION

Pritchardia Forest

The significance of *Pritchardia* palms in the lowland forests of Oʻahu has already been mentioned. It is probably premature, however, to say exactly how these palms were distributed in the lowlands. For example, were the *Pritchardia* palms fairly evenly spread over the lowland landscape among other trees and shrubs, perhaps with different species of *Pritchardia* occupying different elevation zones and different moisture regimes? Or could there have been clumped distributions, with dense *Pritchardia* groves occurring here and there in favorable geographic settings? This would be a situation reminiscent of present-day Nihoa Island (see Fig. 12.4), which is a small uninhabited island 240 km northwest of Kauaʻi in the main Hawaiian group (though numerous archaeological features attest to prehistoric human occupation; see Cleghorn 1988). Here the *Pritchardia* palms are restricted to two valleys, being absent from the more exposed locations (Conant 1985:141).

The Kapunahala Marsh core sediments (pre-Polynesian intervals) contain dense clusters of *Pritchardia* roots in vertical position (roots identified by International Archaeological Research Institute wood anatomist Gail Murakami), suggesting that the samples are from an actual *Pritchardia* forest. The significance of this fortuitous find is that for the first time reference data are available on the percentage contribution of *Pritchardia* pollen within a pristine *Pritchardia* forest, and these data can now be applied to the interpretation of fossil pollen assemblages. Though some reference studies of pollen

Figure 12.4 Photograph of Pritchardia *palms, showing their clumped distribution in the valley on Nihoa Island. Photo by Terry Hunt*

rain in modern communities of natural vegetation in Hawaii have been undertaken (e.g., Hotchkiss and Juvik 1994), the interpretive value of this work suffers from the problem that there are simply no modern lowland pristine plant communities that can be regarded as even remotely similar to what was present in Hawaii during pre-Polynesian times.

The two pollen percentage values for *Pritchardia* obtained from the pre-Polynesian samples at Kapunahala Marsh—43 and 51% of the total pollen and spores (or 69 and 64% of the total pollen)—are on average about double the values found for *Pritchardia* in the pre-Polynesian parts of other O'ahu cores. The other cores are from basins in which the pollen clearly has been secondarily deposited, but it is nevertheless remarkable that the values for *Pritchardia* pollen in these other cores are in fact so high—usually 15 to 25% of the total count for pollen and spores. We infer, therefore, that *Pritchardia* probably did in fact represent an important if not dominant taxon in the lowland forests virtually everywhere on O'ahu. It also seems probable that *Pritchardia* palms would have tended to cluster densely in favorable areas like valley floors or gulches. These conclusions, however, need considerable substantiation and should be regarded for the present as interpretations that remain to be confirmed.

The Ka'au Crater data show that *Pritchardia* palms, while present in the higher elevations of O'ahu to at least 518 m, were not nearly as common as in the lower elevations. We currently have no data for differential distributions or densities of *Pritchardia* palms between leeward and windward areas. It will be quite interesting, however, to see from our Kalaeloa cores (analysis of which is still under way) whether *Pritchardia* is as common on the very arid 'Ewa Plain as it seems to have been elsewhere in the O'ahu lowlands (analysis shows it to be present in some quantity, though percentage figures are not yet available).

Kanaloa kahoolawensis

This plant has one of the fascinating stories of recent Hawaiian botany. It is a leguminous shrub 1 m tall and 2 m across (Fig. 12.5). First identified in 1992 on a nearly inaccessible rocky spire adjacent to the coast of Kaho'olawe Island, only two plants are known to exist in all of Hawaii (Lorence and Wood 1994). Acting on nothing more than a hunch and the suspicion that an unknown but common fossil pollen type (then called "Unknown Tricolporate Type 1") was a legume, we obtained flowers from the live plants to extract the pollen to see if a match could be obtained. Against all odds, the match was perfect. Thus, by lucky coincidence, it was possible to establish the identity of

Figure 12.5 One of two known existing plants of Kanaloa kahoolawensis *on a sea stack just off the south shore of Kaho'olawe Island. Photo by Kenneth R. Wood, National Tropical Botanical Garden, Lawai, Hawaii*

one of the most common under-story plant types in the long-vanished pristine flora of Hawaii.

The pollen core data suggest that *Kanaloa* favors an arid or semiarid environment, occurring from the coastal lowlands to at least the elevation of Ka'au Crater (518 m). In a 1965 geological boring (Stearns and Chamberlain 1967) on the 'Ewa Plain, the driest location on O'ahu, early Pleistocene samples contained an average of 27% *Kanaloa* pollen (interpreted from a pollen study by Bennett 1985; samples are currently being reanalyzed by Ward). With limited information available concerning this core, it is of course difficult to know whether the climate of this location was the same almost two million years ago as it is today. However, *Kanaloa* pollen always occurs in the greatest concentrations from cores taken in the driest environments. At Ka'au Crater, for example, the full to late glacial portion of the core contains an average of 6% *Kanaloa* pollen, which then falls to less than 1% in the early Holocene. This finding exactly mirrors Selling's (1948) finding of a xerophytic pollen spectrum in montane bogs during glacial times, followed by a recovery of mesic pollen types in postglacial times; (upper-elevation areas in Hawaii had a much drier climate than at present during the last glacial period.)

Similarly, cores from the drier coastal settings also produce the greatest amounts of *Kanaloa* pollen. Kawai Nui Marsh and 'Uko'a Pond are two areas that have roughly 1000 mm of rainfall each year. At Kawai Nui Marsh *Kanaloa* makes up 15 to 30% of the prehuman pollen assemblage, while at 'Uko'a Pond *Kanaloa* comprises about 5 to 9% of the pollen record. However, at one point at 'Uko'a Pond, during what is believed to have been a dry climatic episode, *Kanaloa* pollen accounts for as much as 17% of the total pollen amount (including pteridophytes). Other cores on O'ahu typically have less than 1% *Kanaloa* pollen. The most extreme examples for the minimal presence of *Kanaloa* are from the inland settings at Kapunahala Marsh and upper Maunawili Valley, both of which are near the base of the Ko'olau Mountains on the windward side, where rainfall is very high. Here *Kanaloa* pollen registers at 0.1 to 0.7% of the total record, which appears to confirm the plant's preference for drier settings.

FLORISTIC PATTERNS

Important pristine floristic patterns on O'ahu have already been touched upon in my preceding discussions of *Pritchardia* and *Kanaloa*. As noted in our original Kawai Nui and Ft. Shafter–flats pollen studies, there appears to be a difference between leeward and windward lowland plant communities. The leeward forests, dominated by an upper story of *Pritchardia* palms, seem to have had a high diversity of secondary and lower-story species, including *Acacia koa,* Myrtaceae, *Antidesma,* a Chenopodiaceae/Amaranthaceae type (probably *Chenopodium oahuense*), *Dodonaea viscosa, Kanaloa, Erythrina sandwicensis, Colubrina,* and *Gouania.* The windward side at the lowest elevations, in contrast, had a forest characterized by three dominant pollen types, including *Pritchardia, Dodonaea viscosa,* and *Kanaloa.* The secondary and lower-story species, many of which are the same as on the leeward side, do not seem as common in the windward vegetation communities. Moving inland and upward in elevation on both the leeward and windward sides, there is of course substantially increased rainfall. Here the leeward-windward dichotomy loses its significance and the vegetation pattern changes to more of a mixed-dry to mesic forest, with an increasing contribution of mesic to wet forest taxa with elevation.

The Schofield Saddle, an interior plateau averaging about 250 to 350 m in elevation in central Oʻahu between the Koʻolau and Waiʻanae mountain ranges, represents a significant interior land area that encompasses the upper limits of what is here considered the lowlands. Some time ago St. John (1966:12–14) offered the suggestion that *Acacia koa* originally forested the Schofield Saddle. Although evaluation of his proposition is difficult with respect to our coring data—we have no cores from the Schofield Saddle—findings of slightly elevated levels of *Acacia koa* in Pollen Zone B at ʻUkoʻa Pond (3400 to 1000 B.P., Athens et al. 1995:67–90) may suggest pollen transport from the tablelands above ʻUkoʻa Pond, lending some credence to St. John's view. Nevertheless, we are relatively certain that *Pritchardia* would also have been very common, along with a diverse array of other forest taxa. To regard the area as having an *A. koa* forest in the sense of this being the dominant taxon is probably not correct. Whether *A. koa* was more common in the Schofield Saddle than in the coastal forests of Oʻahu remains an open question.

With regard to these floristic inferences, it must be understood that there is nothing like a one-to-one correspondence between the quantity of different pollen types in a core sample and how common a particular species was in a former ecosystem. There are a number of important intervening variables, including taxa that are high pollen producers (wind pollinated) as opposed to low pollen producers (insect or bird pollinated), pollen types that are more resistant to degradation as a result of transport and redeposition, and the effects of the chemical and biological environment of deposition, and the like. Also, there is the fact that most of the pollen observed in the cores could have been transported from almost anywhere within the watershed from which the sample was recovered and thus the pollen spectra may not represent a strictly local flora. For example, the pollen assemblages in most of our leeward core samples almost always contain a small contribution of mesic to wet forest taxa. Because of our botanical knowledge, it is easy to surmise that the pollen of these types was almost certainly transported to the lowlands, presumably by fluvial action. It can be difficult, however, to infer floristic patterns when habitat preferences of the major taxa are not restricted to narrowly defined environmental zones.

CLIMATE CHANGE

The most significant climatic event documented by our coring records concerns the transition from the Pleistocene to the Holocene about 10,000 years ago. As I noted earlier, Selling's (1948) pollen investigations in montane bogs documented arid conditions, and this interpretation seems to hold for our Ka'au Crater core. Aridity was caused by global Pleistocene cooling, which resulted in the depression of the inversion layer above which rain clouds do not form (and sea level was about 100 m lower than at present during the last glacial maximum 18,000 years ago; Shackleton 1987; Dawson 1992:223–34, Stearns 1978:11). Below the inversion layer, rainfall models predict considerably greater precipitation than occurs at present in the lowlands (see Gavenda 1992; Manabe and Hahn 1977; Ruhe 1964).

Although core records have been difficult to obtain for the late glacial period in the Hawaiian lowlands, we do have one likely basal core sample from near the Ma'alaea coastline on the lee side of Maui (Athens et al. 1994). This sample, which could only be of Pleistocene age, shows a predominately mesic-wet forest spectrum, with elevated values for Myrtaceae and *Cheirodendron,* and small contributions from numerous other mesic-wet types. Above this undated zone (Zone A1) the pollen spectrum in a radiocarbon-dated early Holocene zone abruptly changes to one that is typical of leeward lowland environments during the Holocene. This evidence, therefore, confirms a downward shift of vegetation zones in Hawaii during the glacial period, presumably a result of increased rainfall with cooler temperatures, though decreased evapotranspiration could also be part of the explanation.

Regarding climate change during the Holocene, there is now substantial evidence for a drier episode between about 5500 to 3500 or 2200 B.P. (the terminal date is still uncertain). The evidence is from three widely dispersed cores, including two at 'Uko'a Pond on the north shore of O'ahu and one in the Liliha area of urban Honolulu (Athens et al. 1995; Athens and Ward 1995a). Besides a shift in the pollen spectrum to a more xerophytic assemblage in these cores, two of the 'Uko'a Pond cores have crusted red soil horizons at this time, suggesting that the basin dried either seasonally or for longer periods. How widespread or synchronous this climatic shift was in the Pacific remains to be determined, though other investigators have noted dry mid-Holocene episodes on Mangaia and Atiu in the Cook Islands (Kirch et al. 1992:177; Ellison 1994:11–12). Furthermore, a dry mid-Holocene episode has been documented throughout South America, with a corresponding wet episode in Africa (Absy et al. 1991; Ledru 1993; Martin et al. 1993). A global climatic shift during the mid-Holocene, therefore, seems evident, though it was apparently not synchronous at all locations. This episode of climatic change is thought to relate to a long-duration ENSO (El Niño–Southern Oscillation) event in the Pacific (Martin et al. 1993; see also Diaz and Markgraf 1993).

VEGETATION CHANGE

Natural vegetation change due to climatic shifts occurred in Hawaii during relatively recent millennia before human contact, as I pointed out earlier. Although there were profound alterations in the natural environment of Hawaii at these times, it is hard to see that there was any lasting effect. The vegetation communities evidently returned to their prior condition, suggesting great overall long-term stability in the floral ecosystems of Hawaii. Similarly, Steadman (1995:1128–29) has suggested tremendous stability in the avifauna taxa found prior to human contact on many central Pacific islands. This is in spite of global cooling caused by ice ages, major changes in sea levels, tectonic processes, and other natural perturbations. However, by at least A.D. 1000 in Hawaii and usually earlier elsewhere in the Pacific, another major change occurred, but this time the change had a lasting effect on native flora and fauna. This change was instigated by the arrival of humans on previously unoccupied islands.

Pollen diagram after pollen diagram from the coastal lowlands of O'ahu show the same thing. The native forests of the lowlands disappeared in a matter of centuries. By A.D. 1400 to 1500 there was essentially nothing left of the lowland forest. *Pritchardia* palms had all but disappeared, *Kanaloa* did disappear, and the other native trees and shrubs dwindled to negligible numbers, if not outright rarity. By the time Captain Cook arrived in Hawaii in A.D. 1778 the native lowland forest was absolutely gone.

We had previously fixed the date for the start of the forest decline at A.D. 1000 (ca. 950 B.P.) in the coastal lowlands, based primarily on our Kawai Nui Marsh core (Athens et al. 1992:26; Athens and Ward 1993:211–14). From our recent work at 'Uko'a Pond, however, it appears that a case could be made for a slightly earlier date for the north shore area. Figure 12.6 illustrates changing pollen percentages over time for *Pritchardia* and *Kanaloa* at 'Uko'a Pond, Kawai Nui Marsh, and the upper Maunawili Valley (note also the charcoal-particle record for 'Uko'a Pond, discussed below). These are two of the most common and sensitive taxa indicative of a comprehensive decline of native lowland forests and plants. Beginning at about A.D. 800 at 'Uko'a Pond, 1050 at Kawai Nui Marsh, and just after 1200 at Maunawili (Athens and Ward 1994), an unmistakable decline in the two taxa becomes evident (though *Kanaloa* is largely absent from Maunawili, because the core is from a high rainfall area next to the Ko'olau Mountains—as I mentioned earlier). The temporal differences among the three areas appear to be real. Thus, the onset

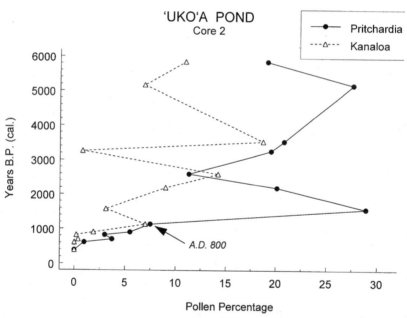

Figure 12.6 Pollen percentages for Pritchardia *and* Kanaloa *in Core 2 of 'Uko'a Pond, Core B of Kawai Nui Marsh, and the upper Maunawili Valley core (original pollen and radiocarbon-dating evidence is presented in Athens and Ward 1991, Athens and Ward 1994, and Athens et al. 1995). The Kawai Nui Marsh dating record presented here is based on Maher's (1992) depth-age program using a cubic spline function with six radiocarbon dates (two marine samples were corrected for a partial reservoir effect using an estimated 40% marine-carbon figure; a third sample was corrected using an estimated 100% marine-carbon figure [all marine samples were calibrated with a Delta R value of 115 ± 50 years]). The Maunawili dates are also based on a cubic spline function.*

Hawaiian Native Lowland Vegetation in Prehistory 263

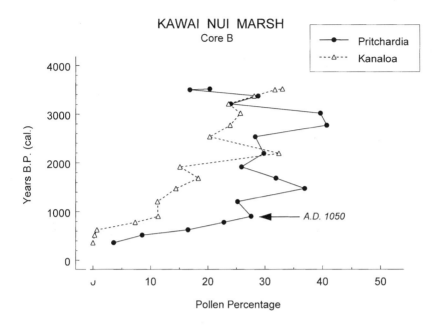

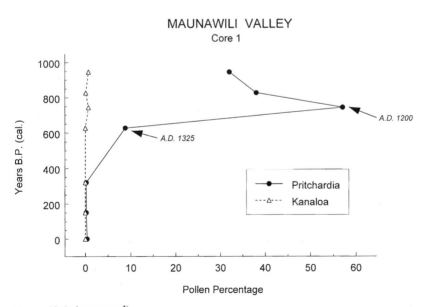

Figure 12.6 (continued)

of processes leading to the demise of the native lowland forest on O'ahu may have begun even earlier than suggested in our original assessment. Obviously, precision in our chronology now requires closer-interval radiocarbon dating and pollen sampling, which will be a goal of future work. The reason for the difference in dates between 'Uko'a and Kawai Nui is unknown at this time, though it may be that the north shore area was settled before the windward area. The later date for Maunawili concerns the inland expansion of settlement and agriculture.

The relatively late date for the decline of the native forest at Maunawili is consistent with Hommon's (1986) model for the inland expansion of population and agriculture once coastal areas were completely occupied. We regard what little contrary evidence exists in the archaeological record (e.g., Allen 1992) as equivocal, because of problems in the radiocarbon dating of charcoal samples that may not be contemporaneous with the archaeological features they purport to date. This may in part be due to an old-wood problem (as has been documented in New Zealand; see Anderson 1991:779–81), though the question of association also may be raised.

We are admittedly uneasy with our position regarding the process of late inland settlement, as thus far it is based on evidence from just one interior lowland coring site. Fortunately we shall soon have the complete data from Kapunahala Marsh, our second inland coring site. An earlier run of a few dating and pollen samples proved this to be an extremely interesting core for both paleoenvironmental and archaeological studies (as I mentioned in my comments about the finding of in situ *Pritchardia* roots). With further radiocarbon dating and sample processing (for pollen and charcoal-particle counts) we shall be in a strong position either to confirm our view on late inland settlement and agriculture or to broaden our understanding of settlement processes on O'ahu if the record of change proves to be earlier than expected.

The reason for the decimation of the native lowland forests of Hawaii after settlement by Polynesians is very much an unresolved issue. Agricultural clearing has often been viewed as a major factor, along with the introduction of new plants and animals (see Cuddihy and Stone 1990; Kirch 1982). With a growing population, more and more land would be devoted to agriculture. It also seems true, however, that many lowland locations around the islands would have been unsuitable for agriculture (narrow gulches, steep and rocky slopes, and so forth). Yet such places do not support relict stands of pristine native lowland forest. For this reason we tend to believe that other factors beyond just land clearing for agriculture may have been involved in the disappearance of the native lowland forest. As previously suggested (Athens and Ward 1993:215), such factors as inadvertently introduced plant diseases and possibly the interruption of complex symbiotic relationships between native species should be considered. While predation on seeds by the introduced Pacific Rat (*Rattus exulans*) has frequently been suggested as the major contributor, it is difficult to see how so many plant species would have been equally susceptible. Clearly, further research is needed.

As a final note on the subject of vegetation change, there is an interesting fact regarding the extirpated *Kanaloa* plant that was once so common in the pristine lowland forests of O'ahu. To date not a single piece of *Kanaloa* wood has been identified from the hundreds of archaeological charcoal samples that have been submitted for wood-species identification (G. Murakami, pers. comm.). Although there could be a variety of explanations for this observation (e.g., there was a cultural prohibition against using the wood, other woods were preferred for cooking fires), we believe the simplest explanation is that *Kanaloa* succumbed very quickly with the onset of Polynesian settlement and from a very early time was not present in sufficient numbers to be readily available for use.

INITIAL POLYNESIAN SETTLEMENT

It is well known among paleoenvironmental investigators conducting wetland and lacustrine sediment coring investigations that microfossil data—usually pollen and particulate charcoal—almost always provide indications for settlement and agriculture earlier than those provided by traditional kinds of archaeological data. We have shown this to be true in our studies in Micronesia (Athens 1995; Athens and Ward 1995b) and Ecuador (Athens n.d.), and other investigators have demonstrated the fact in many other areas of the world (e.g., Burney 1993; Burney et al. 1994; Kirch and Ellison 1994; Piperno 1990; Piperno et al. 1991). While Spriggs and Anderson (1993:211) in the heat of argument make the unfortunate remark that finding "signs of earlier human habitation in pollen cores is fraught with various technical and interpretational problems," such an ill-considered view is not in keeping with data presented in a considerable body of literature (though we acknowledge that great care that must be used in collecting and interpreting such data; see also the rebuttal by Kirch and Ellison 1994).

Archaeologists in Hawaii are divided between those favoring an early time frame for initial Polynesian colonization sometime between the first century B.C. (Beggerly 1990; Hunt and Holsen 1991) and A.D. 300 or 400 (Kirch 1985), and those favoring a time frame between about A.D. 600 or so and 750. This later period has generally been favored by an earlier generation of researchers (Emory 1963; Sinoto 1970, 1983c; see also discussion by Cachola-Abad 1993 and Graves and Addison 1995), though Spriggs and Anderson (1993) have recently flaunted modern convention with their argument that a later settlement date for Hawaii is more in keeping with present information regarding the prehistory of Eastern Polynesia.

What, then, is the very earliest coring evidence we have for the Polynesian presence in Hawaii? The answer is approximately A.D. 800, which is from 'Uko'a Pond. The securest evidence is in the form of microscopic particulate charcoal (Fig. 12.7), though it is also supported by the pollen evidence as discussed above. Because particulate charcoal does not occur in sample intervals predating Polynesian occupation in *any* of our cores on O'ahu, we feel confident that its presence in the 'Uko'a Pond core (and other cores) must be entirely due to anthropogenic causes. The deepest charcoal, therefore, should pertain to initial Polynesian occupation within the watershed.[2] The statement by Spriggs and Anderson (1993:200) that "the case for the rarity of natural fire in Hawaii, particularly in dry, leeward areas, has not been convinc-

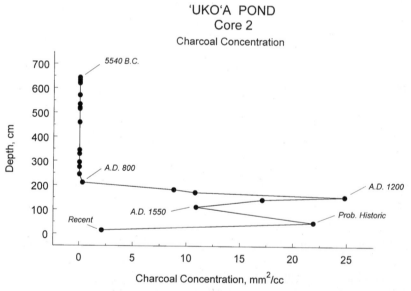

Figure 12.7 Graph showing charcoal-concentration values in Core 2 at 'Uko'a Pond. The dates are derived from the Maher (1992) depth-age computer program (see Athens et al. 1995:56–57). The indicated dates should not be regarded as precise age estimates. Rather, they incorporate an undefined error range, though presumably of a similar magnitude to those of the radiocarbon dates from which the interpolated ages derive (see discussion in note 2).

ingly established" flies in face of our extensive evidence from numerous core samples, including those from Kalaeloa, the driest part of O'ahu.[3]

Although we do not currently regard our data on the early settlement of Hawaii as definitive, at the same time we are reluctant to believe archaeological evidence that places settlement in Hawaii before the late A.D. 600s or 700s at the earliest. The demonstrated value of paleoenvironmental microfossil data for determining questions of initial settlement and land use in other parts of the world is too compelling to dismiss or to make an exception for the Hawaii case.

PARTICULATE CHARCOAL AND AGRICULTURE

Based on the lack of particulate charcoal data from the Kawai Nui Marsh cores, one of our first Hawaii coring locations, we suggested several years ago that fire for agricultural clearing may not have been utilized to a significant extent by prehistoric Polynesian cultivators (Athens and Ward 1993). As our more recent cores are now showing a consistent presence of particulate charcoal, however, this initial assessment must now be modified. Figure 12.7 illustrates the concentration of charcoal particles in Core 2 of 'Uko'a Pond through time, suggesting a fluctuating pattern of burning in the watershed. There is a dramatic increase in charcoal production over a roughly 400-year period until about A.D. 1200. Thereafter, the presence of charcoal declines precipitously in the late prehistoric period, only to climb again during what is probably the historic period when plantation agriculture was initiated. We interpret this data as indicating that much of the native forest was burned during the early part of the Polynesian occupation of O'ahu, and that by about A.D. 1200 much of the primary forest no longer existed. A reduction in the quantity of combustible material, therefore, resulted in a decrease of particulate charcoal entering the depositional record. Presumably the burning was primarily related to the opening of the forest for agricultural plantings. As analysis proceeds with our other cores, we shall be looking closely at the particulate charcoal record to see if the same pattern holds for other areas on O'ahu.

POLYNESIAN PLANT INTRODUCTIONS

One of the interesting aspects of the Polynesian settlement of Hawaii is that at some point many plants (ca. 32 taxa) from the Polynesian homeland were introduced into Hawaii. These were not just food plants, such as taro, breadfruit, sweet potato, and bananas, but also many other plants, some of obvious economic utility and others for which indigenous substitutes were seemingly available (Athens and Ward 1991:20–21; Krauss 1994). Some of these, such as herbs and weeds, apparently arrived in Hawaii hidden in or attached to the purposeful introductions, or perhaps their seeds arrived stuck to clothing or other items.

As the introduced status of most nonfood plants brought by Polynesians has been determined botanically and not archaeologically, it is of interest to verify this assessment with the microfossil evidence. It is also of interest to determine when the introductions first appear in the paleoenvironmental record, as this information may be a primary method for archaeologically testing and dating Hawaiian oral accounts of multiple voyages to and from Hawaii.

Our pollen data regarding the suspected introductions has unfortunately not been as productive as we had initially hoped. This is undoubtedly because many of the introductions are not wind-pollinated, and thus are poor pollen producers and do not disperse well in the environment. Nevertheless, a few comments regarding several species can be made. These tentative interpretations will surely be modified or refined as new information becomes available.

1. *Cocos nucifera* (coconut, *niu*): The coring data has repeatedly indicated the lack of coconut pollen in pre-Polynesian contexts. This is a wind-pollinated tree and a high pollen producer. To date we have not seen coconut in the coring records until after A.D. 1300. *Cocos* appears to be a definite Polynesian introduction.
2. *Aleurites moluccana* (candelnut, *kukui*): There is no evidence for this taxon until after A.D. 1100 (either the pollen or the very hard and usually well preserved nut shell). *Aleurites* appears to be a definite Polynesian introduction.
3. *Hibiscus tiliaceus* (*hau*): This pollen taxon is very rare in the microfossil record, presumably because it does not disperse well. A single grain of pollen was found in the A.D. 350 interval of the 'Uko'a Pond core. This is presumably before Polynesian settlement and thus would indicate an indigenous rather than an introduced status. This conclusion, however, awaits confirmation by further coring data.
4. *Pandanus tectorius* (*hala*): Until recently the introduced or indigenous status of *Pandanus* has been uncertain. Although rare, its pollen has been present in a number of cores in pre-Polynesian pollen levels. A geological macrofossil was also recently reported, which has a minimum age of 500,000 years (TenBruggencate 1993). There can be no doubt about the indigenous status of *Pandanus*.

5. *Ludwigia octivalvis* (primrose willow, *kamole*): Although it is not common in the core records, *Ludwigia* consistently occurs in prehistoric Polynesian levels after about A.D. 1250. *Ludwigia*, therefore, appears to be a Polynesian-introduced species. (Kirch [1994:155] observes that *L. octivalvis* is a common and persistent weed in the irrigated taro pondfields of Futuna Island in Western Polynesia.)
6. *Cordyline fruticosa* (*ki, ti*): The pollen of this taxon is uncommon in the core records. It has been documented only as early as A.D. 1350 and therefore appears to be a Polynesian introduction.

CONCLUSION

Paleoenvironmental research, particularly sediment coring and pollen analysis in wetland basins, has much to offer for an understanding of both natural history and archaeology in Hawaii. The data obtained during the past six years have transformed our knowledge of the pristine natural environment in the Hawaiian lowlands. We have also learned much about the changes that occurred with Polynesian settlement, climate change during the Holocene, the initial date of Polynesian settlement, and other issues. Our work is still in its early stages, and we have much to learn. Experience has shown that each coring location offers unique opportunities and possibilities for addressing important issues. We must be careful to take advantage of these opportunities, building on earlier research, formulating and addressing new questions, and bringing new and more sophisticated techniques to bear on the research. A lost opportunity—or worse, muddled data—can be an incalculable tragedy. Our experience with *Kanaloa* should be a constant reminder of just how fleeting some opportunities can be.

Notes

The pollen data and many of the ideas presented here are the product of a long-standing collaboration with Jerome Ward, a tropical palynologist. I offer sincere thanks and gratitude for his diligent efforts and enthusiasm over the years. I also thank our many sources of financial support and those colleagues who have assisted us along the way and who are named in our various technical reports. In addition, I thank Tom Dye, who provided the original impetus for writing this chapter by asking me to present a summary of our O'ahu paleoenvironmental investigations at the Eighth Annual Hawaiian Archaeology Conference of the Society for Hawaiian Archaeology in Honolulu, April

8–9, 1995. I also thank Terry Hunt and Kenneth Wood for allowing me to use their photographs.

1. The term *lowland vegetation* is used here in a general sense to refer to vegetation found below an elevation of roughly 300 m. It does not refer to a particular classification system and is not intended in any sense to be used as an analytical unit. One of the most useful and detailed modern vegetation classification systems is that developed by Gagné and Cuddihy (1990). It may be possible in the future to develop a classification system based on paleontological data.
2. The date for the earliest particulate charcoal in Core 2 of 'Uko'a Pond was originally reported as A.D. 819 (Athens et al. 1995). This date was determined using the Maher (1992) depth-age computer program (cubic spline function), which does not provide information on confidence intervals. In accordance with the recent admonishment by Bennett (1994), Tom Dye of the Hawaii State Historic Preservation Division wrote a program to do this with a polynomial function using AXUM graphics software. Although various assumptions will lead to slightly different results (whether a polynomial, cubic spline, or linear interpolation function is used), it is clearly beneficial to have some mathematical notion of the degree of error inherent in the derived date. Using Dye's program, a two-sigma calibrated date range of A.D. 663 to 799 was determined for the 211 cm depth, which is where the earliest particulate charcoal was found in Core 2 of 'Uko'a Pond. Nevertheless, it is of interest that a radiocarbon-dated fragment of wood from the 194 to 200 cm interval of this core had an age of A.D. 798 to 1028 (cal. two-sigma range). With an estimated sediment accumulation rate of 0.1052 cm/yr as determined from the Maher program, a sample 11 cm deeper (at 211 cm) should date 105 years earlier. Based on the radiocarbon date, the two-sigma calibrated date range for this depth, therefore, should amount to roughly 700 to 900 B.P. Although this age range for the 211 cm depth interval has little mathematical basis and no elegance at all, it seems to be more in keeping with the radiocarbon date found so close to the interval of interest and within the same sediment layer. In the final analysis, of course, both determinations have considerable of overlap (A.D. 700 to 900 vs. 663 to 799).
3. This statement is only true for O'ahu, however. We have evidence for charcoal particles from throughout the Holocene in our core from Ma'alaea on Maui (Athens et al. 1994). This certainly relates to considerable volcanic activity on that island during the Holocene, and perhaps also on neighboring Hawaii Island.

13
Prehistoric Polynesian Impact on the New Zealand Environment: Te Whenua Hou

Atholl Anderson

Te whenua hou, "the new land," refers to the view that the discovery of New Zealand was so novel within the Polynesian experience that early colonists were compelled to restructure abruptly the pattern of adaptation that had served them and their forebears well over thousands of years of migration in the Pacific. New Zealand was comparatively vast and topographically diverse, but the crucial difference was climatic. In particular, a frost-restricted growing season severely limited establishment of the familiar cultivated environments, or "transported landscapes" (Anderson 1952; Kirch 1982c:2), of tropical Polynesia. These, created by what Crosby (1986:289–93) in an analogous case calls a "portmanteau biota," were reduced in New Zealand to comparatively tiny patches of low diversity and productivity—a sparse terrestrial archipelago of cultivation in a sea of forest and fern. No pig or chicken, no coconut, breadfruit, banana, sugarcane, or *Pandanus* survived, and of those relatively few successful introductions only *kumara*, rat, and dog became widespread (Table 13.1). As a result, in New Zealand Maori economies were necessarily founded on hunting, foraging, and fishing to a significantly greater extent than in Polynesia generally. Understanding of environmental change

Table 13.1 Biota Transported to Prehistoric New Zealand

Prehistoric, introduced at discovery:
  Man (*Homo sapiens*)
Prehistoric, possibly arrived within first generation:
  Dog (*Canis familiaris*)
  Rat (*Rattus exulans*)°
  Kumara (*Ipomoea batatas*)
  Taro (*Colocasia antiquorum*)
  Yam (*Dioscorea* spp.)
  Gourd (*Lagenaria siceraria*)
  Paper mulberry (*Broussonetia papyrifera*)
Probably prehistoric arrival:
  Tropical ti (*Cordyline terminalis*)
Possibly prehistoric arrival:
  Some ferns: *Marattia salicina* (edible large rhizome and stem), *Thelyteris* spp. (in thermal areas), *Nephrolepis* spp. (in thermal areas)
  Sow thistle (*Sonchus* spp.)
  Small-flowered nightshade (*Solanum americanum*)
  Sphinx moth (*Agrius convolvuli*)
  Land snails: *Tornatellinops novoseelandiae, T. subperforata*

SOURCES: G. Mason (pers. comm.), H. Leach (pers. comm.)
°Possibly arrived before human settlement (Anderson n.d.)

has to begin from an appreciation of that point. Three hypotheses have been advanced.

At the Pacific Science Congress in 1961, Cumberland's approach to environmental change in New Zealand prehistory was deeply influenced by Duff's (1956) assertion that the first colonists had been nonhorticultural and had survived for perhaps 600 years before the fourteenth century by hunting and gathering, notably by moa-hunting, only to be supplanted by new Polynesian migrants, the Maori, who introduced gardening. Cumberland proposed that major environmental changes were attributable to profligate hunting and associated behavior by the Moa-hunters, while Maori were essentially conscious conservationists. They induced some environmental changes by incremental patch burning for horticulture and simply through population growth to a level at least ten times greater than when they arrived, but "nowhere did [they] cause a wholesale transformation of the environment or a disastrous disturbance of the ecosystem" (Cumberland 1963:193). The double-migration hypothesis has been widely questioned, and evidence that the early settlers were also horticulturalists, where conditions permitted, vitiates the Duff-Cumberland model.

A second hypothesis, not published authoritatively or in detail but commanding popular support, is achronological in form. Early modern, protohistoric, and pre-European Maori—without distinction—are credited with effective environmental management, expressed through the judicious imposition of *rahui* and other systematic, group-monitored economic devices. Thus, in Orbell and Moon's (1985:26) description of Maori foraging, "all hunting and fishing was carried out according to strict rules. Families and tribal groups were careful to confine their operations to their own territories . . . birds, rats and fish were protected for much of the year by a rahui, or ban, which ensured that they were not disturbed when breeding, that their numbers were not unduly depleted and that they were taken only in the best of condition." Beyond such ethnographic reconstructions, there are now more radical "New Age" assertions that Maori subsistence behavior was actuated by a deep sense of ecological relationships coupled with a mystical reverence for the environment. Describing the alleged revelation several years ago of the all-encompassing, secret knowledge of the "Waitaha Nation" (published by Brailsford 1994), Paterson (1994:34) reverses the Duff-Cumberland model to assert that the earliest people in New Zealand were gardeners who walked "in gentleness with each other and with the land," whose idyll was brutally shattered by the arrival of the later Maori. Conservative, spiritually informed behavior by the first settlers is contrasted with aggressive, exploitative attitudes attributed to later immigrants (e.g., Orbell and Moon 1985:217; Brailsford 1994:294).

The evidential basis of these assertions is highly dubious in all respects. They inflate an indigenous knowledge of some aspects of animal and plant behavior into a mystical understanding of the principles and content of modern ecology. They consistently blur epistemological distinctions among historical observations, ethnographic data, and beliefs about ancestral behavior expressed by modern Maori. And the demonstrable evidence of archaeological and palaeoenvironmental research, especially the widespread evidence of massive environmental changes in prehistory, is basically ignored.

A third hypothesis asserts that there is no evidence to support claims of conservative resource management in prehistory. Rather, people sought resources according to immediate needs or desires, took proportionately more from those that offered the best returns, and, as these were often the most vulnerable, created severe depletion or extinction in many cases before switching their focus to alternative resources or strategies (Anderson 1989a, 1989b; Anderson and McGlone 1991; McGlone 1989; McGlone, Anderson, and

Holdaway 1994; Holdaway 1989). This model describes differential impacts on the environment during the course of prehistory but ascribes them to changes in population (consumer) size and geographical and resource targeting rather than attitudes or behavior. The model is outlined below. McGlone (1989; Anderson and McGlone 1991) has described the prehistoric changes in vegetation patterns and their causes, which are incorporated in the model, and I supplement previous publications on megafaunal extinctions (e.g., Anderson 1989a, 1989b) with some thoughts on the broader pattern of avifaunal loss.

A MODEL OF ENVIRONMENTAL AND CULTURAL CHANGE

An initial assumption of this model is that habitation of New Zealand began relatively late and was more or less contemporaneous with the widespread horizon of extensive deforestation that appears quite suddenly early in the second millennium A.D., at about the same time as the first direct evidence of horticulture and the first sites of big-game butchery—indeed the first unequivocal archaeological evidence of any kind (Anderson 1991; McGlone, Anderson, and Holdaway 1994; but rats may have reached New Zealand before human settlement [see Anderson n.d.]). The advent of these phenomena is otherwise attributed to increasing archaeological visibility of a human population that had reached New Zealand perhaps a millennium earlier and had slowly been growing in areas where subsistence was substantially by gardening (Sutton 1987; Bulmer 1989). But it is difficult to see how archaeological visibility could be attained coincidentally throughout much of New Zealand when the activities represented in it, and almost certainly the population densities, were so regionally variable. Either a general change in behavior toward the environment and its major resources can be inferred or the early archaeological evidence actually reflects the period of initial colonization. An important consequence of this second conclusion is that a narrow band of acceptable radiocarbon dates for earliest sites from throughout the country and a similar pattern of dates for deforestation, among other data, suggest colonization by a relatively large number of people, probably some hundreds, which in turn argues for a period of return voyaging early in the settlement sequence.

The model outlined here attributes chronological differences in the nature and scale of human impact on the environment to different regional patterns of economic opportunity. A broad arrangement in the distribution of wild resources and cultivation opportunities according to latitude is the usual way of describing these patterns, but an additional perspective is gained—and New

Zealand is drawn into a common Polynesian discourse about environmental change—by proposing a model in which common processes are divided according to their effects in windward and leeward provinces that lie on either side of the main axial ranges (Fig. 13.1). The emphasis is on the terrestrial environment where changes are most apparent. There were important changes in the marine environment, including retreat of the seal-breeding margin from northern to southern New Zealand (Smith 1988). Various instances of

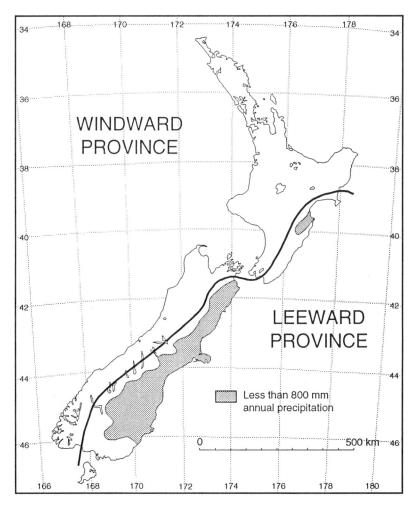

Figure 13.1 Approximate division of windward and leeward provinces in New Zealand

decline in the size-frequency distributions of fish and shellfish have been inferred from measurement of midden remains (e.g., Leach and Anderson 1979; Anderson 1981; Anderson and McGlone 1991), but these declines were common to both provinces, and most were probably localized.

The windward province is comparatively warm and humid. It contains nearly all of the area's fertile volcanic soils, was under heavy closed forest at the time people arrived, and has most of the large bays, harbors, and lagoons in New Zealand. The leeward province is comparatively cool and dry. Its zonal soils are mostly infertile, and they developed on graywacke and other old sedimentary rocks. The vegetation was more open and comprised mainly beech (*Nothofagus* spp.)–dominated forests and various associations of forest and scrub ranging down to grasslands in the driest inland areas upon the arrival of people. The 800 mm isohyet (Fig. 13.1) marks the approximate threshold of beech-forest regeneration, so forest fires in drier areas would usually result in replacement by smaller trees, such as *Sophora* sp., and scrub species.

The dynamics of the model (Fig. 13.2) incorporate this broad provincial division in an argument of palaeoeconomic optimization—exploitation of habitats and resources occurs in order of perceived energy efficiency and

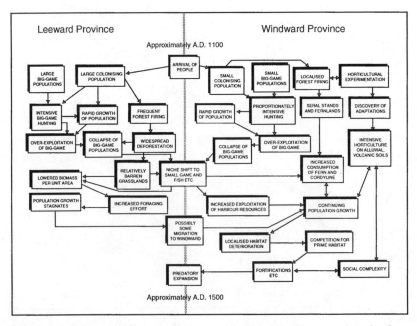

Figure 13.2 Model of environmental and related cultural changes in New Zealand prehistory

desirability. Thus the first colonists exploit selectively the large packages of readily available game, most of which are concentrated in the leeward province. Those taxa, particularly moa and seal, were highly vulnerable to exploitation because of conservative life histories and naive behavior, with the result that the human population could expand in conditions of low-cost access to high-quality food.

Optimal returns among the colonists were maintained by group dispersal and high mobility throughout the leeward province, and further afield wherever there were smaller pockets of comparable habitat and resources (e.g., northern Northland and Coromandel). The rate of expansion was archaeologically instantaneous because optimal exploitation requires dispersal at low densities, rather than near carrying capacity as in the Caughley (1988) model of subsistence and colonization (see Anderson and McGovern-Wilson 1990).

The limiting factor in this situation is the availability of plant foods. Because horticulture was generally marginal or impossible in the leeward province, except in its northern district, there was a premium on creating and maintaining optimal conditions for the growth of the main native plant foods, *Cordyline australis* (cabbage tree, of which the stem and root in young specimens produced a sweet starch) and bracken fern (*Pteridium esculentum*) rhizomes. Both plants thrive best in forest-edge conditions created by repeated burning. Up to a point, then, such reduction of the forest actually enhanced the resource environment (McGlone et al. 1994). The New Zealand rainforest (podocarp as well as beech-dominated)—much more than the tropical rainforest, in which hunting and gathering was so marginal (Bailey, Jenike, and Rechtman 1991)—was virtually barren of economically obtainable resources, and most of those found there occurred in greater abundance on its margins.

The more open and productive leeward forests, however, proved unusually vulnerable to fire and were rapidly pushed below optimal levels for both vegetable foods and game as the human population continued to expand. Big-game taxa became substantially depleted or extinct, and there was a niche shift into small game, particularly colonial or schooling taxa of coastal birds and fish (Anderson and McGlone 1991). These, notably harbor species of fish and shellfish (e.g., snapper [*Chrysophrys auratus*], schooling sharks, and the shellfish *pipi* [*Paphies australe*] and cockle [*Austrovenus stutchburyi*], among others), were more abundant in the windward province. Forest firing proved less disastrous there and more able to maintain the most productive mosaic character in which there is a high proportion of forest edge. In addition, because fern root and *Cordyline* were more productive in fertile soils and warm conditions, which also provided the best situation for horticulture, the balance of economic optimization swung toward the windward province. Leeward populations stagnated as the economy settled into a broad-spectrum foraging mode, which demanded high levels of effort in seasonal movement and food preservation, while windward populations expanded through a succession of horticulturally favorable habitats, which in most cases were associated with major harbors or estuaries (Fig. 13.2).

At this point we might see two contrasting strategies developing. The leeward strategy was switching from an intensive, specialist mode to an extensive, generalist mode, while the windward strategy was moving in the opposite direction, toward specialized economic behavior in high-quality habitats.

I should like to make two additional points about this model. First, the same general hypothesis explains why the pattern of windward-leeward development in New Zealand was opposite to that proposed in Hawaii and other Polynesian archipelagoes. In those, indigenous resources were at best little more productive in leeward districts than to windward, while horticulture, conversely, was immediately and significantly more productive in windward situations, at least in conditions of relatively low population density. It has been a mistake, I think, to regard New Zealand environments as appealing to early settlers in the same order of priority as in the tropical Pacific merely because horticulture was also possible there.

Second, it is apparent that this model falls within a broader theory of island habitation. There is a set of similar hypotheses of behavioral optimization that bind together ideas about why people migrate between islands, how they colonize newly discovered islands, and what they do to exploit new arrays of resource niches. Migration models emphasize push-pull causality, the perception that conditions are declining at source below levels expected at destination (Anthony 1990), which states in another way what colonization models call autocatalysis, and subsistence hypotheses call economic optimization (Diamond and Case 1986; Keegan and Diamond 1987). The argument can be carried down to the microenvironmental level of, for instance, rocky-shore shellfishing (Anderson 1981). The principle is the same, and is merely expressed at different scales of time and space and with respect to different levels of existence. Similarly, the argument that optimal exploitation will follow a first-choice seam of resources rather than expand into broad-spectrum foraging in the first instance is analogous to the observation that migration is commonly channelized rather than on a broad front, and this is consistent with Irwin's (1989) sequence of prudent exploration. In both cases a perceived optimal strategy has been adopted. It has been argued, as well, that rapid Polynesian migration at low population-density levels may have been impelled by repeated discoveries, and thus expectations, of rich avifaunal colonies (cf. Groube's [1971] strandlooper model), and that it was observation of migratory shearwaters and petrels that attracted exploration in the direction of New Zealand (McGlone et al. 1994), by A.D. 1000 probably the largest and most populous seabird colony remaining outside the polar regions.

FAUNAL EXTINCTIONS

There were various kinds of changes in the indigenous fauna of New Zealand during the prehistoric era (meaning here the period after the arrival of Polynesians and until European discovery): some introductions; some restrictions of range, notably among seals, lizards, and flightless birds; various localized changes in faunal composition and diversity; and changes in the size-frequency distribution of fish and shellfish populations. However, I shall concentrate here on extinctions.

These were relatively common in New Zealand, primarily because the lack of cursorial mammals and snakes enabled birds, in particular, to develop energy-efficient, ground-dwelling habits and structural adaptations to a degree that subsequently rendered them extraordinarily susceptible to mammalian predation and catastrophic deforestation by burning. Much the same was true of the invertebrate megafauna. Very large flightless crickets (wetas: Stenopelmatidae) and weevils (Curculionidae) shared, with semiterrestrial bats, the niche that small mammals might otherwise have occupied, and they were highly vulnerable to predation by introduced rats. The degree of impact is unknown, but new species have turned up on the few occasions that late subfossil deposits of insect remains have been found (Worthy 1983; Kuschel 1987), and it is assumed that extinctions were widespread among the invertebrate megafauna (Harris, pers. comm.), including land snails. Evidence is hard to find, according to Mason (pers. comm.), because calcium is relatively scarce in the forest environment and invertebrate exoskeletons, including land-snail shells, were consumed by the living.

Rat predation, along with deforestation, probably accounts for a number of extinctions among small vertebrates. Half of the frog species disappeared, and there were almost certainly more extinctions than have been shown among the lizards (Fig. 13.3). Systematic work on subfossil deposits is now underway in New Zealand, and evidence of further extinctions can be anticipated.

Severe depletion of invertebrates and small vertebrates by rats, which also consumed seeds and fruits, may have presented a significant problem to some cursorial birds, including moas, which, as breeding birds and chicks, probably relied on these foods as rich sources of proteins and fats. It is speculated, as

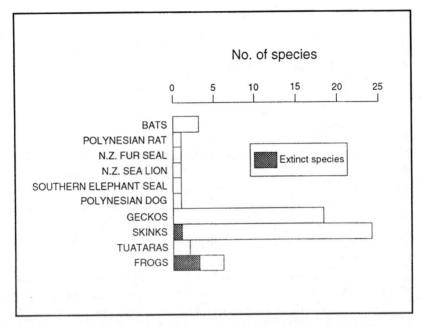

Figure 13.3 Pre-European extinction rate among non-avian terrestrial and shore-breeding vertebrates

well, that rats destroyed huge colonies of petrels and shearwaters. Until the arrival of rats there was nothing to prevent those birds from breeding safely on the ground throughout the main islands of New Zealand.

There are no known extinctions among marine birds, however. The impact of extinction was uneven across the avifauna as a whole, although it affected 11 families and resulted in the loss of 37 species and subspecies during the prehistoric era (Fig. 13.4). Most of these were structurally or facultatively flightless, and many species that became extinct had large body sizes. In fact, all 17 species over 10 kg disappeared: 11 moas, one pelican, two adzebills, two geese, and an eagle (Fig. 13.5). In the landscape, the really lethal biotope for avian extinctions was the open leeward forest and associated waterways (Fig. 13.6). The leeward forest was more completely destroyed by fire than any windward forest, which must have played a role in avian extinctions. However, three-quarters of the 300 moa-hunting sites are located in or near the leeward forest, and these contain sufficient remains to support over-hunting as the primary factor of extinction in the case of Dinornithiformes (Anderson 1989a, 1989b). So far as other taxa are concerned, there are no compelling archaeological data. Pelican and adzebill bones are rare in archaeological sites,

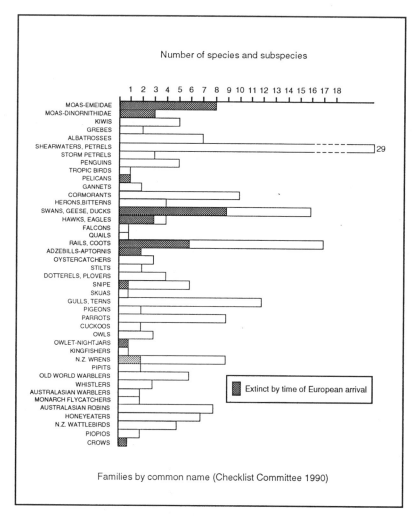

Figure 13.4 Pre-European extinction rate among New Zealand avifauna

goose and eagle occur in only one or two, and there are no archaeological remains at all in the cases of the Pink-eared, Blue-billed, Musk, and Chatham's Ducks, the owlet-nightjar, sea-eagle, and Stout-legged Wren. They occur in late subfossil sites and are assumed to have disappeared within the human era. At any rate, most of them shared with species that did become extinct at that time various factors of vulnerability to people, dogs, and rats.

Still, it is fair to make the point that avian extinctions are not all of the same status. It is really only in the instance of moas that we can be certain of extinc-

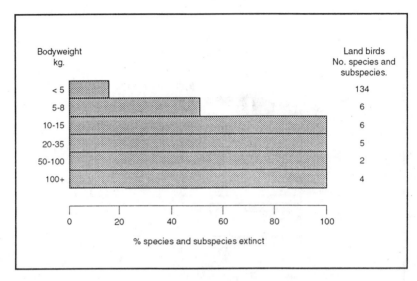

Figure 13.5 Extinction rate of avifauna by body weight

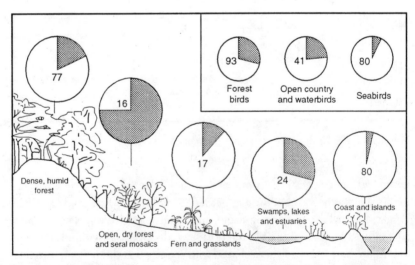

Figure 13.6 Extinction rate (shaded proportion) of avifauna by habitat type; inset., summary for major environments; and figures in all cases are total numbers of species and subspecies before prehistoric extinction events

tion by cultural behavior (Anderson 1989b), and even there the argument is one of analogy rather than weight of evidence in the case of the windward province as a whole. That is, given that moas were forced into extinction in the leeward province and disappeared at the same time to windward, given that for ecological reasons their windward populations were substantially smaller, and given that there is no evidence of an environmental catastrophe, then over-exploitation seems most probable. Whatever the case, the point that I want to make here is that the early archaeological evidence discloses a single-minded pursuit of one out of 22 orders of birds represented in New Zealand, with no real broadening of the search image until moas were substantially depleted.

CONCLUSION

The hypothesis developed here argues that an optimizing approach to the New Zealand environment by early colonists would have involved an initial emphasis on mining the easily accessible and protein-rich seam of big-game resources that outcropped repeatedly through the leeward province in particular. Perhaps similar strategies were followed elsewhere in Polynesia. Once these game sources declined significantly, the relative economic value of other resources and other modes of production increased attractively. An economic-optimization model of this kind (and it need not be optimization of food supplies as such, other resources or commodities might provide a better fit) offers a useful framework for understanding environmental change in a cultural context, not only in Oceania but also more widely.

Note

My thanks to the symposium organizers, Patrick Kirch and Terry Hunt, and for useful information and comment to Helen Leach, Graeme Mason, Ian Smith, and Richard Walter (University of Otago), Tony Harris (Otago Museum), and Matt McGlone (Landcare Research, Christchurch).

Epilogue
Islands as Microcosms of Global Change?

Patrick V. Kirch

As the twentieth century draws rapidly to a close, the attention not only of natural and social scientists but also of planners and policymakers is increasingly focused on the accelerating pace of "global change." The exponential consumption of both finite and renewable resources, spiraling population growth, increased pollution, and global warming are rightly viewed as problems that threaten the persistence of civilized life as we have known it since the advent of the Industrial Revolution (Meadows et al. 1992). There is an understandable tendency to view these problems as new developments on the time scale of the human career, developments that can be fully understood within a temporal frame of the past 300 years or less (e.g., Turner II et al., eds., 1990). However, as Robert McC. Adams—whose own work on long-term demographic cycles and human impacts in the Near East has been seminal—stresses, "only with the lengthy time perspective of historically ordered change can the complex considerations [of global change] be disentangled" (1990:x). Although research on recent and contemporary aspects of the human dimensions of global change must doubtless receive priority, it would be a mistake to ignore the insights and lessons to be derived from longer-term historical perspectives (see Silver and DeFries 1990:45–47).

The various contributors to this volume have demonstrated the value of a historical ecological approach to understanding the complex interactions between natural and cultural processes in island ecosystems, over time spans ranging from as short as 1,000 years to several tens of thousands of years. They show that the disentangling of natural inputs (e.g., climatic change, rising sea levels) from cultural effects (e.g., forest clearance, hunting of birds, translocation of animals and plants) is not always a straightforward task and requires carefully formulated research strategies and interdisciplinary cooperation. They also demonstrate that the island landscapes first documented by European explorers and naturalists in the eighteenth century were far from being "pristine" and were, indeed, often highly anthropogenic. As Steadman (Chapter 4; see also Steadman 1995) has remarked of Pacific Island avifaunas, the "biodiversity crisis" was something that had already occurred before the first systematic ornithological collections were initiated! These discoveries have important implications for current research on global change: establishing the "baseline" against which the postindustrial changes of the past 300 years have occurred requires the evidence of longer-term historical ecological research.

As has been argued elsewhere many times, islands are useful venues for studying both natural and cultural change—and the interactions between people and environment—because they are relatively small-scale, bounded systems. Paul Bahn and John Flenley have gone so far as to suggest that the ecological history of one such isolated island, Rapanui (Easter Island), can be taken as "a model for the whole planet" (1992:212). They compare the environmental history of Rapanui since Polynesian colonization with the global computer model of the Club of Rome, and find disturbing congruence between the two (1992:213–14, Figs. 192, 193). Some may find their comparisons overdrawn, and certainly Rapanui could never be more than a grossly simplified analogue. Nonetheless, precisely because in the Rapanui case the variables can be more readily defined and the geographic scale is so "microcosmic," the historical lessons are not entirely irrelevant to understanding the contemporary human dimensions of global change.

The advances that have been made in our understanding of the historical ecology of Pacific islands, and especially of the role of humans in transforming insular ecosystems, are in large part due to the heightened interdisciplinary collaboration between archaeologists and natural scientists, collaboration exemplified well by the contributors to this volume. Such interdisciplinary collaboration has indeed been identified by the National Research Council's Committee on the Human Dimensions of Global Change (Stern et al., eds., 1992:168–72) as one of the highest priorities for theory and method advancement in the study of global change. The Pacific research summarized here nicely underscores this point, for before the interdisciplinary collaboration of the past decade or two, independent research by anthropologists, archaeologists, biogeographers, and ecologists had led to little historical understanding of the long-term dynamism of island environments or of the critical role of human populations in shaping island landscapes.

One area that still requires much research, however, is that of prehistoric human population growth on islands, and the relation between demographic transitions and the anthropogenic transformation of island ecosystems. Gordon Wolman writes that "the absence of satisfactory historical information relating both land and population change to the many factors that influence both suggests the obvious need for comparative studies combining demography, land use, and environmental change" (1993:27). Again, because of the various advantages that islands offer, it may be that such comparative studies will be more readily carried out in island contexts. For example, paleodemographic research in the Hawaiian Islands is producing increasingly fine-grained reconstructions of the rates of prehistoric population growth, reconstructions that can be measured against paleoenvironmental evidence for human impacts through agricultural expansion and intensification (see Chapters 11 and 12), among other forms of human impact. Moreover, the rich ethnographic record of Pacific Island societies allows us to explore the social and ideational correlates of demographic change, such as various forms of population regulation, whether voluntary or coercive.

The past few years have been an exciting and stimulating period in Pacific Islands archaeology, paleoecology, and historical ecology, thanks to the heightened levels of cooperation and collaboration among scientists from both the natural and social sciences. We fully anticipate that the pace of research will accelerate, and we hope that future advances will contribute as much to contemporary efforts to comprehend the human dimensions of global change as to the historical understanding of how island landscapes came to be as we see them today.

References

Absy, M. L., A. Cleef, M. Fournier, L. Martin, M. Servant, A. Sifeddine, M. Ferreira da Silva, F. Soubies, K. Suguio, B. Turcq, and T. Van Der Hammen. 1991. Mise en évidence de quatre phases d'ouverture de la forêt dense dans le sud-est de l'Amazonie au cours des 60,000 dernières années: première comparaison avec d'autres régions tropicales. *Comptes Rendus de l'Académie des Sciences*, 2d ser., 312:673–78.

Adams, R. McC. 1990. Foreword: The relativity of time and transformation. In B. L. Turner II, W. C. Clark, R. W. Kates, J. F. Richards, J. T. Mathews, and W. B. Meyer, eds., *The Earth as Transformed by Human Action: Global and Regional Changes in the Biosphere over the Past 300 Years*, vii–x. Cambridge: Cambridge University Press.

Adamson, A. 1939. *Review of the Fauna of the Marquesas Islands and Discussion of Its Origin*. Bernice P. Bishop Museum Bulletin 159, Honolulu.

Allen, B. 1971. Wet-field taro terraces on Mangaia, Cook Islands. *Journal of the Polynesian Society* 80:371–78.

Allen, B., and R. Crittenden. 1987. Degradation and a pre-capitalist political economy. In P. Blaikie and H. Brookfield, eds., *Land Degradation and Society*, 145–56. London: Methuen.

Allen, Jane. 1987a. Preliminary report, archaeological investigations at sites 50-Oa-G6-44 and 50-Oa-G6-45, Maunawili, Kailua, O'ahu. Department of Anthropology, B. P. Bishop Museum, and Hawaii State Department of Land and Natural Resources, Honolulu.

———. 1987b. Preliminary report, archaeological investigations at sites 50-Oa-G6-46 through 50-Oa-G6-62, Mauka Golf Course Project Area, Maunawili, Kailua, Ko'olaupoko, O'ahu. Department of Anthropology, B. P. Bishop Museum, and Hawaii State Department of Land and Natural Resources, Honolulu.

———. 1988. Trade, transportation, and tributaries: Exchange, agriculture, and settlement distribution in early Historic-Period Kedah, Malaysia. Ph.D. dissertation, University of Hawaii, Honolulu. Ann Arbor: University Microfilms.

———. 1989. Preliminary report, archaeological investigations at sites 50-Oa-G6-17 and G6-68 through G6-71, Royal Hawaiian Country Club, Inc., Makai Golf Course Project Area, Maunawili, Kailua, Ko'olaupoko, O'ahu. Department of Anthropology, Bishop Museum, and Hawaii State Department of Land and Natural Resources, Honolulu.

———. 1990. Agriculture, hydraulics, and urbanism at Satingpra. *Asian Perspectives* 28(2):163–77.

———. 1991a. Review, J. Stephen Athens and Jerome V. Ward, Paleoenvironmental and archaeological investigations, Kawainui Marsh Flood Control Project, O'ahu Island, Hawaii. U.S. Army Engineer Division, Pacific Ocean Environmental Section, Honolulu.

———. 1991b. The role of agriculture in the evolution of the pre-contact Hawaiian state. *Asian Perspectives* 30:117–32.

———. 1992. Farming in Hawaii from colonization to contact: Radiocarbon chronology and implications for cultural change. *New Zealand Journal of Archaeology* 14:45–66.

Allen, Jane, ed. 1994. *Kula* and *kahawai*: Geoarchaeological and historical investigations in middle Maunawili Valley, Kailua, Ko'olau Poko, O'ahu. Manuscript on file, Department of Anthropology, B. P. Bishop Museum, Honolulu.

Allen, Jane, M. Riford, T. Bennett, Murakami, and M. Kelly. 1987. Five Upland 'Ili: Archaeological and Historical Investigations in the Kane'ohe Interchange, Interstate Highway H-3, Island of O'ahu. Report prepared for the Department of Transportation, State of Hawaii. Department of Anthropology, B. P. Bishop Museum, Honolulu.

Allen, Jim. 1989. When did humans first colonize Australia? *Search* 20 (5):149–55.

Allen, Jim, C. Gosden, R. Jones, and J. White. 1988. Pleistocene dates for the human occupation of New Ireland, northern Melanesia. *Nature* 331:707–9.

Allen, Jim, C. Gosden, and J. White. 1989. Human Pleistocene adaptations in the tropical island Pacific: Recent evidence from New Ireland, a Greater Autsralian outlier. *Antiquity* 63:548–61.

Allen, M. S. 1992a. Dynamic landscapes and human subsistence: Archaeological investigations on Aitutaki Island, southern Cook Islands. Ph.D. dissertation, University of Washington. Ann Arbor: University Microfilms.

———. 1992b. Temporal variation in Polynesian fishing strategies: The southern Cook Islands in regional perspective. *Asian Perspectives* 31:183–204.
Allen, M. S., and S. Schubel. 1990. Recent archaeological research on Aitutaki, Southern Cooks: The Moturakau Rockshelter. *Journal of the Polynesian Society* 99:265–95.
Allen, M. S., and D. W. Steadman. 1990. Excavations at the Ureia site, Aitutaki, southern Cook Islands: Preliminary results. *Archaeology in Oceania* 25:24–37.
Allen-Wheeler, J. 1981. Archaeological excavations in Kawainui Marsh, Island of O'ahu. Report prepared for Department of Planning and Economic Development, Department of Anthropology, B. P. Bishop Museum, Honolulu.
Amerson, A., Jr., W. Whistler, and T. Schwaner. 1982a. *Wildlife and Wildlife Habitat of American Samoa*. Vol. 1. U.S. Department of the Interior, Fish and Wildlife Service, Washington, D.C.
———. 1982b. *Wildlife and Wildlife Habitat of American Samoa*. Vol. 2. U.S. Department of the Interior, Fish and Wildlife Service, Washington, D.C.
Anderson, A. J. 1981. A model of prehistoric collecting on the rocky shore. *Journal of Archaeological Science* 8:109–20.
———. 1984. The extinction of moa in southern New Zealand. In P. Martin and R. Klein, eds., *Quaternary Extinctions*, 728–40. Tucson: University of Arizona Press.
———. 1989a. Mechanics of overkill in the extinction of New Zealand moas. *Journal of Archaeological Science* 16:137–51.
———. 1989b. *Prodigious Birds: Moas and Moa-Hunting in Prehistoric New Zealand*. Cambridge: Cambridge University Press.
———. 1991. The chronology of colonization in New Zealand. *Antiquity* 65:767–95.
———. n.d. Rat voyaging and Polynesian colonization. Paper in preparation.
Anderson, A. J., and M. S. McGlone. 1992. Prehistoric land and people in New Zealand. In J. Dodson, ed., *The Naive Lands: Human-Environmental Interactions in Australia and Oceania*, 199–241. Sydney: Longman Cheshire.
Anderson, A. J., and R. McGovern-Wilson. 1990. The pattern of prehistoric Polynesian colonisation in New Zealand. *Journal of the Royal Society of New Zealand* 20:41–63.
Anderson, E. 1952. *Plants, Man, and Life*. Berkeley: University of California Press.
Anderson, I. 1991. First Australian headed south in haste. Supplement to *New Scientist* no. 1796. *Science and Educaton in Australia and New Zealand* 2:3.
Andrews, E., and I. Andrews. 1944. *A Comparative Dictionary of the Tahitian Language*. Chicago: Chicago Academy of Sciences.
Anthony, D. W. 1990. Migration in archaeology: The baby and the bathwater. *American Anthropologist* 92:895–914.
Ash, J. 1987. Holocene sea levels in northern Viti Levu, Fiji. *New Zealand Journal of Geology and Geophysics* 30:431–35.
Athens, J. 1982. Prehistoric pondfield agriculture in Hawaii: Archaeological investigations at the Hanalei National Wildlife Refuge, Kaua'i. Manuscript, Department of Anthropology, B. P. Bishop Museum, Honolulu.
———. 1988. Archaeological survey and testing for airfield perimeter fence project,

Bellows Air Force Station, Oʻahu, Hawaii. International Archaeological Research Institute, Inc., and U.S. Army Engineer District, Honolulu.

———. 1995. Landscape archaeology: Prehistoric settlement, subsistence, and environment of Kosrae, Eastern Caroline Islands, Micronesia. Report prepared for Kosrae State Government, F.S.M. International Archaeological Research Institute, Inc., Honolulu.

———. n.d. Early maize agriculture in northern highland Ecuador. Manuscript submitted for publication.

Athens, J., and J. V. Ward. 1991. Paleoenvironmental and archaeological investigations, Kawainui Marsh flood control project, Oʻahu Island, Hawaii. In *Report Prepared for the U.S. Army Engineer Division, Pacific Ocean, Ft. Shafter, Hawaii.* Micronesian Archaeological Services, Guam.

———. 1993. Environmental change and prehistoric Polynesian settlement in Hawaii. *Asian Perspectives* 32:205–23.

———. 1994. Prehistoric inland expansion of settlement and agriculture, Oʻahu, Hawaii. Paper presented at the Seventh Annual Hawaiian Archaeology Conference, Hilo, Hawaii. 1–3 April 1994.

———. 1995a. Liliha paleoenvironmental study, Honolulu, Hawaii. Draft report prepared for Scientific Consultant Services, Inc., Kaneʻohe, Hawaii. International Archaeological Research Institute, Inc., Honolulu.

———. 1995b. Paleoenvironment of the Orote Peninsula, Guam. *Micronesica* 28:51–76.

Athens, J., J. V. Ward, and D. W. Blinn. 1995. Paleoenvironmental investigations at ʻUkoʻa Pond, Kawailoa Ahupuaʻa, Oʻahu, Hawaii. Report prepared for Engineers Surveyors Hawaii, Inc., Honolulu. International Archaeological Research Institute, Inc., Honolulu.

Athens, J., J. V. Ward, and M. Tomonari-Tuggle. 1994. Archaeological inventory survey and paleoenvironmental investigations at Maʻalaea Power Plant, Maui Electric Co., Ltd., Waikapu Ahupuaʻa, Maʻalaea, Maui. Report prepared for Belt Collins Hawaii, Ltd., Honolulu. International Archaeological Research Institute, Inc., Honolulu.

Athens, J., J. Ward, and S. Wickler. 1992. Late Holocene lowland vegetation, Oʻahu, Hawaii. *New Zealand Journal of Archaeology* 14:9–34.

Atkinson, I. 1985. The spread of commensal species of *Rattus* to oceanic islands and their effects on island avifaunas. *International Council for Bird Preservation Technical Publication* 3:35–81.

Bahn, P., and J. Flenley. 1992. *Easter Island: Earth Island.* London: Thames and Hudson.

Bailey, R., M. Jenike, and R. Rechtman. 1991. Reply to Colivaux and Bush. *American Anthropologist* 93:160–62.

Balean, C. 1989. Caves as refuse sites: An analysis of shell material from Buang Merabak, New Ireland. B.A. honors thesis, Department of Prehistory and Anthropology, Australian National University, Canberra.

Baldwin, E. D. 1917. Map of Waimanalo Plantation, Hawaii. Registered map 2681, Survey Office, State of Hawaii, Honolulu.

Balme, J., and J. Hope. 1990. Radiocarbon dates from midden sites in the lower Darling River area of western New South Wales. *Archaeology in Oceania* 25 (3):85–101.

Balouet, J., and S. Olson. 1987. A new extinct species of giant pigeon (Columbidae: *Ducula*) from archaeological deposits on Wallis ('Uvea) Island, South Pacific. *Proceedings of the Biolological Society of Washington* 100:769–75.

———. 1989. Fossil birds from late Quaternary deposits in New Caledonia. *Smithsonian Contributions to Zoology* 469:1–38.

Banks, M., E. Colhoun, and G. van de Geer. 1976. Late Quaternary *Palorchestes azael* (Mammalia, Diprotodontidae) from northwestern Tasmania. *Alcheringa* 1:159–66.

Barrera, W., Jr. 1984. Archaeological services during installation of five replacement antennas at Bellow Air Force Station, O'ahu, Hawaii. Manuscript on file, U.S. Army Engineer District, Honolulu.

Bates, M. 1963. Nature's effect on and control of man. In F. Fosberg, ed., *Man's Place in the Island Ecosystem: A Symposium,* 101–13. Honolulu: Bishop Museum Press.

Bayliss-Smith, T. 1988. The role of hurricanes in the development of reef islands, Ontong Java, Solomon Islands. *Geographical Journal* 154:377–91.

Beaglehole, E., and P. Beaglehole. 1938. *Ethnology of Pukapuka.* Bernice P. Bishop Museum Bulletin 150, Honolulu.

Beaglehole, J. C., ed. 1967. *The Journals of Captain James Cook on His Voyages of Discovery: The Voyage of the* Resolution *and* Discovery, *1776–1780.* Cambridge: Hakluyt Society.

———. 1968. *The Journals of Captain James Cook on His Voyages of Discovery: The Voyage of the* Endeavour, *1768–1771.* Cambridge: Hakluyt Society.

Beggerly, P. 1990. Kahana Valley, Hawaii, a geomorphic artifact: A study of the interrelationships among geomorphic structures, natural processes, and ancient Hawaiian technology, land use, and settlement patterns. Ph.D. dissertation, University of Hawaii, Honolulu. Ann Arbor: University Microfilms.

Bellwood, P. 1978a. *Man's Conquest of the Pacific.* Auckland: W. Collins.

———. 1978b. *Archaeological Research in the Cook Islands.* Pacific Anthropological Records 27. B. P. Bishop Museum, Honolulu.

Bennett, K. D. 1994. Confidence intervals for age estimates and deposition times in late-Quaternary sediment sequences. *The Holocene* 4:337–48.

Bennett, T. M. 1985. Palynology of selected horizons from the Ewa coastal plain, Oahu, Hawaii. M.S. thesis, Department of Botany, University of Hawaii, Honolulu.

Benninghoff, W. 1963. Calculations of pollen and spore density in sediments by the addition of exotic pollen in known quantities. *Pollen et Spores* 4:323–24.

Birdsell, J. 1957. Some population problems involving Pleistocene man. *Cold Spring Harbor Symposia on Quantitative Biology* 22:47–69.

———. 1977. The recalibration of a paradigm for the first peopling of Greater Australia. In J. Allen, J. Golson, and R. Jones, eds., *Sahul and Sunda,* 113–67. London: Academic Press.

Birkeland, P. W. 1974. *Pedology, Weathering and Geomorphological Research.* New York: Oxford University Press.

Blaikie, P., and H. Brookfield. 1987a. Defining and debating the problem. In P. Blaikie and H. Brookfield, eds., *Land Degradation and Society*, 1–26. London: Methuen.

———. 1987b. Decision making in land management. In P. Blaikie and H. Brookfield, eds., *Land Degradation and Society*, 64–83. London: Methuen.

Blaikic, P., and H. Brookfield, eds. 1987. *Land Degradation and Society*. London: Methuen.

Blom W. M. 1988. Late Quaternary sediments and sea-levels in Bass Basin, Southeastern Australia: A preliminary report. *Search* 19:94–96.

Blong, R., and R. Gillespie. 1978. Fluvially transported charcoal gives erroneous ^{14}C ages for recent deposits. *Nature* 271:739–41.

Bloom, A. L. 1978. *Geomorphology: A Systematic Analysis of Late Cenozoic Landforms*. Englewood Cliffs, N.J.: Prentice-Hall.

———. 1980. Late Quaternary sea level change on South Pacific coasts: A study in tectonic diversity. In N.-A. Morner, ed., *Earth Rheology, Istostasy and Eustasy*, 505–16. New York: John Wiley and Sons.

———. 1983. Sea level and coastal changes. In H. Wright, ed., *Late Quaternary Environments of the United States: The Holocene*, 42–53. Minneapolis: University of Minnesota Press.

Bowdler, S. 1977. The coastal colonisation of Australia. In J. Allen, J. Golson, and R. Jones, eds., *Sahul and Sunda*, 205–46. London: Academic Press.

Bowers, N. 1968. The ascending grasslands: An anthropological study of ecological succession in a high mountain valley of New Guinea. Ph.D. dissertation, Columbia University, New York. Ann Arbor: University Microfilms.

Brailsford, B. J., ed. 1994. *Song of Waitaha: The Histories of a Nation*. Christchurch: Ngatapuwae Trust.

Braudel, F. 1980. *On History*. Chicago: University of Chicago Press.

Bridgman, H. 1983. Could climate change have had an influence on the Polynesian migrations? *Palaeogeography, Palaeoclimatology, and Palaeoecology* 41:193–206.

Brookfield, H. 1972. Intensification and disintensification in Pacific agriculture: A theoretical approach. *Pacific Viewpoint* 13:30–48.

———. 1980. *Population-Environment Relations in Tropical Islands: The Case of Eastern Fiji*. MAB Technical Notes 13. Paris: UNESCO.

———. 1989. Frost and drought through time and space, part III: What were conditions like when the high valleys were first settled? *Mountain Research and Development* 9:306–21.

———. 1991. Research in the mountains of the island of New Guinea. *Mountain Research and Development* 11:203–11.

Brookfield, H., ed. 1979. Lakeba: Environmental change, population dynamics and resource use. In *UNESCO/UNFPA Population and Environment Project in the Eastern Islands of Fiji, Island Reports* 5, Canberra.

Brown, F. 1931. *Flora of Southeastern Polynesia*. Bernice P. Bishop Museum Bulletin 84, Honolulu.

Bruner, P. 1972. *Birds of French Polynesia, Tahiti, Pacific Scientific Information Center*. Honolulu: B. P. Bishop Museum.

Brunton, R. 1991. *Aborigines and Environmental Myths: Apocalypse in Kakadu*. IPA Environmental Backgrounder 4. Canberra: Institute of Public Affairs.

Bryan, W. 1915. *Natural History of Hawaii*. Honolulu: Hawaiian Gazette.

Bub, H. 1991. *Bird Trapping and Bird Banding: A Handbook for Trapping Methods All over the World*. Ithaca: Cornell University Press.

Buck, P. [Te Rangi Hiroa]. 1927. *The Material Culture of the Cook Islands (Aitutaki): Memoirs of the Board of Maori Ethnological Research*. New Plymouth, N.Z.

———. 1934. *Mangaian Society*. Bernice P. Bishop Museum Bulletin 122, Honolulu.

———. 1944. *Arts and Crafts of the Cook Islands*. Bernice P. Bishop Museum Bulletin 179, Honolulu.

Bulmer, R. 1968. The strategies of hunting in New Guinea. *Oceania* 38:302–18.

Bulmer, S. 1989. Gardens in the south: Diversity and change in prehistoric Maori agriculture. In D. Harris and G. Hillman, eds., *Foraging and Farming: The Evolution of Plant Exploitation*, 688–705. London: Unwin Hyman.

Burney, D. A. 1993. Late Holocene environmental changes in arid southwestern Madagascar. *Quaternary Research* 40:98–106.

Burney, D. A., L. P. Burney, and R. D. E. MacPhee. 1994. Holocene charcoal stratigraphy from Laguna Tortuguero, Puerto Rico, and the timing of human arrival on the island. *Journal of Archaeological Science* 21:273–81.

Butzer, K. 1974. Accelerated soil erosion: A problem of man-land relationships. In I. Manners and M. Mikesell, eds., *Perspectives on Environment*, 47–78. Association of American Geographers Publication 13, Washington, D.C.

Buxton, P. 1930. Description of the environment. In *Insects of Samoa*, part 9, fasc. 1. London: British Museum.

Cachola-Abad, C. K. 1993. Evaluating the orthodox dual settlement model for the Hawaiian Islands: An analysis of artefact distribution and Hawaiian oral traditions. In M. Graves and R. C. Green, eds., *The Evolution and Organization of Prehistoric Society in Polynesia*, 13–32. New Zealand Archaeological Association Monograph 19. Auckland.

Carlquist, S. 1970. *Hawaii: A Natural History*, 2d edition. Pacific Tropical Botanical Garden, Lawai, Hawaii. Honolulu: S. B. Printers.

Cassels, R. 1984. Faunal extinction and prehistoric man in New Zealand and the Pacific islands. In P. Martin and R. Klein, eds., *Quaternary Extinctions: A Prehistoric Revolution*, 741–67. Tucson: University of Arizona Press.

Caughley, G. 1988. The colonisation of New Zealand by the Polynesians. *Journal of the Royal Society of New Zealand* 18:245–70.

Chappell, J. 1982. Sea levels and sediments: Some features of the context of coastal archaeological sites in the tropics. *Archaeology in Oceania* 17:69–78.

Chazine, J. 1978. Contribution à l'étude des anciennes structures d'habitat dans une vallée de Tahiti (vallée de la Papeno'o). M.A. thesis, University of Paris.

Chikamori, M. 1987. Archaeology on Pukapuka atoll. *Man and Culture in Oceania* 3 (special issue):105–15.

Christensen, C. 1981. Preliminary analysis of non-marine mollusks from the Fa'ahia archaeological site. Manuscript on file, B. P. Bishop Museum, Honolulu.

Christensen, C., and P. Kirch. 1981. Nonmarine mollusks from archaeological sites on Tikopia, Southeastern Solomon Islands. *Pacific Science* 35:75–88.

———. 1986. Nonmollusks and ecological change at Barbers Point, O'ahu, Hawaii. *Bishop Museum Occasional Papers* 26:52–80.

Christensen, N. L. 1989. Landscape history and ecological change. *Journal of Forest History* 33:116–24.

Clark, J., and C. Lingle. 1979. Predicted sea-level changes (18,000 years B.P. to present) caused by late-glacial retreat of the Antarctic ice sheet. *Quaternary Research* 11:279–98.

Clark, J. T. 1980. Phase I archaeological survey of the slopes around the Kawainui Marsh, O'ahu. In M. Kelly and J. Clark, eds., *Kawainui Marsh, O'ahu: Historical and Archaeological Studies*, 29–73. Departmental Report Series 80–3. Department of Anthropology, B. P. Bishop Museum, Honolulu.

———. 1981. Early settlement and man-land relationships at Kawainui Marsh, O'ahu Island, Hawaii. *New Zealand Archaeological Association Newsletter* 24:30–37.

Clark, R. 1982. Proto-Polynesian birds. *Transactions of the Finnish Anthropological Society* 11:121–43.

Clarke, W. 1977. A change in subsistence staple in prehistoric New Guinea. In C. L. A. Leakey, ed., *Proceedings of the Third Symposium of the International Society for Tropical Root Crops*, 159–63. Ibadan: International Society for Tropical Root Crops in Collaboration with International Institute for Tropical Agriculture.

Cleghorn, P. L. 1988. The settlement and abandonment of two Hawaiian outposts: Nihoa and Necker Islands. *Bishop Museum Occasional Papers* 28:35–49.

Cline, M. G., A. S. Ayres, W. Crosby, P. F. Philipp, R. Elliott, O. G. Magistad, J. C. Ripperton, E. Y. Hosaka, M. Takahashi, G. D. Sherman, and C. K. Wentworth. 1955. *Soil Survey of the Territory of Hawaii*. U.S. Department of Agriculture Soil Survey Series 1939 No. 25.

Conklin, H. 1980. *An Ethnographic Atlas of Ifugao*. New Haven: Yale University Press.

Conant, S. 1985. Recent observations on the plants of Nihoa Island, northwestern Hawaiian Islands. *Pacific Science* 39:135–49.

Cooke, C. M., Jr. 1926. Notes on Pacific island land snails. *Proceedings of the Third Pan-Pacific Science Congress* 2276–84. Tokyo.

Cordy, R. 1977. A cultural resources study for the city and county of Honolulu's permit request: Kawainui Marsh Sewerline (O'ahu). Archaeological reconnaissance and pre-1850 literature search. Manuscript on file, U.S. Army Engineer District, Honolulu.

———. 1978. Test excavations, site 7, Kawainui Marsh, Kailua Ahupua'a, O'ahu. Manuscript on file, U.S. Army Engineer District, Honolulu.

Cordy, R., and H. D. Tuggle. 1976. Bellows, O'ahu, Hawaiian Islands: New work and new interpretations. *Archaeology and Physical Anthropology in Oceania* 11:207–35.

Cosgrove, R. 1991. The illusion of riches: Issues of scale, resolution and explanation of Pleistocene human behaviour. Ph.D. dissertation, La Trobe University, Bundoora, Australia.

Cosgrove, R., J. Allen, and B. Marshall. 1990. Palaeo-ecology and Pleistocene human occupation in south central Tasmania. *Antiquity* 64 (242):59–78.

Coulter, J. 1941. *Land Utilization in American Samoa*. Bernice P. Bishop Museum Bulletin 170, Honolulu.

Cox, P., T. Elmquist, E. Pierson, and W. Rainey. 1991. Flying foxes as strong interactors in South Pacific island ecosystems: A conservation hypothesis. *Conservation Biology* 5:448–54.

Crosby, A. 1986. *Ecological Imperialism: The Biological Expansion of Europe, A.D. 900–1900*. Cambridge: Cambridge University Press.

Crumley, C. L. 1994. Historical ecology: A multidimensional ecological orientation. In C. L. Crumley, ed., *Historical Ecology: Cultural Knowledge and Changing Landscapes*, 1–16. Santa Fe: School of American Research Press.

Crumley, C. L., ed. 1994. *Historical Ecology: Cultural Knowledge and Changing Landscapes*. Santa Fe: School of American Research Press.

Cuddihy, L. W., and C. P. Stone. 1990. *Alteration of Native Hawaiian Vegetation: Effects of Humans, Their Activities and Introductions*. Honolulu: University of Hawaii Press.

Cumberland, K. 1962. Moas and men: New Zealand about A.D. 1250. *Geographical Review* 52:151–73.

———. 1963. Man's role in modifying island environments in the southwest Pacific, with special reference to New Zealand. In F. R. Fosberg, ed., *Man's Place in the Island Ecosystem: A Symposium*, 187–206. Honolulu: Bishop Museum Press.

Curr, E. 1883. *Recollections of Squatting in Victoria*. Melbourne: George Robertson.

Dames and Moore, Consultants, Inc. 1961. Preliminary site investigation, proposed land development, Kawainui Marsh, Kailua, Hawaii (with appendices). In Report prepared by Dames and Moore, consultants for the Trousdale Construction Co. Manuscript on file, International Archaeological Research Institute, Inc., Honolulu.

Darwin, C. R. 1859. *On the Origin of Species by Means of Natural Selection; or, The Preservation of Favoured Races in the Struggle for Life*. London: John Murray.

———. 1957 [1845]. *Journal of Researches into the Natural History and Geology of the Countries Visited during the Voyage of H.M.S. Beagle*. Norwalk, Conn.: Easton Press.

Dawson, A. G. 1992. *Ice Age Earth: Late Quaternary Geology and Climate*. New York: Routledge.

Dawson, S. 1990. A chemical and mineralogical study of a sediment core from Lake Tiriara, Mangaia, southern Cook Islands, with special reference to the impact of early man. B.Sc. dissertation, School of Geography and Earth Resources, Hull University.

Dening, G. 1963. The geographic knowledge of the Polynesians and the nature of inter-island contact. In J. Golson, ed., *Polynesian Navigation*, Polynesian Society Memoir 34, 102–53. Wellington: Polynesian Society.

Diaz, H. F., and V. Markgraf, eds. 1993. *El Niño: Historical and Paleoclimatic Aspects of the Southern Oscillation*. New York: Cambridge University Press.

Diamond, J. M. 1986. The environmentalist myth. *Nature* 324:19–20.

Diamond, J., and T. J. Case. 1986. Overview: Introductions, extinctions, exterminations, and invasions. In J. Diamond and T. J. Case, eds., *Community Ecology*, 65–79. New York: Harper and Row.

Dicks, A. M., A. E. Haun, and P. H. Rosendahl. 1987. Archaeological reconnaissance survey for environmental impact statement, West Loch Estates Golf Course and Parks, Land of Honouliuli, 'Ewa District, Island of O'ahu. Manuscript on file, Hawaii State Department of Land and Natural Resources, Honolulu.

Di Piazza, A. 1990. Les jardins enfouis de Futuna: Une ethnoarchéologie de l'horticulture. *Journal de la Société des Océanistes* 91:151–62.

Di Piazza, A., and D. Frimigacci. 1991. A thousand years of gardening: A history of subsistence on Futuna. *Indo-Pacific Prehistory Association Bulletin* 11:124–40.

Dodson, J., ed. 1992. *The Naive Lands: Prehistory and Environmental Change in Australia and the Southwest Pacific*. Melbourne: Longman Cheshire.

Dodson, J., R. Fullager, and L. Head. 1992. Dynamics of environment and people in the forested crescents of temperate Australia. In J. Dodson, ed., *The Naive Lands*, 115–59. Melbourne: Longman Cheshire.

Donn, J. M. 1902. Map of Kailua, O'ahu. Hawaii Territorial Survey, Honolulu.

Douglas, I. 1977. *Humid Landforms: An Introduction to Systematic Geomorphology*. Cambridge: MIT Press.

Driessen, M. 1988. Age structure and reproduction of Bennetts and Rufous Wallabies in Tasmania. Report (phase 1) to the Australian National Parks and Wildlife Service, Canberra.

Drigot, D. 1982. *Ho'ona'auao No Kawai Nui (Educating about Kawai Nui): A Multimedia Educational Guide*. Honolulu: Environmental Center.

Dubois, J. 1991. Contribution de l'histoire biogéographique du rat à la connaissance des peuplements humains en Polynésie. M.A. thesis, University of Paris.

Dudal, R., and F. R. Moorman. 1964. Major soils of South-east Asia. *Journal of Tropical Geography* 18:54–80.

Duff, R. 1956. *The Moa-Hunter Period in Maori Culture*. Wellington: Government Printer.

Dwyer, P. 1982. Wildlife conservation and tradition in the Highlands of Papua New Guinea. In L. Morauta, J. Pernetta, and W. Heaney, eds., *Traditional Conservation in Papua New Guinea: Implications for Today*, 173–89. Boroko: Institute of Applied Social and Economic Research.

Dye, T., and D. Steadman. 1990. Polynesian ancestors and their animal world. *American Scientist* 78:207–15.

Ellison, J. 1994. Palaeo-lake and swamp stratigraphic records of Holocene vegetation and sea-level changes, Mangaia, Cook Islands. *Pacific Science* 48:1–15.

Emory, K. P. 1933. *Stone Remains in the Society Islands*. Bernice P. Bishop Museum Bulletin 116, Honolulu.

———. 1963. East Polynesian relationships: Settlement pattern and time involved as indicated by vocabulary agreements. *Journal of the Polynesian Society* 72:78–100.

Emory, K. P., and Y. Sinoto. 1961. *Hawaiian Archaeology: Oahu Excavations*. Bernice P. Bishop Museum Special Publication 49, Honolulu.

———. 1964. Eastern Polynesian burials at Maupiti. *Journal of the Polynesian Society* 73:143–60.

———. 1965. Preliminary report on the archaeological investigations in Polynesia. Manuscript on file, Department of Anthropology, B. P. Bishop Museum, Honolulu.

Evans J. 1972. *Land Snails in Archaeology*. London: Seminar Press.

Faegri, K., and J. Iversen. 1975. *Textbook of Pollen Analysis*. Oxford: Blackwell Scientific.

Fairbridge, R. 1961. Eustatic changes in the sea level. *Physics and Chemistry of the Earth* 4:99–185.

Feil, D. 1987. *The Evolution of Highland Papua New Guinea Societies*. Cambridge: Cambridge University Press.

Feinberg, R. 1981. *Anuta: Social Structure of a Polynesian Island*. Institute of Polynesian Studies. Provo: Brigham Young University Press.

Finney, B. R. 1979. Voyaging. In J. D. Jennings, ed., *The Prehistory of Polynesia*, 323–51. Cambridge: Harvard University Press.

Firth, R. 1936. *We, the Tikopia*. London: George Allen and Unwin.

Flannery, T. 1990a. *Mammals of New Guinea*. Carina: Robert Brown.

———. 1990b. Pleistocene faunal loss: Implications of the aftershock for Australia's past and future. *Archaeology in Oceania* 25:45–67.

Flannery, T., P. Kirch, J. Specht, and M. Spriggs. 1988. Holocene mammal faunas from archaeological sites in island Melanesia. *Archaeology in Oceania* 23:89–94.

Flenley, J. n.d. Man-vegetation interactions in the Pacific. Report to the British Academy.

Flenley, J., and S. King. 1984. Late Quaternary pollen records from Easter Island. *Nature* 307:47–50.

Flenley, J., S. King, J. Jackson, C. Chew, J. Teller, and M. Prentice. 1991. The Late quaternary vegetational and climatic history of Easter Island. *Journal of Quaternary Science* 6:85–115.

Flood, J. 1983. *Archaeology of the Dreamtime*. Honolulu: University of Hawaii Press.

Florence, J. 1982. *Recherches Botaniques en Polynesie Française*. Papeete: ORSTOM.

Foote, D. E., E. L. Hill, S. Nakamura, and F. Stephens. 1972. *Soil Survey of the Islands of Kaua'i, O'ahu, Maui, Molokai, and Lana'i, State of Hawaii*. Edited by U.S. Department of Agriculture Soil Conservation Service. Washington, D.C.: Government Printing Office.

Fosberg, F. 1963a. The island ecosystem. In F. R. Fosberg, ed., *Man's Place in the Island Ecosystem: A Symposium*, 1–7. Honolulu: Bishop Museum Press.

———. 1963b. Disturbance in island ecosystems. In J. L. Gressitt, ed., *Pacific Basin Biogeography*, 557–61. Honolulu: Bishop Museum Press.

Fosberg, F., ed. 1963. *Man's Place in the Island Ecosystem: A Symposium*. Honolulu: Bishop Museum Press.

Franklin, J., and M. Merlin. 1992. Species-environment patterns of forest vegetation on the uplifted reef limestone of Atiu, Mitiaro, and Ma'uke, Cook Islands. *Journal of Vegetation Science* 3:3–14.

Franklin, J., and D. Steadman. 1991. The potential for conservation of Polynesian birds through habitat mapping and species translocation. *Conservation Biology* 5:506–521.

Friedman, J. 1981. Notes on structure and history in Oceania. *Folk* 23:275–95.

———. 1982. Catastrophe and continuity in social evolution. In C. Renfrew, M. Rowlands, and B. Seagraves, eds., *Theory and Explanation in Archaeology*, 175–96. New York: Academic Press.

Frimigacci, D. 1990. *Aux Temps de la Terre Noire: Ethnoarchéologie des Iles Futuna et Alofi*. Langues et Cultures du Pacifique 7. Paris: Peeters.

Fujita, M., and M. Tuttle. 1991. Flying foxes (Chiroptera: Pteropodidae): Threatened animals of key ecological importance. *Conservation Biology* 5:455–63.

Gagné, W. C., and L. W. Cuddihy. 1990. Vegetation. In W. L. Wagner, D. R. Herbst, and S. H. Sohmer, eds., *Manual of the Flowering Plants of Hawaii* 2 vols., 45–114. Honolulu: University of Hawaii Press and Bishop Museum Press.

Galzin, R., and J. Pointier. 1985. Moorea Island, Society Archipelago. In Delesalle B., Galzin R. and B. Salvat, eds., *Proceedings of the Fifth International Coral Reef Congress, Tahiti* 1985, vol. 1, 73–102. Tahiti.

Gavenda, R. T. 1992. Hawaiian Quaternary paleoenvironments: A review of geological, pedological, and botanical evidence. *Pacific Science* 46:295–307.

Gill, W. 1856. *Gems from the Coral Islands*. London: Ward.

———. 1885. *Jottings from the Pacific*. New York: American Trust Society.

———. 1894. *From Darkness to Light in Polynesia*. London: Religious Tract Society.

Gillespie, R., and P. Swadling. 1979. Marine shells give reliable radiocarbon ages for middens. *Search* 10:92–3.

Gillieson, D., and M.-J. Mountain. 1983. Environmental history of Nombe Rockshelter, Papua New Guinea Highlands. *Archaeology in Oceania* 18:53–62.

Glacken, C. 1967. *Traces on the Rhodian Shore*. Berkeley: University of California Press.

Gladwin, T. 1970. *East Is a Big Bird: Navigation and Logic on Puluwat Atoll*. Cambridge: Harvard University Press.

Goede, A., and J. Bada. 1985. Electron spin resonance dating of Quaternary bone material from Tasmanian caves: A comparison with ages determined by aspartic acid racemization and C^{14}. *Australian Journal of Earth Sciences* 32:155–62.

Goede, A., and P. Murray. 1977. Pleistocene man in south central Tasmania: Evidence from a cave site in the Florentine Valley. *Mankind* 11:2–10.

———. 1979. Late Pleistocene bone deposits from a cave in the Florentine Valley, Tasmania. *Papers and Proceedings of the Royal Society of Tasmania* 113:39–52.

Goldman, I. 1970. *Ancient Polynesian Society*. Chicago: University of Chicago Press.

Golson, J. 1971. Both sides of the Wallace Line: Australia, New Guinea, and Asian prehistory. *Archaeology and Physical Anthropology in Oceania* 6:124–44.

———. 1977. No room at the top: Agricultural intensification in the New Guinea Highlands. In J. Allen, J. Golson, and R. Jones, eds., *Sunda and Sahul: Prehistoric*

Studies in Southeast Asia, Melanesia and Australia, 601–38. London: Academic Press.

———. 1981. Agricultural technology in New Guinea. In D. Denoon and C. Snowden, eds., *A Time to Plant and a Time to Uproot: A History of Agriculture in Papua New Guinea*, 43–53. Boroko: Institute of Papua New Guinea Studies.

———. 1982. The Ipomoean Revolution revisited: Society and the sweet potato in the upper Wahgi Valley. In A. Strathern, ed., *Inequality in New Guinea Highlands Societies*, 109–36. Cambridge: Cambridge University Press.

———. 1988. The origins and development of New Guinea agriculture. In D. Harris and G. Hillman, eds., *Foraging and Farming: The Evolution of Plant Exploitation*, 678–87. London: Unwin Hyman.

———. 1991a. Introduction: Transitions to agriculture in the Pacific region. *Bulletin of the Indo-Pacific Prehistory Association* 11:48–53.

———. 1991b. The New Guinea Highlands on the eve of agriculture. *Bulletin of the Indo-Pacific Prehistory Association* 11:82–91.

Golson, J., and D. Gardner. 1990. Agriculture and sociopolitical organization in New Guinea Highlands prehistory. *Annual Review of Anthropology* 19:395–417.

Golson, J., and A. Steensberg. 1985. The tools of agricultural intensification in the New Guinea Highlands. In I. Farrington, ed., *Prehistoric Intensive Agriculture in the Tropics, part 1*, BAR International Series 232. Oxford: British Archaeological Reports.

Gorecki, P. 1986. Human occupation and agricultural development in the Papua New Guinea Highlands. *Mountain Research and Development* 6:159–66.

Gorecki, P., M. Mabin, and Campbell. 1991. Archaeology and geomorphology of the Vanimo Coast, Papua New Guinea: Preliminary results. *Archaeology in Oceania* 26:119–22.

Gosden, C., and N. Robertson. 1991. Models for Matenkupkum: Interpreting a late Pleistocene site from southern New Ireland, Papua New Guinea. In J. Allen and C. Gosden, eds., *Report of the Lapita Homeland Project*, 20–45. Occasional Paper 20, Department of Prehistory, Research School of Pacific Studies. Canberra: Australian National University.

Grange, L., and J. Fox. 1953. *Soils of the Lower Cook Group*. DSIR Soil Bureau Bulletin 8. Wellington: Government Printer.

Grant, P. 1985. Major periods of erosion and alluvial sedimentation in New Zealand during the Late Holocene. *Journal of the Royal Society of New Zealand* 15:67–121.

Graves, M. W., and D. J. Addison. 1995. The Polynesian settlement of the Hawaiian archipelago: Integrating models and methods in archaeological interpretation. *World Archaeology* 26:380–99.

Grayson, D. 1984. Explaining Pleistocene extinctions: Thoughts on the structure of a debate. In P. Martin and R. Klein, eds., *Quaternary Extinctions: A Prehistoric Revolution*, 807–23. Tucson: University of Arizona Press.

Green, R. 1974. Excavations of the prehistoric occupation of SU-SA-3. In R. Green and J. Davidson, eds., *Archaeology in Western Samoa*, vol. 2, 108–54. Auckland: Auckland Institute and Museum.

———. 1986. Lapita fishing: The evidence of Site SE-RF-2 from the main Reef islands, Santa Cruz Group, Solomons. In A. Anderson, ed., *Traditional Fishing in the Pacific*, 119–35. Pacific Anthropological Records 37, Department of Anthropology, B. P. Bishop Museum, Honolulu.

———. 1988. Those mysterious mounds are for the birds. *Archaeology in New Zealand* 31:153–58.

Groube, L. 1971. Tonga, Lapita pottery and Polynesian origins. *Journal of the Polynesian Society* 80:278–316.

———. 1986. Waisted axes of Asia, Melanesia and Australia. In G. Ward, ed., *Archaeology at ANZAAS Canberra*, 168–77. Canberra: Canberra Archaeological Society.

———. 1988. The taming of the rain forests: A model for late Pleistocene forest exploitation in New Guinea. In D. Harris and G. Hillman, eds., *Foraging and Farming: The Evolution of Plant Exploitation*, 293–317. London: Unwin Hyman.

Gunn, W. 1911. News from Aneityum. *New Hebrides Magazine* 40 (April):6–8.

Haberle, S., G. Hope, and Y. DeFretes. 1991. Environmental change in the Bliem Valley, montane Irian Jaya, Republic of Indonesia. *Journal of Biogeography* 18:25–40.

Hammatt, H., D. Shideler, R. Chiogioji, and R. Scoville. 1990. Sediment coring in Kawainui Marsh, Kailua, O'ahu, Ko'olaupoko. In Report prepared for M & E Pacific, Inc., Honolulu. Cultural Surveys Hawaii, Kailua.

Handy, E. 1923. *The Native Culture of the Marquesas.* Bernice P. Bishop Museum Bulletin 9, Honolulu.

Hather, J., and P. Kirch. 1991. Prehistoric sweet potato (*Ipomoea batatas*) from Mangaia Island, central Polynesia. *Antiquity* 65:887–93.

Hay, R. 1986. Bird conservation in the Pacific Islands. In *International Council for Bird Preservation Study Report No. 7.* Cambridge: International Council for Bird Preservation.

Henry, T. 1951. *Tahiti aux Temps Anciens.* Paris: Société des Océanistes.

Heyen, G. 1963. Primitive navigation in the Pacific. In J. Golson, ed., *Polynesian Navigation*, vol. 1, 64–80. Polynesian Society Memoir 34. Wellington: Polynesian Society.

Heyerdahl, T. 1961. Surface artifacts. In T. Heyerdahl and E. Ferdon, eds., *Archaeology of Easter Island*, vol. 1, 397–489. Monograph 24, School of American Research, Musuem of New Mexico, Santa Fe.

Hilder, B. 1963. Primitive navigation in the Pacific. In J. Golson, ed., *Polynesian Navigation*, vol. 1, 81–97. Polynesian Society Memoir 34. Wellington: Polynesian Society.

Hoare, M. E., ed. 1982. *The* Resolution *Journal of Johann Reinhold Forster, 1772–1775*, 4 vols., 2d ser., nos. 152–55. London: Hakluyt Society.

Hocking, G., and E. Guiler. 1982. The mammals of the Lower Gordon River Region, South-West Tasmania. *Australian Wildlife Research* 10:1–23.

Holdaway, R. N. 1989. New Zealand's pre-human avifauna and its vulnerability. *New Zealand Journal of Ecology* 12 (suppl.):11–25.

Holyoak, D. 1974. Undescribed land birds from the Cook Islands. *Bulletin of the British Ornithologists' Club* 94:145–50.

———. 1980. *Guide to Cook Island Birds*. Private copyright, n.p.
Holyoak, D., and J. Thibault. 1978. Notes on the biology and systematics of Polynesian swiftlets, *Aerodramus*. *Bulletin of the British Ornithologists' Club* 98:59–65.
———. 1984. Contribution à l'étude des oiseaux de Polynésie orientale. *Mémoires du Museum National d'Histoire Naturelle*, n.s., ser. A, Zoologie 127:1–209.
Hommon, R. J. 1986. Social evolution in ancient Hawaii. In P. V. Kirch, ed., *Island Societies: Archaeological Approaches to Evolution and Transformation*, 55–68. Cambridge: Cambridge University Press.
Hommon, R. J., and W. J. Barrera. 1971. Archaeological survey of *Kahana* Valley, Ko'olauloa District, Island of O'ahu. Departmental Report Series 71-3. Department of Anthropology, B. P. Bishop Museum, Honolulu.
Hommon, R., and R. F. Bevacqua. 1973. Excavations in Kahana Valley, O'ahu. 1972. Department of Anthropology, B. P. Bishop Museum, Honolulu.
Hope, G. S. 1976. The vegetational history of Mt. Wilhelm, Papua New Guinea. *Journal of Ecology* 64:627–64.
———. 1982. Pollen from archaeological sites: A comparison of swamp and open archaeological site pollen spectra at Kosipe Mission, Papua New Guinea. In W. Ambrose and P. Duerden, eds., *Archaeometry: An Australian Perspective*, 211–19. Department of Prehistory, Research School of Pacific Studies. Canberra: Australian National University.
Hope, G., J. Golson, and J. Allen. 1983. Palaeoecology and prehistory in New Guinea. *Journal of Human Evolution* 12:37–60.
Hope, G., and M. Spriggs. 1982. A preliminary pollen sequence from Aneityum Island, Southern Vanuatu. *Indo-Pacific Prehistory Association Bulletin* 3:88–94.
Hope, J., and G. Hope. 1976. Palaeoenvironments for man in New Guinea. In R. Kirk and A. Thorne, eds., *The Origin of the Australians*, 28–54. Human Biology Series 6. Canberra: Australian Institure of Aboriginal Studies.
Hornell, J. 1946. The role of birds in early navigation. *Antiquity* 20:142–49.
Horton, D. 1980. A review of the extinction debate. *Archaeology and Physical Anthropology in Oceania* 15:86–97.
———. 1981. Water and woodland: The peopling of Australia. *Australian Institute of Aboriginal Studies Newsletter* 16:21–27.
———. 1982. The burning question: Aborigines, fire and Australian ecosystems. *Mankind* 13:237–51.
Hotchkiss, S., and J. O. Juvik. 1994. Pollen record from Ka'au Crater, Oahu, Hawaii. *Program and Abstracts of the Thirteenth Biennial Meeting of the American Quaternary Association*, 96. University of Minnesota, Minneapolis, 19–22 June 1994.
Hughes, P. 1985. Prehistoric man-induced soil erosion: Examples from Melanesia. In I. Farrington, ed., *Prehistoric Intensive Agriculture in the Tropics*, part 1, 393–408. BAR International Series 232. Oxford: British Archaeological Reports.
Hughes, P., G. Hope, M. Latham, and M. Brookfield. 1979. Prehistoric man-induced degradation of the Lakeba landscape: Evidence from two inland swamps. In H. Brookfield, ed., *Lakeba: Environmental Change, Population Dynamics, and Resource Use*, 93–110. Paris: UNESCO.

Hughes, P., M. Sullivan, and D. Yok. 1991. Human-induced erosion in a Highlands catchment in Papua New Guinea: The prehistoric and contemporary records. *Zeitschrift für Geomorphologie* 83(suppl.):227–39.

Hunt, T. L., and C. Erkelens. 1993. The To'aga ceramics. In P. V. Kirch and T. L. Hunt, eds., *The To'aga Site: Three Millennia of Polynesian Occupation in the Manu'a Islands, American Samoa*, 123–49. Contributions of the University of California Archaeological Research Facility 51. Berkeley: University of California.

Hunt, T. L., and R. M. Holsen. 1991. An early radiocarbon chronology for the Hawaiian Islands: A preliminary analysis. *Asian Perspectives* 30:147–61.

Hunt, T. L., and P. V. Kirch. 1988. An archaeological survey of the Manu'a Islands, American Samoa. *Journal of the Polynesian Society* 97:153–83.

Intes, A. 1982. La nacre de Polynésie Française: *Pinctada margaritifera* Linne, Mollusca, Bivalvia: Evolution des stockes naturels et de leru exploitation. *Oceanographie: Notes et Documents* 16. Papeete: ORSTOM.

Irwin, G. 1981. How Lapita lost its pots: The question of continuity in the colonization of Polynesia. *Journal of the Polynesian Society* 90:481–94.

———. 1989. Against, across and down the wind: The first exploration of the Pacific islands. *Journal of the Polynesian Society* 98:167–206.

———. 1991. Pleistocene voyaging and the settlement of Greater Australia and its near Oceanic neighbours. In J. Allen and C. Gosden, eds., *Report of the Lapita Homeland Project*, 9–19. Occasional Papers in Prehistory 20, Department of Prehistory, Australian National University, Canberra.

James, H., and S. Olson. 1991. Descriptions of thirty-two species of birds from the Hawaiian Islands: Part 2. *Ornithological Monogographs* 46:1–88.

James, H., T. Stafford, W. D. Steadman, S. Olson, P. Martin, A. Jull, and P. McCoy. 1987. Radiocarbon dates on bones of extinct birds from Hawaii. *Proceedings of the National Academy of Science, USA* 84:2350–2354.

Jarrard, R., and D. Turner. 1979. Comments on "Lithospheric flexure and uplifted atolls" by M. McNutt and H. W. Menard. *Journal of Geophysical Research* 84:5691–5694.

Jennings, J. 1974. The Ferry Berth Site, Mulifanua District, Upolu. In R. C. Green and J. Davidson, eds., *Archaeology in Western Samoa*, vol. 2, 176–78. Bulletin of the Auckland Institute and Museum.

Jennings, J., ed. 1979. *The Prehistory of Polynesia*. Cambridge: Harvard University Press.

Johnson, C. 1987. Macropod studies at Wallaby Creek IV: Home range and movements of the red-necked wallaby. *Australian Wildlife Research* 14:125–32.

Jones, A. T. 1992. Holocene coral reef on Kauai, Hawaii: Evidence for a sea-level highstand in the Central Pacific. In C. H. Fletcher III and J. F. Wehmiller, eds., *Quaternary Coasts of the United States: Marine and Lacustrine Systems*, 267–71. SEPM Special Publication 48. Tulsa: Society for Sedimentary Geology.

Jones, R. 1969. Firestick farming. *Australian Natural History* 16:224–28.

———. 1989. Island occupation. *Nature* 337:605–6.

Jordon, D., and A. Seale. 1906. *The Fishes of Samoa*. Bureau of Fisheries Bulletin 25, 173–455. Washington, D.C.: Government Printing Office.

Kautai, N., et al., eds. 1984. *Atiu: An Island Community.* Suva: University of the South Pacific.

Keegan, W. F., and J. M. Diamond. 1987. Colonization of islands by humans: A biogeographical perspective. *Advances in Archaeological Method and Theory* 10:49–92. New York: Academic Press.

Kelly, M., and J. T. Clark. 1980. *Kawainui Marsh, O'ahu: Historical and Archaeological Studies.* Departmental Report Series 80–83, B. P. Bishop Museum, Honolulu.

Kelly, R. 1988. Etoro suidology: A reassessment of the pig's role in the prehistory and comparative ethnology of New Guinea. In J. Weiner, ed., *Mountain Papuans: Historical and Comparative Perspectives from New Guinea Fringe Highlands Societies,* 116–86. Ann Arbor: University of Michigan Press.

Kiernan, K., R. Jones, and D. Ranson. 1983. New evidence from Fraser Cave for glacial age man in Southwest Tasmania. *Nature* 301:28–32.

Kikuchi, W. 1976. Prehistoric Hawaiian fishponds. *Science* 193:295–99.

Kimber, R. 1983. Black lightning: Aborigines and fire in Central Australia and the Western Desert. *Archaeology in Oceania* 18:38–45.

Kirch, P. 1974. The chronology of early Hawaiian settlement. *Archaeology and Physical Anthropology in Oceania* 9:110–19.

———. 1975a. Cultural adaptation and ecology in Western Polynesia: An ethnoarchaeological study. Ph.D dissertation, Yale University, New Haven. Ann Arbor: University Microfilms.

———. 1975b. Excavations at sites A1–3 and A1–4: Early settlement and ecology in the Halawa Valley. In P. V. Kirch and M. Kelly, eds., *Prehistory and Ecology in a Windward Hawaiian Valley: Halawa Valley, Molokai,* 17–78. Pacific Anthropological Records 24. Department of Anthropology, B. P. Bishop Museum, Honolulu.

———. 1976. Ethnoarchaeological investigations in Futuna and 'Uvea (Western Polynesia). *Journal of the Polynesian Society* 85:27–69.

———. 1977. Valley agricultural systems in prehistoric Hawaii: An archaeological consideration. *Asian Perspectives* 20:246–80.

———. 1981. Lapitoid settlements of Futuna and Alofi, Western Polynesia. *Archaeology in Oceania* 16:127–43.

———. 1982a. Advances in Polynesian prehistory: Three decades in review. *Advances in World Archaeology* 1:51–97.

———. 1982b. Transported landscapes. *Natural History* 91 (December):32–35.

———. 1982c. The impact of the prehistoric Polynesians on the Hawaiian ecosystem. *Pacific Science* 36:1–14.

———. 1983. Man's role in modifying tropical and subtropical Polynesian ecosystems. *Archaeology in Oceania* 18:26–31.

———. 1984. *The Evolution of the Polynesian Chiefdoms.* Cambridge: Cambridge University Press.

———. 1985. *Feathered Gods and Fishhooks: An Introduction to Hawaiian Archaeology and Prehistory.* Honolulu: University of Hawaii Press.

———. 1986. Rethinking East Polynesian prehistory. *Journal of the Polynesian Society* 95:9–40.

———. 1988. *Niuatoputapu: The Prehistory of a Polynesian Chiefdom.* Thomas Burke Memorial Washington State Museum Monograph 5. Seattle: Burke Museum.

———. 1993a. Ofu Island and the To'aga site: Dynamics of the natural and cultural environment. In P. V. Kirch and T. L. Hunt, eds., *The To'aga Site: Three Millennia of Polynesian Occupation in the Manu'a Islands, American Samoa*, 9–22. Contributions of the University of California Archaeological Research Facility 51. Berkeley: University of California.

———. 1993b. The To'aga site: Modelling the morphodynamics of the land-sea interface. In P. V. Kirch and T. L. Hunt, eds., *The To'aga Site: Three Millennia of Polynesian Occupation in the Manu'a Islands, American Samoa*, 31–42. Contributions of the University of California Archaeological Research Facility 51. Berkeley: University of California.

———. 1993c. Radiocarbon chronology of the To'aga site. In P. V. Kirch and T. L. Hunt, eds., *The To'aga Site: Three Millennia of Polynesian Occupation in the Manu'a Islands, American Samoa*, 85–92. Contributions of the University of California Archaeological Research Facility 51. Berkeley: University of California.

———. 1993d. Non-marine molluscs from the To'aga site sediments and their implications for environmental change. In P. V. Kirch and T. L. Hunt, eds., *The To'aga Site: Three Millennia of Polynesian Occupation in the Manu'a Islands, American Samoa*, 115–21. Contributions of the University of California Archaeological Research Facility 51. Berkeley: University of California.

———. 1994. *The Wet and the Dry: Irrigation and Agricultural Intensification in Polynesia.* Chicago: University of Chicago Press.

Kirch, P., C. Christensen, and D. Steadman. n.d. Extinct achatinellid land snails from Easter Island: biogeographic, ecological, and archaeological implications.

Kirch, P., and J. Ellsion. 1994. Palaeoenvironmental evidence for human colonization of remote Oceanic islands. *Antiquity* 68:310–21.

Kirch, P., J. Flenley, and D. Steadman. 1991. A radiocarbon chronology for human-induced environmental change on Mangaia, Southern Cook Islands, Polynesia. *Radiocarbon* 33:217–28.

Kirch, P., J. Flenley, D. Steadman, F. Lamont, and S. Dawson. 1992. Ancient environmental degradation. *National Geographic Research and Exploration* 8:166–79.

Kirch, P., and R. C. Green. 1987. History, phylogeny, and evolution in Polynesia. *Current Anthropology* 28:431–56.

Kirch, P., and T. L. Hunt. 1988. The spatial and temporal boundaries of Lapita. In P. V. Kirch and T. L. Hunt, eds., *Archaeology of the Lapita Cultural Complex: A Critical Review*, 9–32. Thomas Burke Memorial Washington State Museum Research Report 5. Seattle: Burke Museum.

———. 1993a. Excavations at the To'aga site. In P. V. Kirch and T. L. Hunt, eds., *The To'aga Site: Three Millennia of Polynesian Occupation in the Manu'a Islands, American Samoa*, 43–84. Contributions of the University of California Archaeological Research Facility 51. Berkeley: University of California.

———. 1993b. Synthesis and interpretations. In P. V. Kirch and T. L. Hunt, eds., *The To'aga Site: Three Millennia of Polynesian Occupation in the Manu'a Islands,*

American Samoa, 229–48. Contributions of the University of California Archaeological Research Facility 51. Berkeley: University of California.

Kirch, P., and T. L. Hunt, eds. 1988. *Archaeology of the Lapita Cultural Complex: A Critical Review.* Thomas Burke Memorial Washington State Museum Research Report 5. Seattle: Burke Museum.

———. 1993. *The To'aga Site: Three Millennia of Polynesian Occupation in the Manu'a Islands, American Samoa.* Contributions of the University of California Archaeological Research Facility 51. Berkeley: University of California.

Kirch, P., and M. Kelly, eds. 1975. *Prehistory and Ecology in a Windward Hawaiian Valley: Halawa Valley, Molokai.* Pacific Anthropological Records 24. Department of Anthropology, B. P. Bishop Museum, Honolulu.

Kirch, P., and M. Sahlins. 1992. *Anahulu: The Anthropology of History in the Kingdom of Hawaii*, 2 vols. Chicago: University of Chicago Press.

Kirch, P. V., D. W. Steadman, V. L. Butler, J. Hather, and W. I. Weisler. 1995. Prehistory and human ecology in Eastern Polynesia: Excavations at Tangatatau Rockshelter, Mangaia, Cook Islands. *Archaeology in Oceania* 30:47–65.

Kirch, P., and D. Yen. 1982. *Tikopia: Prehistory and Ecology of a Polynesian Outlier.* Bernice P. Bishop Museum Bulletin 238, Honolulu.

Koopman, K. F., and D. W. Steadman. 1995. Extinction and biogeography of bats on 'Eua, Kingdom of Tonga. *American Museum Novitates* 3125. New York.

Kraft, C. 1980a. Letter report to Mr. Susumo Ono regarding corings in Kawai Nui Marsh, 15 July 1980. On file, Hawaii State Historic Preservation Office, Honolulu.

———. 1980b. Letter report to Mr. Ed Marcus regarding corings in Kawai Nui Marsh, 18 December 1980. On file, Hawaii State Historic Preservation Office, Honolulu.

———. 1989. Letter to Ms. Joyce Bath, Hawaii State Historic Preservation Office regarding archaeological finds at Kailua Beach, O'ahu, 18 April 1989. On file, Hawaii State Historic Preservation Office, Honolulu.

Krauss, B. 1993. *Plants in Hawaiian Culture.* Honolulu: University of Hawaii Press.

Kuschel, G. 1987. The subfamily Molytinae (Coleoptera: Curculionidae): General notes and descriptions of new taxa from New Zealand and Chile. *New Zealand Entomologist* 9:11–29.

Lamont, F. 1990. A 6,000 year pollen record from Mangaia, Cook Islands, South Pacific: Evidence for early human impact. B.Sc. dissertation, School of Geography and Earth Resources, Hull University.

Layard, E. 1876. Notes on the birds of the Navigators' and Friendly Islands, with some additions to the ornithology of Fiji. *Proceedings of the Zoological Society of London* 46:490–506.

Leach, B. F., and A. J. Anderson. 1979. Prehistoric exploitation of crayfish in New Zealand. In A. J. Anderson, ed., *Birds of a Feather: Osteological and Archaeological Papers from the South Pacific in Honour of R. J. Scarlett*, vol. 62, 141–64. Oxford: British Archaeological Reports.

Leach, H., and R. C. Green. 1989. New information for the Ferry Berth Site, Mulifauna, Western Samoa. *Journal of the Polynesian Society* 98:319–29.

Ledru, M.-P. 1993. Late Quaternary environmental and climatic changes in central Brazil. *Quaternary Research* 39:90–98.

Lee, G. 1986. The birdman motif of Easter Island. *Journal of New World Archaeology* 7:39–49.

Leidemann, H., and P. L. Cleghorn. 1983. Archaeological monitoring of vegetation clearance on antenna fields at Bellows Air Force Station, O'ahu, Hawaii. Manuscript on file, Hawaii State Historic Preservation Office, Honolulu.

Leopold, A. 1991. *Round River: From the Journals of Aldo Leopold.* Edited by L. B. Leopold. Minocqua: Northword Press.

Lewis, D. 1964. Polynesian navigational methods. *Journal of the Polynesian Society* 73:364–74.

———. 1972. *We, the Navigators.* Honolulu: University of Hawaii Press.

Loeb, E. 1926. *History and Traditions of Niue.* Bernice P. Bishop Museum Bulletin 32, Honolulu.

Loope, L. L., and D. Mueller-Dombois. 1989. Characteristics of invaded islands, with special reference to Hawaii. In J. A. Drake, ed., *Biological Invasions: A Global Perspective,* 257–81. New York: John Wiley.

Lorence, D. H., and K. R. Wood. 1994. *Kanaloa,* a new genus of Fabaceae (Mimosoideae) from Hawaii. *Novon* 4 (2):137–45.

Lowe, R. 1974. Environmental requirements and pollution tolerance of freshwater diatoms. *Environmental Monitoring Series, Program Element No. 1BA027.* Cincinnati: Environmental Protection Agency.

Lyons, C. J. 1903. A history of the Hawaiian Government Survey with notes on land matters in Hawaii. Appendixes 3 and 4 to *Surveyor's Report for 1902.* Honolulu.

Macdonald, G. A., and A. T. Abbott. 1983. *Volcanoes in the Sea: The Geology of Hawaii,* 2d edition. Honolulu: University of Hawaii Press.

Maher, L. J., Jr. 1992. Depth-age conversion of pollen data. INQUA Commission for the Study of the Holocene, Working Group on Data-Handling Methods, *INQUA Newsletter* 7:13–17.

Manabe, S., and D. G. Hahn. 1977. Simulation of the tropical climate of an ice age. *Journal of Geophysical Research* 82:3889–3911.

Marshall, B., and J. Allen. 1991. Excavations in Panakiwuk Cave, New Ireland. In J. Allen and C. Gosden, eds., *Report of the Lapita Homeland Project,* 59–91. Occasional Papers in Prehistory 20, Department of Prehistory, Australian National University, Canberra.

Marshall, P. 1927. *Geology of Mangaia.* Bernice P. Bishop Museum Bulletin 36, Honolulu.

———. 1930. *Geology of Rarotonga and Atiu.* Bernice P. Bishop Museum Bulletin 72, Honolulu.

Martin, L., M. Fournier, P. Mourguiart, A. Sifeddine, B. Turcq, M. L. Absy, and J.-M. Flexor. 1993. Southern Oscillation signal in South American palaeoclimatic data of the last 7000 years. *Quaternary Research* 39:338–46.

Martin, P. S. 1984. Prehistoric overkill: The global model. In P. S. Martin and R. G. Klein, eds., *Quaternary Extinctions,* 354–403. Tucson: University of Arizona Press.

———. 1990. 40,000 years of extinctions of the "planet of doom." *Palaeogeography, Palaeoclimatology, Palaeoecology* 82:187–201.

Martin, P. S., and R. Klein, eds. 1984. *Quaternary Extinctions: A Prehistoric Revolution.* Tucson: University of Arizona Press.

McCormack, G., and J. Kunzle. 1990a. *Kakerori: Rarotonga's Endangered Flycatcher.* Rarotonga: Cook Island Conservation Service.

———. 1990b. *Rarotonga's Cloud Forest.* Rarotonga: Cook Island Conservation Service.

McCoy, P. C. 1976. Intensive survey: Archaeological investigations along upper Kamo'oali'i Stream, Kane'ohe, O'ahu. In P. H. Rosendahl, ed., *Archaeological Investigations in Upland Kane'ohe,* 4-ii to 4–28. Departmental Report Series 76–1, Anthropology Department, B. P. Bishop Museum, Honolulu.

McDougal, I. 1985. Age and evolution of the volcanoes of Tutuila, American Samoa. *Pacific Science* 39:311–20.

McFadgen, B. 1985. Late Holocene stratigraphy of coastal deposits between Auckland and Dunedin, New Zealand. *Journal of the Royal Society of New Zealand* 15:27–65.

McGlone, M. S. 1983. Polynesian deforestation of New Zealand: A preliminary synthesis. *Archaeology in Oceania* 18:11–25.

———. 1989. The Polynesian settlement of New Zealand in relation to environmental and biotic changes. *New Zealand Journal of Ecology* 12 (suppl.):115–30.

McGlone, M. S., A. J. Anderson, and R. Holdaway. 1994. An ecological approach to the Polynesian settlement of New Zealand. In D. Sutton, ed., *The Origins of the First New Zealanders,* 136–63. Auckland: Auckland University Press.

McKern, W. 1929. *Archaeology of Tonga.* Bernice P. Bishop Museum Bulletin 60, Honolulu.

McLean, R. 1980. The land-sea interface of small tropical islands: Morphodynamics and man. In H. Brookfield, ed., *Population-Environment Relations in Tropical Islands: The Case of Eastern Fiji,* 125–30. MAB Technical Notes 13. Paris: UNESCO.

McNiven, I., B. Marshall, J. Allen, N. Stern, and R. Cosgrove. 1993. The southern forests archaeological project. In M. Smith, M. Spriggs, and B. Fankhauser, eds., *Sahul in Review: Pleistocene Archaeology in Australia, New Guinea, and Island Melanesia,* 213–24. Occasional Papers in Prehistory 24, Department of Prehistory, Australian National University, Canberra.

McNutt, M., and H. Menard. 1978. Lithospheric flexure and uplifted atolls. *Journal of Geophysical Research* 83:1206–12.

Meadows, D. H., D. L. Meadows, and J. Randers. 1992. *Beyond the Limits: Global Collapse or a Sustainable Future.* London: Earthscan Publications.

Menzies, J. 1991. *A Handbook of New Guinea Marsupials and Monotremes.* Madang: Kristen Press.

Menzies, J., and J. Pernetta. 1986. A taxonomic revision of cuscuses allied to *Phalanger orientalis* (Marsupalia: Phalangeridae). *Journal of Zoology* (B)1:551–618.

Merlin, M. 1985. Woody vegetation in the upland region of Rarotonga, Cook Islands. *Pacific Science* 39:81–99.

———. 1991. Woody vegetation on the raised coral limestone of Mangaia, southern Cook Islands. *Pacific Science* 45:131–151.

Metraux, A. 1940. *Ethnology of Easter Island.* Bernice P. Bishop Museum Bulletin 160, Honolulu.

Miller, L. 1989. Archaeological monitoring of the Tinker Road Bridge repair project, Bellows Air Force Station, Waimanalo, O'ahu Island, Hawaii. Manuscript on file, Hawaii State Historic Preservation Office, Honolulu.

Milner, G. 1966. *Samoan Dictionary.* London: Oxford University Press.

Moberly, R., Jr., D. C. Cox, T. Chamberlain, F. W. McCoy, Jr., and J. F. Campbell. 1963. Coastal geology of Hawaii. Appendix. In *Hawaii Shoreline,* Hawaii Institute of Geophysics Report 41. Honolulu: University of Hawaii.

Montaggioni, L., and P. Pirazzoli. 1984. The significance of exposed coral conglomerates from French Polynesia (Pacific Ocean) as indicators of recent relative sea-level change. *Coral Reefs* 3:29–42.

Morse, K. 1988. Mandu Mandu Creek Rockshelter: Pleistocene human coastal occupation of North West Cape, Western Australia. *Archaeology in Oceania* 23:82–88.

Mosimann, J., and P. Martin. 1975. Simulating overkill by paleoindians. *American Scientist* 63:304–13.

Mountain, M.-J. 1983. Preliminary report on excavations at Nombe Rockshelter, Simbu Province, Papua New Guinea. *Indo-Pacific Prehistory Association Bulletin* 4:84–89.

———. 1991. Highland New Guinea hunter-gatherers from the Pleistocene: Nombe rockshelter, Simbu. Ph.D. dissertation, Australian National University, Research School of Pacific Studies, Department of Prehistory, Canberra.

Murdock, G. P. 1965. Human influences on the ecosystem of high islands of the tropical Pacific. In F. R. Fosberg, ed., *Man's Place in the Island Ecosystem: A Symposium,* 145–52. Honolulu: Bishop Museum Press.

Murray, P., A. Goede, and J. Bada. 1980. Pleistocene human occupation at Beginners-Luck Cave, Florentine Valley, Tasmania. *Archaeology and Physical Anthropology in Oceania* 15:142–52.

Muse, C., and S. Muse. 1982. *The Birds and Birdlore of Samoa.* Walla Walla: Pioneer Press.

Nagaoka, L. 1993. Faunal assemblages from the To'aga site. In P. V. Kirch and T. L. Hunt, eds., *The To'aga Site: Three Millennia of Polynesian Occupation in the Manu'a Islands, American Samoa,* 189–216. Contributions of the University of California Archaeological Research Facility 51. Berkeley: University of California.

Nagata, K. M. 1985. Early plant introductions in Hawaii. *Hawaiian Journal of History* 19:35–61.

Nakamura, S. 1984. *Soil Survey of American Samoa.* Soil Conservation Service, U.S. Department of Agriculture. Washington, D.C.: Government Printing Office.

Norrish, K., and J. Hutton. 1969. An accurate X-ray spectrographic method for the analysis of a wide range of geological samples. *Geochim, Cosmochim* 33:431–453.

Nunn, P. 1990. Recent environmental changes on Pacific islands. *Geographical Journal* 156:125–40.

———. 1991. *Keimami sa vakila na liga ni Kalou* (Feeling the Hand of God): Human and nonhuman impacts on Pacific island environments. Occasional Papers of the East-West Environment and Policy Institute 13. Honolulu: East-West Center.

———. 1994. *Oceanic Islands*. Oxford: Blackwell.

O'Connor, S. 1989. New radiocarbon dates from Koolan Island, West Kimberley, W.A. *Australian Archaeology* 28:92–104.

Oldfield, F. 1977. Lake sediments and human activities in prehistoric Papua New Guinea. In J. H. Winslow, ed., *The Melanesian Environment*, 57–60. Canberra: Australian National University Press.

Oldfield, F., P. G. Appleby, and J. Thompson. 1980. Paleoecological studies of lakes in the highlands of New Guinea: I. The chronology of sedimentation. *Journal of Ecology* 68:457–77.

Olson, S. L. and H. F. James. 1982a. Fossil birds from the Hawaiian Islands: Evidence for wholesale extinction by man before Western contact. *Science* 217:633–35.

———. 1982b. Prodromus of the fossil avifauna of the Hawaiian Islands. *Smithsonian Contributions to Zoology* 365:1–59.

———. 1984. The role of Polynesians in the extinction of the avifauna of the Hawaiian Islands. In P. S. Martin and R. L. Klein, eds., *Quaternary Extinctions: A Prehistoric Revolution*, 768–80. Tucson: University of Arizona Press.

———. 1991. Descriptions of thirty-two species of birds from the Hawaiian Islands: Part I. *Ornithology Monograph* 45:1–88.

Orbell, M., and G. Moon. 1985. *The Natural World of the Maori*. Auckland: Collins.

Orliac, M. 1981. Etude préliminaire des rats de l'abri-sous-roche de la terre Putoa, vallée de Papeno'o, Tahiti. Laboratoire d'Ethnologie Préhistorique, URA 275 du CNRS, Paris.

———. 1984a. Variation récentes du climat et abandon de sites en Polynésie. Séminaire sur les structures d'habitat. Circulation et échanges. Le déplacement et le séjour. Laboratoire d'Ethnologie Préhistorique, URA 275 du CNRS, Paris.

———. 1984b. Niveaux archéologiques enfouis de la vallée de la Papeno'o. Eléments pour l'étude du peuplement de la vallée et des variations récentes du climat à Tahiti. Laboratoire d'Ethnologie Préhistorique, URA 275 du CNRS, Paris.

———. 1986. Reconstitution du cadastre ancien de la vallée de Papeno'o (inédit). Laboratoire d'Ethnologie Préhistorique, URA 275 du CNRS, Paris.

———. 1990. Tahiti-Papeno'o, Tahinu 1989, rapport préliminaire sur les fouilles de sauvetage de la Tahinu (vallée de Papeno'o, Tahiti). Laboratoire d'Ethnologie Préhistorique, URA 275 du CNRS, Paris.

———. 1991. Etude du fonctionnement de fours expérimentaux à Tahiti, rapport mutigraphié. Laboratoire d'Ethnologie Préhistorique, URA 275 du CNRS, Paris.

Orliac, C., and J. Wattez. 1989. Un four polynésien et son interprétation archéologique, Actes du colloque de Nemours, 1987: Nature et fonctions des foyers préhistoriques. *Mémoires du Musée de Préhistoire d'Ile-de-France* 2:69–75.

Ottino, P. 1985. Un site ancien aux Iles Marquises: L'Abri-sous-rouche d'Anapua, à Ua Pou. *Journal de la Société des Océanistes* 80:33–37.

Parkes, A. 1991. Impact of prehistoric Polynesian colonization: Pollen records for central Polynesia. Paper presented at Seventeenth Pacific Congress, 27 May to 2 June 1991, Honolulu.

Parkes, A., and J. Flenley. 1990. Hull University Mo'orea Expedition, 1985. School of Geography and Earth Resources Miscellaneous Series 37, University of Hull.

Parkes, A., J. Flenley, and M. Johnston. n.d. Environmental change in the Pacific: 1986 Preliminary report. Unpublished manuscript.

Parsonson, G. 1963. The settlement of Oceania: An examination of the accidental voyage theory. In J. Golson, ed., *Polynesian Navigation,* Polynesian Society Memoir 34. Wellington: Polynesian Society.

Paterson, K. 1994. Waka of dreams. *New Spirit* (August):33–35.

Pawley, A., and R. Green. 1973. Dating the dispersal of the Oceanic languages. *Oceanic Linguistics* 12:1–67.

Pearce, R. H., and M. Barbetti. 1981. A 38,000-year-old archaeological site at Upper Swan, Western Australia. *Archaeology in Oceania* 16:173–78.

Pearson, R., P. V. Kirch, and M. Pietrusewsky. 1971. An early prehistoric site at Bellows Beach, Waimanalo, Oahu, Hawaiian Islands. *Archaeology and Physical Anthropology in Oceania* 6:204–34.

Piperno, D. R. 1990. Aboriginal agriculture and land usage in the Amazon Basin, Ecuador. *Journal of Archaeological Science* 17:665–77.

Piperno, D. R., M. B. Bush, and P. A. Colinvaux. 1991. Paleoecological perspectives on human adaptation in central Panama: II. The Holocene. *Geoarchaeology* 6:227–50.

Pirazzoli, P., R. Brousse, G. Delibrias, L. Montaggioni, M. Sachet, B. Salvat, and Y. Sinoto. 1985. Leeward Islands: Maupiti, Tupai, Bora Bora, Huanhine (Society Archipelago). In *French Polynesian Coral Reefs, Reef Knowledge and Field Guides,* 17–72. Fifth International Coral Reef Congress, Tahiti.

Pirazzoli, P., and L. Montaggioni. 1986. Late Holocene sea-level changes in the northwest Tuamotu Islands, French Polynesia. *Quaternary Research* 25:350–68.

Pirazzoli, P., and L. Montaggioni, B. Salvat, and G. Faure. 1988. Late Holocene sea-level indicators from twelve atolls in the central and eastern Tuamotus (Pacific Ocean). *Coral Reefs* 7:57–68.

Powell, J. 1970. The impact of man on the vegetation of the Mt. Hagen Region, New Guinea. Ph.D. dissertation, Australian National University, Canberra.

———. 1980. Studies of New Guinea vegetation history. *International Palynological Conference, Lucknow* (1976–77) 3:11–20.

———. 1982. History of plant use and man's impact on the vegetation. In J. L. Gressitt, ed., *Biogeography and Ecology of New Guinea,* vol. 1, 207–27. W. Junk, Monographiae Biologicae 42. The Hague: Junk.

Pratt, H., P. Bruner, and D. Berrett. 1987. *A Field Guide to the Birds of Hawaii and the Tropical Pacific.* Princeton: Princeton University Press.

Pregill, G., and T. Dye. 1989. Prehistoric extinction of giant iguanas in Tonga. *Copeia* 1989:505–8.

Pukui, M. K., and S. H. Elbert. 1986. *Hawaiian Dictionary.* Honolulu: University of Hawaii Press.

Pukui, M. K., S. H. Elbert, and E. T. Mookini. 1974. *Place Names of Hawaii*. Honolulu: University of Hawaii Press.

Rappaport, R. A. 1965. Aspects of man's influence upon island ecosystems: Alteration and control. In F. R. Fosberg, ed., *Man's Place in the Island Ecosystem: A Symposium*, 155–70. Honolulu: Bishop Musuem Press.

———. 1968. *Pigs for the Ancestors: Ritual and Ecology of a New Guinea People*. New Haven: Yale University Press.

Raynal, M. 1980–1981. "*Koau*," l'oiseau insaisissable des îles Marquises. *Bulletin de la Société d'Etude des Sciences Naturelles de Béziers*, n.s., 8:20–26.

Raynal, M., and M. Dethier. 1990. Lézards géants des Maoris et oiseaux énigmatiques des Marquisiens: La vérité derrière la légende. *Bulletin Mensuel de la Société Linnéenne de Lyon* 59:85–91.

Reinecke, H., and I. Singh. 1980. *Depositional Sedimentary Environments*. Berlin: Springer.

Rinke, D. 1989. The relationship and taxonomy of the Fijian parrot genus *Prosopeia*. *Bulletin of the British Ornithologists' Club* 109:185–95.

Roberts, R., R. Jones, and M. Smith. 1990a. Thermoluminescence dating of a 50,000-year-old-human occupation site in northern Australia. *Nature* 345:153–56.

———. 1990b. Early dates at Malakunanja II: A reply to Bowdler. *Australian Archaeology* 31:94–97.

Ruhe, R. V. 1964. An estimate of paleoclimate in Oahu, Hawaii. *American Journal of Science* 262:1098–15.

St. John, H. 1966. *Monograph of* Cyrtandra *(Gesneriaceae) on Oahu, Hawaiian Islands*. Bernice P. Bishop Museum Bulletin 229, Honolulu.

Sabels, B. 1966. Climatic variation in the tropical Pacific as evidenced by trace element analysis of soils. In D. Blumenstock, ed., *Pleistocene and Post-Pleistocene Variations in the Pacific Area*, 131–51. Honolulu: Bishop Museum Press.

Sahlins, M. 1958. *Social Stratification in Polynesia*. Seattle: American Ethnological Society.

Sanchez, W., and J. Kutzbach. 1974. Climate of the American tropics and subtropics in the 1960s and possible comparisons with climatic variations of the last millennium. *Quaternary Research* 4:128–35.

Sanderson, M., ed. 1993. *Prevailing Trade Winds: Weather and Climate in Hawaii*. Honolulu: University of Hawaii Press.

Savage, S. 1980. *A Dictionary of the Maori Language of Rarotonga*. Suva: University of the South Pacific.

Schilt, A. 1980. Archaeological investigations in specified areas of the Hanalei Wildlife Refuge, Hanalei Valley, Kaua'i. Manuscript on file, Department of Anthropology, Bernice P. Bishop Museum, Honolulu.

———. 1984. *Subsistence and Conflict in Kona, Hawai'i: An Archaeological Study of the Kuakini Highway Realignment Corridor*. Departmental Report Series 84–1, Anthropology Department, B. P. Bishop Museum, Honolulu.

Schofield, J. 1970. Notes on late Quaternary sea levels, Fiji and Rarotonga. *New Zealand Journal of Geology and Geophysics* 13:199–206.

Schubel, S., and D. Steadman. 1989. More bird bones from Polynesian archaeological sites on Henderson Island, Pitcairn Group, South Pacific. *Atoll Research Bulletin* 325:1–18.

Selling, O. 1948. *Studies in Hawaiian Pollen Statistics: Part III. On the Late Quaternary History of the Hawaiian Vegetation.* Bernice P. Bishop Museum Special Publication 39, Honolulu.

Semah, F., H. Ouwen, and M. Charleux. 1978. Fouilles archéologiques sur Raiatea: Vaihi. Report, Centre ORSTOM et Musée de Tahiti et des Iles, Papeete.

Shackleton, N. J. 1987. Oxygen isotopes, ice volume and sea level. *Quaternary Science Reviews* 6:183–90.

Sharp, A. 1964. *Ancient Voyagers in Polynesia.* Berkeley: University of California Press.

Sheppard, F. 1963. Thirty-five thousand years of sea level. In T. Clements, ed., *Essays in Marine Geology in Honor of K. O. Emery,* 1–10. Los Angeles: University of California Press.

Shun, K., P. Price-Beggerly, and J. S. Athens. 1987. Archaeological inventory survey of an inland parcel, Kaneohe-Kailua, O'ahu, Hawaii. Manuscript on file, Hawaii State Historic Preservation Office, Honolulu.

Silver, C. S., with R. S. DeFries. 1990. *One Earth, One Future: Our Changing Global Environment.* Washington, D.C.: National Academy Press.

Sim, R. 1989. Flinders Island prehistoric land use survey. A report to the National Estate Grants Programme on behalf of the Tasmanian Archaeological Society.

Singh, G., and E. Geissler. 1985. Late Cainozoic history of vegetation, fire, lake levels and climate, at Lake George, New South Wales, Australia. *Philosophical Transactions of the Royal Society of London* (B) 311:379–47.

Singh, G., A. Kershaw, and R. Clark. 1981. Quaternary vegetation and fire history in Australia. In A. Gill, A. Groves, and I. Noble, eds., *Fire and Australian Biota,* 23–54. Canberra: Academy of Sciences.

Sinoto, Y. H. 1970. An archaeologically based assessment of the Marquesas Islands as a dispersal center in East Polynesia. In R. C. Green and M. Kelly, eds., *Studies in Oceanic Culture History* vol. 1, 105–32. Pacific Anthropological Records 11, Department of Anthropology, B. P. Bishop Museum, Honolulu.

———. 1979. Excavation on Huahine, French Polynesia. *Pacific Studies* 3:1–40.

———. 1983a. Archaeological excavations of the Vaito'otia and Fa'ahia sites on Huahine Island, French Polynesia. *National Geographic Society Research Reports* 15:583–99.

———. 1983b. The Huahine excavation: Discovery of an ancient Polynesian canoe. *Archaeology* 36:10–15.

———. 1983c. An analysis of Polynesian migrations based on the archaeological assessments. *Journal de la Société des Océanistes* 39:57–67.

Sinoto, Y. H., H. Kurashina, and R. Stevenson. 1988. The discovery of a Melanesian potsherd from the southern Cook Islands. News release, Department of Anthropology, B. P. Bishop Museum, Honolulu.

Sinoto, Y. H., and P. C. McCoy. 1975. Report on the preliminary excavation of an early

habitation site on Huahine, Society Islands. *Journal de la Société des Océanistes* 31:143–86.
Skottsberg, C. 1920. *The Natural History of Juan Fernandez and Easter Island.* Uppsala: Almqvist and Wiksells.
Smith, B. 1985. *European Vision and the South Pacific.* New Haven: Yale University Press.
———. 1992. *Imagining the Pacific: In the Wake of Cook's Voyages.* New Haven: Yale University Press.
Smith, I. W. G. 1988. Maori impact on the marine megafauna: Pre-European distributions of New Zealand sea mammals. In D. Sutton, ed., *Saying So Doesn't Make It So: Papers in Honour of B. Foss Leach,* 76–108. New Zealand Archaeological Association Monograph 17, Auckland.
Solem, A. 1959. Systematics and zoogeography of the land and fresh-water mollusca of the New-Hebrides. *Fieldiana: Zoology* 43. Chicago: Field Museum of Natural History.
———. 1964. New records of New-Caledonian non-marine mollusks and an analysis of the introduced mollusks. *Pacific Science* 18:130–37.
Southwell, C. 1987. Macropod studies at Wallaby Creek II: Density and distribution of macropod species in relation to environmental variables. *Australian Wildlife Research* 14:15–33.
Spores, R. 1969. Settlement, farming technology, and environment in the Nochixtlan Valley. *Science* 166:557–69.
Spencer, J. E., and G. A. Hale. 1961. The origin, nature, and distribution of agricultural terracing. *Pacific Viewpoint* 2:1–40.
Spriggs, M. 1981. Vegetable kingdoms: Taro irrigation and Pacific prehistory. Ph.D. dissertation, Australian National University, Canberra.
———. 1984. Early coconut remains from the South Pacific. *Journal of the Polynesian Society* 93:71–76.
———. 1985. Prehistoric human-induced landscape enhancement in the Pacific: Examples and implications. In I. Farrington, ed., *Prehistoric Intensive Agriculture in the Tropics,* part 1, 409–34. BAR International Series 232. Oxford: British Archaeological Reports.
———. 1986. Landscape, land use and political transformation in southern Melanesia. In P. V. Kirch, ed., *Island Societies: Archaeological Approaches to Evolution and Transformation,* 6–19. Cambridge: Cambridge University Press.
———. 1990. Why irrigation matters in Pacific prehistory. In D. E. Yen and J. Mummery, eds., *Pacific Production Systems,* 174–89. Occasional Papers in Prehistory 18, Department of Prehistory. Australian National University, Canberra.
———. 1991. "Preceded by forest": Changing interpretations of landscape change on Kaho'olawe. *Asian Perspectives* 30:71–116.
———. In press a. Adornment or degradation? Prehistoric human impact on the Hawaiian landscape. In P. Griffin and M. Spriggs, eds., *New Directions in Hawaiian Archaeology: Papers in Honor of Kenneth Emory.* Honolulu: Social Science Research Institute.

———. In press b. The Solomon Islands as bridge and barrier in the settlement of the Pacific. In W. Ayres, ed., *Southeast Asia and Pacific Archaeology.* Pullman: Washington State University Press.

Spriggs, M., and A. Anderson. 1993. Late colonization of East Polynesia. *Antiquity* 67:200–17.

Stargardt, J. 1983. *The Environmental and Economic Archaeology of South Thailand.* Studies in Southeast Asian Archaeology 1. BAR International Series 158. Oxford: British Archaeological Reports.

Steadman, D. W. 1985. Fossil birds from Mangaia, Southern Cook Islands. *Bulletin of the British Ornithologists' Club* 105:58–66.

———. 1986. Holocene vertebrate fossils from Isla Floreana, Galápagos. *Smithsonian Contributions in Zoology* 413:1–103.

———. 1987. Two new species of rails (Aves: Rallidae) from Mangaia, Southern Cook Islands. *Pacific Science* 40:27–43.

———. 1988. A new species of *Porphyrio* (Aves: Rallidae) from archaeological sites in the Marquesas Islands. *Proceedings of the Biological Society of Washington* 101:162–70.

———. 1989a. Extinction of birds in Eastern Polynesia: A review of the record, and comparisons with other Pacific island groups. *Journal of Archaeological Science* 16:177–205.

———. 1989b. New species and records of birds (Aves): Lifuka, Tonga. *Proceedings of the Biological Society of Washington* 102:537–52.

———. 1991a. Extinction of species: Past, present, and future. In R. Wyman, ed., *Global Climate Change and Life on Earth,* 156–69. New York: Routledge, Chapman, and Hall.

———. 1991b. Extinct and extirpated birds from Aitutaki and Atiu, Southern Cook Islands. *Pacific Science* 45:325–47.

———. 1991c. The identity and taxonomic status of *Megapodius stairi* and *M. burnabyi* (Aves: Megapodiiae). *Proceedings of the Biological Society of Washington* 104:870–77.

———. 1992a. Extinct and extirpated birds from Rota, Mariana Islands. *Micronesica* 25:71–84.

———. 1992b. New species of *Gallicolumba* and *Macropygia* (Aves: Columbidae) from archaeological sites in Polynesia. *Los Angeles County Museum of Natural History, Science Series* 36:329–48.

———. 1993a. Biogeography of Tongan birds before and after human impact. *Proceedings of the National Academy of Science* 90 (3):818–22.

———. 1993b. Birds from the To'aga Site, Ofu, American Samoa: Prehistoric loss of seabirds and megapodes. In P. V. Kirch and T. L. Hunt, eds., *The To'aga Site: Three Millennia of Polynesian Occupation in the Manu'a Islands, American Samoa,* 217–28. Contributions of the University of California Archaeological Research Facility 51. Berkeley: University of California.

———. 1995. Prehistoric extinctions of Pacific island birds: Biodiversity meets zooarchaeology. *Science* 267:1123–30.

———. In press. Biogeography and prehistoric exploitation of birds in the Mussau Islands, Papua New Guinea. In P. V. Kirch, ed., *Archaeology of the Lapita Cultural Complex on Mussau, Bismarck Islands, Papua New Guinea*. Berkeley: Archaeological Research Facility, University of California.

Steadman, D. W., E. Greiner, and C. Wood. 1990. Absence of blood parasites in indigenous and introduced birds from the Cook Islands, South Pacific. *Conservation Biology* 4:398–404.

Steadman, D. W., and P. V. Kirch. 1990. Prehistoric extinction of birds on Mangaia, Cook Islands, Polyneisa. *Proceedings of the National Academy of Science, USA* 87:9605–9.

Steadman, D. W., and S. Olson. 1985. Bird remains from an archaeological site on Henderson Island, South Pacific: Man-caused extinctions on an "uninhabited" island. *Proceedings of the National Academy of Science, USA* 82:6191–95.

Steadman, D. W., and D. S. Pahlavan. 1992. Prehistoric exploitation and extinction of birds on Huahine, Society Islands, French Polynesia. *Geoarchaeology* 7:449–83.

Steadman, D. W., D. Pahlavan, and P. V. Kirch. 1990. Extinction, biogeography, and human exploitation of birds on Tikopia and Anuta, Polynesian outliers in the Solomon Islands. *Bishop Museum Occasional Papers* 30:118–53.

Steadman, D. W., S. Schubel, and D. Pahlavan. 1988. A new subspecies and new records of *Papasula abbotti* (Aves: Sulidae) from archaeological sites in the tropical Pacific. *Proceedings of the Biological Society of Washington* 101:487–95.

Steadman, D. W., T. Stafford, D. Donahue, and S. Jull. 1991. Chronology of Holocene vertebrate extinction in the Galápagos Islands. *Quaternary Research* 36:126–33.

Steadman, D. W., C. Vargas, and F. Cristino. 1994. Stratigraphy, chronology, and cultural context of an early faunal assemblage from Easter Island. *Asian Perspectives* 33:79–96.

Steadman, D. W., J. White, and J. Allen. n.d. Extinction, biogeography, and human exploitation of birds from New Ireland, Papua New Guinea. Unpublished manuscript.

Steadman, D. W., and M. Zarriello. 1987. Two new species of parrots (Aves: Psittacidae) from archaeological sites in the Marquesas Islands. *Proceedings of the Biological Society of Washington* 100:518–28.

Stearns, H. T. 1978. *Quaternary Shorelines in the Hawaiian Islands*. Bernice P. Bishop Museum Bulletin 237, Honolulu.

Stearns, H. T., and T. K. Chamberlain. 1967. Deep cores of Oahu, Hawaii, and their bearing on the geologic history of the Central Pacific Basin. *Pacific Science* 21:153–65.

Stern, P. C., O. R. Young, and D. Druckman, eds. 1992. *Global Environmental Change: Understanding the Human Dimensions*. Washington, D.C.: National Academy Press.

Stice, G., and F. McCoy, Jr. 1968. The geology of the Manu'a Islands, Samoa. *Pacific Science* 22:427–57.

Stoddart, D. 1975a. Mainland vegetation of Aitutaki. In D. Stoddart and P. Gibbs, eds., *The Almost-atoll of Aitutaki: Reef Studies in the Cook Islands, South Pacific*, 117–22. Atoll Research Bulletin 190. Washington, D.C.: Smithsonian Institution.

———. 1975b. Reef Islands of Aitutaki. In D. Stoddart and P. Gibbs, eds., *The Almost-atoll of Aitutaki: Reef Studies in the Cook Islands, South Pacific*, 59–72. Atoll Research Bulletin 190. Washington, D.C.: Smithsonian Institution.

———. 1975c. Almost-atoll of Aitutaki: Geomorphology of reefs and islands. In D. Stoddart and P. Gibbs, eds., *The Almost-atoll of Aitutaki: Reef Studies in the Cook Islands, South Pacific*, 31–58. Atoll Research Bulletin 190. Washington, D.C.: Smithsonian Institution.

———. 1975d. Scientific studies in the southern Cook Islands: Background and bibliography. In D. Stoddart and P. Gibbs, eds., *The Almost-atoll of Aitutaki: Reef Studies in the Cook Islands, South Pacific*, 1–30. Atoll Research Bulletin 190. Washington, D.C.: Smithsonian Institution.

Stoddart, D., T. Spencer, and T. Scoffin. 1985. Reef growth and karst erosion on Mangaia, Cook Islands: A reinterpretation. *Zeitschrift für Geomorphologie, n.s.*, 57:121–40.

Stoddart, D., C. Woodroffe, and T. Spencer. 1990. *Mauke, Mitiaro and Atiu: Geomorphology of Makatea Islands in the Southern Cooks*. Atoll Research Bulletin 341. Washington, D.C.: Smithsonian Institution.

Stuiver, M., and G. Pearson. 1986. High-precision calibration of the radiocarbon time scale, A.D. 1950–500 B.C. *Radiocarbon* 28:805–38.

Stuiver, M., and P. Reimer. 1986. A computer program for radiocarbon age calibration. *Radiocarbon* 28 (2B):1022–30.

Summerhayes, C. 1971. Lagoonal sedimentation at Aitutaki and Manuae in the Cook Islands: A reconnaissance survey. *New Zealand Journal of Geology and Geophysics* 14:351–63.

Summerhayes, G., and J. Allen. 1993. The transport of Mopir obsidian to Pleistocene New Ireland. *Archaeology in Oceania* 28:144–52.

Survey Department, Rarotonga. 1983. *Maps of the Cook Islands*. Rarotonga: Government Survey Office.

Sutton, D. G. 1987. A paradigmatic shift in Polynesian prehistory: Implications for New Zealand. *New Zealand Journal of Archaeology* 9:135–55.

Suggs, R. C. 1961. *The Archaeology of Nuku Hiva, Marquesas Islands, French Polynesia*. Anthropological Papers of the American Museum of Natural History 49 (1), New York.

TenBruggencate, J. 1993. Fossil suggests hala preceded Hawaiians. *Honolulu Advertiser*, 27 September 1993, A1–2, Honolulu.

Thom, B., and P. Roy. 1985. Relative sea levels and coastal sedimentation in Southeast Australia in the Holocene. *Journal of Sedimentary Petrology* 55:257–64.

Thomas, N. 1990. *Marquesan Societies: Inequality and Political Transformation in Eastern Polynesia*. Oxford: Clarendon Press.

Thompson, C. 1986. *The Climate and Weather of the Southern Cook Islands*. New Zealand Meterological Service Miscellaneous Publications 188 (2), Wellington.

Thompson, R., and F. Oldfield. 1986. *Environmental Magnetism.* London: Allen and Unwin.

Tomlinson, P. 1984. Ultrasonic filtration as an aid in pollen analysis of archaeological deposits. *Circaea* 2:139–40.

Townsend, C., and A. Wetmore. 1919. Reports on the scientific results of the expedition to the tropical Pacific on the Albatross. 1899–1900. XXI. The birds. *Bulletin of the Museum of Comparative Zoology, Harvard University* 63:151–225.

Trask, H.-K. 1993. *From a Native Daughter: Colonialism and Sovereignty in Hawaii.* Monroe, Maine: Common Courage Press.

Trigger, B. 1989. *A History of Archaeological Thought.* Cambridge: Cambridge University Press.

Trotter, M., and B. McCulloch. 1984. Moas, men, and middens. In P. Martin and R. Klein, eds., *Quaternary Extinctions,* 708–27. Tucson: University of Arizona Press.

Tuggle, H. D., R. H. Cordy, and M. Child. 1978. Volcanic-glass hydration-rind age determinations for Bellows Dune, Hawaii. *New Zealand Archaeological Association Newsletter* 21:58–77.

Turner, B. L., II, W. C. Clark, R. W. Kates, J. F. Richards, J. T. Mathews, and W. B. Meyer, eds. 1990. *The Earth as Transformed by Human Action: Global and Regional Changes in the Biosphere over the Past 300 Years.* Cambridge: Cambridge University Press.

Turner, D., and R. Jarrard. 1982. K-Ar Dating of the Cook-Austral island chain: A test of the hot-spot hypothesis. *Journal of Volcanology and Geothermal Research* 12:187–220.

U.S. Geological Survey. n.d. Aerial photographs, Kane'ohe and Kawai Nui Marsh, O'ahu. Ca. 1926 to 1930. On file, U.S. Geological Survey, Water Resources Division, Honolulu.

Vayda, A. 1961. Expansion and warfare among swidden agriculturalists. *American Anthropologist* 63:346–58.

Vial, L. 1940. Stone axes of Mount Hagen, New Guinea. *Oceania* 11:158–63.

Visher, S. 1925. *Tropical Cyclones of the Pacific.* Bernice P. Bishop Museum Bulletin 20, Honolulu.

von Haast, H. 1948. *The Life and Times of Sir Julius von Haast, Explorer, Geologist, Museum Builder.*

Wagner, W. C., D. R. Herbst, and S. H. Sohmer. 1990. *Manual of the Fowering Plants of Hawaii,* 2 vols. Bishop Museum Special Publication 83. Honolulu: University of Hawaii Press and Bishop Museum Press.

Walker, D. 1964. A modified Vallentyne mud sampler. *Ecology* 45:642–44.

Walker, D., and G. Hope. 1982. Late Quaternary vegetation history. In J. L. Gressitt, ed., *Biogeography and Ecology of New Guinea,* vol. 1, 263–85. Monographiae Biologicae 42. The Hague: W. Junk.

Walter, R. 1990. The southern Cook Islands in eastern Polynesian prehistory. Ph.D. dissertation, University of Auckland,

Ward, J. V. 1981. Palynological analysis of samples from trenches B and C, Kawainui Marsh, O'ahu. Appendix A. In J. Allen-Wheeler, Archaeological excavations in

Kawainui Marsh, Island of O'ahu. Manuscript on file, Hawaii State Historic Preservation Office, Honolulu.

———. 1990. Pollen analysis of Core 6, Kawainui Marsh, O'ahu. Appendix I. In H. H. Hammatt, D. W. Shideler, R. Chiogioji, and R. Scoville, Sediment coring in Kawainui Marsh, Kailua, Ko'olaupoko, O'ahu. Manuscript on file, Hawaii State Historic Preservation Office, Honolulu.

Watling, D. 1982. *Birds of Fiji, Tonga, and Samoa*. Wellington: Millwood Press.

Weisler, M. 1993. Long-distance interaction in prehistoric Polynesia: Three case studies. Ph.D. dissertation, University of California, Berkeley. Ann Arbor: University Microfilms.

Wentworth, C. K. 1943. Soil avalanches on Oahu, Hawaii. *Bulletin of the Geological Society of America* 54:53–64.

Whistler, W. 1980. *Coastal Flowers of the Tropical Pacific: A Guide to Widespread Seashore Plants of the Pacific Islands*. Honolulu: University of Hawaii Press.

———. 1988. The unique flowers of Polynesia: The Cook Islands Bulletin. *Pacific Tropical Garden* 18:89–98.

———. 1990. The ethnobotany of the Cook Islands: The plants, their Maori names, and their uses. *Allertonia* 5:347–419.

White, J. P. 1972. *Ol Tumbuna: Archaeological Excavations in the eastern Central Highlands, Papua New Guinea*. Department of Prehistory, Terra Australis 2. Canberra: Australian National University.

White, J. P., and T. Flannery 1991. The impact of people on the Pacific world. In J. Dodson, ed., *The Naive Lands*, 1–8. Melbourne: Longman Cheshire.

White, J. P., T. Flannery, R. O'Brien, R. Hancock, and L. Pavlish. 1991. The Balof shelters, New Ireland. In J. Allen and C. Gosden, eds., *Report of the Lapita Homeland Project*, 46–58. Department of Prehistory, Occasional Papers 20. Canberra: Australian National University.

White, J. P., and J. O'Connell. 1982. *A Prehistory of Australia, New Guinea, and Sahul*. Sydney: Academic Press.

Wickler, S., J. S. Athens, and J. V. Ward. 1991. Vegetation and landscape change in a leeward coastal environment: Paleoenvironmental and archaeological investigations, Fort Shafter flats sewerline project, Honolulu, Hawaii. Report prepared for U.S. Army Engineer District, Pacific Ocean Division, Ft. Shafter, Hawaii. International Archaeological Research Institute, Honolulu.

Wickler, S., and M. Spriggs. 1988. Pleistocene human occupation of the Solomon Islands, Melanesia. *Antiquity* 62:703–6.

Wiens, H. 1962. *Atoll Environment and Ecology*. New Haven: Yale University Press.

Wild, S. 1985. Voyaging to Australia: 30,000 years ago. Unpublished paper delivered at Ausgraph 85, Third Australasian Conference on Computer Graphics, Brisbane.

Wilder, G. 1931. *Flora of Rarotonga*. Bernice P. Bishop Museum Bulletin 86, Honolulu.

Wiles, G., T. Lemke, and N. Payne. 1989. Population estimates of fruit bats (*Pteropus mariannus*) in the Mariana Islands. *Conservation Biology* 3:66–76.

Wiles, G., and N. Payne. 1986. The trade in fruit bats *Pteropus* spp. on Guam and other Pacific islands. *Biological Conservation* 38:143–61.

Williams, J. 1838. *A Narrative of Missionary Enterprises in the South Sea Islands.* London: John Snow.

Williams, S. S. 1989. A preliminary report of test excavations on sites 50-Oa-G5-106 and G5-110, Luluku, Kane'ohe, Ko'olau Poko, O'ahu. Manuscript on file, Anthropology Department, Bishop Museum, Honolulu.

———. 1992. Early inland settlement expansion and the effect of geomorphological change on the archaeological record in Kane'ohe, O'ahu. *New Zealand Journal of Archaeology* 14:67–78.

Winterhalder, B. P. 1994. Concepts in historical ecology: The view from evolutionary theory. In C. L. Crumley, ed., *Historical Ecology: Cultural Knowledge and Changing Landscapes,* 17–42. Santa Fe: School of American Research Press.

Witter, D. 1978. Late Pleistocene extinctions: A global perspective. *The Artefact* 3:51–65.

Wolman, W. G. 1993. Population, land use, and environment: A long history. In C. L. Jolly and B. B. Torrey, eds., *Population and Land Use in Developing Countries,* 15–29. Washington, D. C.: National Academy Press.

Wood, A. 1985. *The Stability and Permanence of Huli Agriculture.* Occasional Paper 5. University of Papua New Guinea and Southern Highlands Rural Development Project, Department of Geography, Port Moresby and Mendi.

Wood, B., and R. Hay. 1970. *Geology of the Cook Islands.* New Zealand Geological Survey Bulletin 82. Wellington: New Zealand Department of Scientific and Industrial Research.

Worthy, T. H. 1983. Subfossil insects: A key to past diversity. *Weta* 6:42–43.

Wright, R. 1986a. New Light on the extinction of the megafauna. *Proceedings of the Linnean Society of New South Wales* 109:1–9.

———. 1986b. How old is Zone F at Lake George? *Archaeology in Oceania* 21:138–39.

Wurm, S. 1975. Eastern Central Trans-New Guinea Phylum languages. In S. Wurm, ed., *New Guinea Area Languages and Language Study,* 461–526. Pacific Linguistics Series C, no. 38, vol. 1: *Papuan Languages and the New Guinea Linguistic Scene.* Canberra: Australian National University.

Yen, D. E. 1974. *The Sweet Potato and Oceania: An Essay in Ethnobotany.* Bernice P. Bishop Museum Bulletin 236, Honolulu.

Yen, D. E., P. V. Kirch, P. H. Rosendahl, and T. Riley. 1972. Prehistoric agriculture in the upper valley of Makaha, O'ahu. In E. Ladd and D. E. Yen, eds., *Makaha Valley Historical Project: Interim Report* 3, 59–94. Department of Anthropology, B. P. Bishop Museum, Honolulu.

Yonekura, N., T. Ishii, Y. Saito, Y. Maeda, Y. Matsushima, E. Matsumoto, and H. Kayanne. 1988. Holocene fringing reefs and sea-level change in Mangaia Island, southern Cook Islands. *Palaeogeography, Palaeoclimatology, Palaeoecology* 68:177–88.

Yonekura, N., T. Ishii, Y. Saito, E. Matsumoto, and H. Kayanne. 1986. Geologic and geomorphic development of Holocene fringing reef of Mangaia Island, the South Cooks. In *HIPAC Team, Sea-Level Changes Tectonics in the Middle Pacific, Report HIPAC Project, 1984, 1985,* 44–67. Kobe: Kobe University.

Young, A. 1976. *Tropical Soils and Soil Survey.* Cambridge: Cambridge University Press.

Yunker, T. 1945. *Plants of the Manua Islands.* Bernice P. Bishop Museum Bulletin 184, Honolulu.

Ziegler, A. 1977. Evolution of New Guinea's marsupial fauna in response to a forested environment. In B. Stonehouse and D. Gilmore, eds., *The Biology of Marsupials,* 117–38. London: Macmillan.

Zimmerman, E. C. 1948. *Insects of Hawaii,* vol. 1: *Introduction.* Honolulu: University of Hawaii Press.

Index

Aborigines, Australian, 22, 31, 102
Acacia koa, 239, 259
Acalypha sp., 180
Achatinellidae, 211
Achyranthes aspera, 195
Acridotheres tristis, 60
Acrocephalus kerearako, 60, 76
Acrostichum aureum, 176, 178, 194, 195, 196, 197
Adams, Robert McC., 284
Adamson, A. M., 7
Ageratum conyzoides, 180, 195
Agriculture: charcoal and, 267; effect on forests, 81; intensification of, 49, 150, 163, 195; New Guinea Highlands, 44–47; Polynesian, 7, 234
Ahu Naunau site, 72
Aibura site, 43

Aitutaki Is., 13, 14, 60, 71, 124–46, 149. *See also* Cook Is.
Aleurites moluccana, 65, 159, 234, 239, 268
Allen, Jim, 15
Allen, M. S., 13, 14
Allophylus vitiensis, 176
Alluvial sequences, 86–98. *See also* Sediments
Alocasia, 209
Alphitonia zizyphoides, 210, 214
Alstonia costata, 193, 194, 210
Amomun cevuga, 203
Anakena site, 72
Anas superciliosa, 60, 75
Anderson, Atholl, 18, 20, 265–66
Anderson, Edgar, 271
Aneityum Is., 12, 83, 84, 85, 91–98, 181

Angiopteris evecta, 195
Anous stolidus, 69
Anthropogenic impacts, on islands, 15. See also Ecosystems
Anthropology, 1, 19, 22, 286
Antidesma, 259
Aplonis tabuensis, 108
Araucaria, 32
Archaeology, 8, 9, 10, 18, 40, 48, 52, 65, 74, 77, 141–43, 185, 188, 204, 230, 235, 246, 265, 274, 283, 286
Artocarpus altilis, 5, 65, 128, 210, 214
Ascarina polystachya, 192, 194
Asipani site, 87–88
Asplenium nidus, 75
Assiminea nitida, 121
Astronia fraterna, 193, 194
Athens, J. Stephen, 17, 91, 134, 235, 236
Atiu Is., 16, 71, 76, 126, 134, 149, 168–81, 197, 261
Aunu'u Is., 192
Austral Is., 71, 145

Bahn, Paul, 285
Baliem Valley, 41, 42, 50
Bamboo, 160
Banks, Sir Joseph, 4
Barringtonia sp., 209
Barringtonia asiatica, 181, 193, 195
Bats, 279
Bellows Air Force Base, 232–34
Bellwood, Peter, 127, 136, 162
Bennett's Wallaby, 26–27
Bidens paniculata, 195
Bidens pilosa, 180, 195
Biodiversity crisis, 285
Biogeography, 3, 13, 286
Biota: changes in, 117–21; vulnerability of, 149
Birds: on Aitutaki, 129; association with deities, 76; bones of, 72–73, 120, 158–59, 219–21; coastal, 277; conservation of, 78; decline in, 158; drepaniid, 4, 16; eggs, 64; ethnographic information on, 51, 70; extinction of, 4, 11, 16, 82, 158–59, 280–83; fauna at Balof Cave, 36; feathers, 71–72; fishing and, 69–70; flightlessness in, 60, 74, 279, 280; human impact on, 53–67, 158–59, 164; hunting of, 60–66; imagery of, 73–77; impact on humans, 67–77; legends of, 73–77; names of, 73–77; navigation and, 69–70; parasites, 67; pets, 77; predation on, 51, 159; survival of, 66. See also Seabirds
Bishop Museum, Bernice Pauahi, 7, 8
Bismarck Archipelago, 15, 20, 23, 33, 34–37
Boobies, 69. See also *Sula*
Bradybaena similaris, 211
Breadfruit. See *Artocarpus altilis*
Brookfield, Harold C., 50, 99–100
Brown, Forest, 7
Brushtail Possum, 30
Bryan, W. A., 6
Buang Merabak site, 35
Bush Wallaby, 43

Caesalpinia major, 160
Calophyllum sp., 176
Calophyllum inophyllum, 128, 175, 194, 197
Canarium, 37
Candlenut. See *Aleurites moluccana*
Cannibalism, 159, 165
Capra hircus, 66
Caroline Is., 69
Casuarina, 31, 32, 42, 45, 47, 155
Casuarina equisetifolia, 178, 181, 194, 195
Celtis sp., 176
Cerbera manghas, 210
Charcoal: analyses, 167, 173, 209–10; assemblages, 128; flecking in soils,

87, 93, 96, 116, 239; fragments in cores, 236; influxes, 18, 31; particles, 129, 153, 266, 267, 270; in Putoa rockshelter, 206, 209–10; radiocarbon dating and, 85–86
Charmosyna amabilis, 71–72
Chazine, J.-M., 201
Cheirodendron, 260
Chenopodium oahuense, 259
Chickens, 58, 64, 72, 82, 117, 157, 219
Chiefdoms, 147, 165
Christensen, Carl C., 213
Cibotium, 236, 239
Climate, change in, 14, 26, 31, 40, 80, 124, 196, 242–44, 260–61, 269, 285; current conditions, 130; Holocene trends in, 130–33; seasonality, 130; variability, 39. *See also* Little Climatic Optimum; Little Ice Age
Club of Rome, 285
Coconut. See *Cocos nucifera*
Cocos nucifera, 159, 168, 170–71, 175, 178, 180–81, 185, 192, 194, 195, 197, 198, 268
Collocalia sawtelli, 75
Collocalia spodiopygia, 75
Colocasia esculenta, 6, 40, 84, 126, 149, 160, 194, 195, 234, 239
Colonization: birds and, 68–69; of Greater Australia, 23–24; of islands, 17–18; of Oceania, 51; of Polynesia, 68–69, 197; of Remote Oceania, 35
Colubrina, 259
Columbidae, 56
Cook, Captain James, 4, 250, 261
Cook Is., 53, 58, 71, 72, 75, 114, 124–46, 166, 168–81, 261. *See also* Aitutaki Is.; Atiu Is.; Mangaia Is.; Rarotonga Is.
Cooke, C. M., 8
Corals, 116, 119, 134
Cordyline australis, 278
Cordyline fruticosa, 269

Cordyline terminalis, 6, 160, 194, 202, 239, 240
Coring, 252–54
Cosgrove, R., 27, 30
Crickets, 279
Crosby, A., 271
Cumberland, K., 105, 272–73
Curculionidae, 279
Cyanoramphus ulietanus, 71
Cyanoramphus zealandicus, 71
Cyathea affinis, 193, 194
Cyathea horrida, 193, 194
Cyclones, 116, 185, 224, 225, 227, 228. *See also* Hurricanes
Cyclosorus, 193
Cyperus pennatus, 178
Cyrtosperma chamissonis, 160

Dacrydium, 32
Darwin, Charles, 1, 6, 78, 249
Deforestation, 46, 236, 274. *See also* Forests
Degradation, of land, 49, 50
Diamond, Jared, 78
Diatoms, 173, 180, 197
Dicranopteris linearis, 7, 152, 155, 163, 168, 176, 178, 181, 193, 194, 195, 196, 197
Dingo, 29
Dinornithiformes, 280. *See also* Moa
Dioscorea spp., 6, 40
Diprotodon, 30
Disease, effect of, 151, 199
Dodonaea viscosa, 259
Dogs, 51, 65, 82, 207, 222–23, 246, 271
Domesticated animals, 58
Dorcopsulus vanheurni, 43
Doves, 54, 60, 65, 159
Drepaniidae, 16
Ducks, 75. See also *Anas superciliosa*
Ducula spp., 72
Ducula aurorae, 54, 159

Ducula galeata, 159
Ducula pacifica, 65, 108

Earthquakes, 80
Easter Is., 12, 20, 53, 72, 75, 76, 100, 148, 228, 285
Eastern Polynesia, 125, 129, 133, 134, 145, 160
Eclectus roratus, 72
Ecosystems, island, 2; concept of, 9; fragility of, 19–20, 106; impact on, 3, 13, 14; isolation of, 9; resiliancy of, 19–20; transformation of, 286; vulnerability of, 20, 249
Eels, 218
Egretta sacra, 73
Elaeocarpaceae, 195
Elaeocarpus tonganus, 176, 197
Elephantopus mollis, 180, 195
Ellison, Joanna, 18, 129, 153, 261, 265
El Niño–Southern Oscillation, 261
Emory, Kenneth P., 8, 201
Endemism, 249
Endodontidae, 212, 214
Environmental change: and cultural change, 274–79; and society, 18–19
Erosion, 11, 12, 50, 51, 81, 82, 83, 99, 130, 136, 163, 166, 173, 180, 181, 197, 198, 228, 240, 243, 245, 246
Erythrina sandwicensis, 259
'Eua Is., 74, 77
Eucalyptus, 31, 32
Eudynamis taitensis, 76
Eugenia sp., 193
Eugenia rariflora, 192, 195
Exchange networks, 19, 44, 50, 148
Extinction: background levels of, 53; of birds, 4, 11, 16, 68, 159; blitzkrieg model of, 29, 36, 69; chronologies for, 29–30; faunal, 279–83; kill sites and, 29; megafaunal, 30, 274; Pleistocene faunal, 25, 28–30

Extirpation, of birds, 53
Fa'ahia site, 56
Faunal remains, 214–23. *See also* Birds; Fish; Mollusks; Rats
Feathers, 71–72
Felis catus, 66
Fernlands, 152, 163, 171. See also *Dicranopteris linearis*
Ferns, 180, 277
Ficus sp., 176
Ficus tinctoria, 178, 192
Fiji Is., 12, 71, 74, 113, 243
Fimbristylis cymosa, 193, 194
Fire, 30–32, 81, 105, 166, 181, 236, 243, 266, 277, 280
Firestick farming, 31
Firth, Raymond, 78
Fish, 109, 117, 119, 143, 217–19, 276, 277
Fishing, 69–70, 99, 132, 143, 155, 271
Fishponds, 99
Fitchia, 193
Flannery, Tim, 28
Flenley, John R., 12, 153, 285
Flooding, 82, 89, 96, 225, 228, 240
Flora. *See* Forests; Vegetation
Flying foxes, 67. *See also* Fruit bats
Forests: birds and, 67; clearance of, 33, 41–42, 82, 90, 117, 124, 163, 178, 198, 285; cloud, 193; decline of, 161, 264; destruction of, 11; disturbance of, 193; firing of, 277; leeward, 277–78; lowlands, 261, 264; montane, 193;
Forster, Georg, 5
Forster, Johann Reinhold, 5
Fosberg, Raymond, 2, 9, 20, 105, 230, 249
Foulehaio carunculata, 108
Frecinetia impavida, 193
Fregata spp., 69
Fregata ariel, 73
Fregata minor, 70

Frigatebirds, 69, 70, 73, 75, 77. See also *Fregata*
Frimigacci, Daniel, 87–88
Frogs, 279
Fruit bats, 67, 81, 158, 164
Fruit doves. *See* Doves
Fulica americana, 73
Futuna Is., 12, 84, 86–89, 269

Galápagos Is., 53, 132
Gallicolumba erythroptera, 159
Gallicolumba stairii, 120
Gallinula chloropus, 73
Gallirallus philippensis, 64, 108
Gallirallus ripleyi, 159
Gallus gallus. *See* Chickens
Gambier Is., 145. *See also* Mangareva
Gardenia taitensis, 185, 195
Gastrocopta pediculus, 121, 128
Gastropods, terrestrial. *See* Land snails
Geckonidae, 108
Geochemistry, of lake sediments, 170–73
Geomorphology, 12, 110, 122, 125, 149
Gleichenia. See *Dicranopteris linearis*
Global change, 284–86
Goldman, Irving, 147–48, 165
Golson, Jack, 12, 15, 19, 34
Gouania, 259
Grasslands, 42, 46, 49, 82, 171, 276
Gray Kangaroo, 27, 30
Greater Australia, 23–24
Green, Roger C., 11
Gregory, Herbert E., 7
Groube, Les, 34
Guava. See *Psidium guajava*
Guetarda speciosa, 210
Gygis candida, 69, 75
Gygis microrhyncha, 159

Habitat alteration, 68, 120
Halcyon chloris manuae, 108
Halcyon tuta, 73
Handy, Edward S. C., 7

Hane site, 54, 56, 185
Hawaii Is., 82
Hawaiian Is., 6, 10, 16, 20, 52, 82, 84, 89–91, 99, 101, 134, 147, 230–46, 248–69, 278, 286
Helicarionidae, 211
Henderson Is., 3, 53
Hernandia sp., 160
Herons, 73
Herpestes sp., 66
Hibiscus, 209, 210
Hibiscus tiliaceus, 170, 178, 181, 195, 268
Highlands, New Guinea, 12, 15, 19, 33, 34, 37, 39–50, 100, 130. *See also* New Guinea
Historical ecology, 1–2, 285–86
Huahine Is., 56
Hunters-and-gatherers, 22–38, 81
Hunting, 60–66, 271
Hurricanes, 80, 107, 169–70

Ilex, 236
Industrial Revolution, 284
Inocarpus edulis, 202
Inocarpus fagiferus, 5, 181
Insect remains, 279
Introduction: of land snails, 211, 214; of plants, 178, 267–69
Ipomoea batatas, 15, 39, 47–50, 160, 178
Irrigation, 5, 19, 33, 84, 89–91, 95, 97, 98, 150–51, 161–62, 234
Irwin, Geoff, 23–25
Islands: colonization of, 17–18; human impact on, 3, 15, 18; isolation of, 2; size of, 20
Isolation, 1, 9, 249

James, Helen, 11, 12
Jones, R., 31

Ka'au Crater, 242, 256, 258, 260
Kahana Valley, 238–41

Kaho'olawe Is., 104, 256
Kalaeloa (Barbers Point), 252, 254
Kamapuk site, 43
Kamo'oali'i, 236–38
Kanaloa kahoolawensis, 17, 256–58, 261, 265, 269
Kane'ohe, 236–38
Kapunahala Marsh, 252, 254, 258, 264
Kaua'i Is., 91, 254
Kaugel Valley, 47
Kawai Nui Marsh, 234–36, 250, 252, 258, 262, 264
Kilu site, 37
Kingfishers, 60, 73
Kirch, Patrick V., 11, 12, 13, 16, 18, 68, 73, 75, 84, 85, 86–87, 89, 99, 100, 109, 112, 116, 119, 121, 129, 132, 133, 142, 145, 178, 213, 234, 245, 249, 254, 261, 264, 265, 269, 271
Kosipe site, 34
Kuk site, 44–46

Lack, David, 78
Lake George, 31–32
Lake Temae, 167, 181–96
Lake Te Roto, 167, 168–81, 197
Lake Tiriara, 152–53
Lamellaxis gracilis, 121, 128, 211, 213, 214
Lamellidea oblonga, 128, 211, 212
Lamellidea pusilla, 121, 211, 212
Landscapes: change in, 1, 112–17, 232–41, 241–46; degradation of, 19, 49, 98, 99–101; "enhancement" of, 19, 80–102; evolution of, 114; Hawaiian, 232–41; interaction and feedback, 244–46; on Mangaia, 152–55; morphodynamic model of, 113; New Guinea Highland, 47; of To'aga site, 112–17; transformation of, 165, 214; "transported," 108, 271
Landslides, 243

Land snails, 8, 11, 82, 108, 109, 120, 121, 122, 128, 206, 210–14, 279
Lapita Culture, 15, 37, 68, 87, 125, 127, 129, 132
Leopold, Aldo, 78
Liardetia samoensis, 121
Libera bursatella, 211
Libera dubiosa, 211
Libera incognata, 211
Libera micrasoma, 211
Little Climatic Optimum, 14, 130, 145, 242–43
Little Ice Age, 14, 130, 132, 144, 145, 242, 243–44
Lizards, 108, 279
Lorikeets, 54, 71, 72, 77
Ludwigia octivalis, 239, 240, 269
Luluku site, 89–91, 236–38
Lynch's Crater, 32

Macaranga tahitensis, 192, 194
MacArthur, Robert, 1, 3, 78
Macropiper latifolium, 193
Macropus giganteus, 27
Macropus rufogriseus, 26–27
Macropus titan, 27
Makatea: islands, 126; on Mangaia, 149; resources of, 169; vegetation on, 166
Malaria, avian, 52
Malinowski, Bronislaw, 78
MAN-44, site. *See* Tangatatau rockshelter
Mangaia Is., 13, 16, 18, 19, 20, 58, 60, 61, 69, 71, 72, 75, 76, 77, 109, 126, 128, 129, 133, 141, 142, 143, 145, 147–65, 166, 178, 261
Mangareva, 148
Mangroves, 181
Manu'a Is., 106. *See also* Ofu Is.
Manus Is., 36
Maori, 272–73. *See also* New Zealand
Mariana Is., 52
Marine resources, 99

Marquesas Is., 8, 53, 71, 129, 130, 148, 185
Matenbek site, 35, 36
Matenkupkum site, 35, 36
Maui Is., 260, 270
Ma'uke Is., 60, 126, 132, 134, 143, 149
Maunawili Valley, 262, 264
Mauthodonta parvidens, 211
Mayr, Ernst, 78
McGlone, Matt, 274
Mead, Margaret, 78
Megafauna: in Greater Australia, 28; in New Guinea, 12; in New Zealand, 274; Pleistocene, 38
Megapodes, 17, 52, 74, 120
Megapodius spp., 74, 120
Megapodius alimentum, 74
Megapodius molistructor, 74
Megapodius pritchardii, 74
Melanesia, 32–34, 37, 52
Melania incisa, 192
Merrill, E. D., 8
Messerschmidia, 193
Metrosideros, 181, 236
Metrosideros collina, 192–93, 194
Miconia calvescens, 181, 195, 196
Micronesia, 52, 265
Mikania micrantha, 180
Mimosa pudica, 195
Miscanthus, 49
Mitiaro Is., 126, 134, 176, 192
Moa, 15, 272, 277, 279–83
Moa-hunters, 272
Mollusks, 109, 188, 192, 206, 210–14, 216–17, 223. *See also* Land snails
Mo'orea Is., 16, 166, 167, 181–96, 198
Moturakau Is., 127, 128, 134. *See also* Aitutaki Is.
Mulifanua site, 114, 125
Murdock, George P., 10, 15
Musa spp., 6, 40, 160
Mussau Is., 52

Myrsine affinis, 193
Myrtaceae, 259, 260

Navigation, 69–70
Near Oceania, 17, 20
Neonauclea, 209, 210
Neonauclea forsteri, 192, 194
Nephrolepis, 239
Nesodiscus cretaceus, 211
Nesofregata fuliginosa, 159
New Britain, 23, 35
New Caledonia, 17, 52
New Guinea, 9, 24, 26, 35, 100, 130. *See also* Highlands, New Guinea
New Ireland, 15, 23, 24, 35, 52
New Zealand, 11, 15, 20, 52, 101, 105, 130, 264, 271–83
Nihoa Is., 254
"Noble Savage," 5, 6, 22, 101
Noddies, 69, 70. See also *Anous stolidus; Procelsterna cerulea*
Nombe site, 34, 43
Norikori swamp, 43, 46
Nothofagus spp., 276
Nukuhiva Is., 8
Nunn, Patrick D., 13, 81, 97, 106, 113, 130, 134, 135, 138, 144, 242, 243, 244

O18 site, 232–34
O'ahu Is., 11, 14, 89–91, 230–46, 248–69
Ofu Is., 105–22
Olosega Is., 106
Olson, Storrs, 11, 12
Opeas pumilum, 211
Oral traditions, 151, 160, 227
Orliac, Michel, 17
Outliers, Polynesian, 53. *See also* Tikopia Is.

Paleoecology, 8, 19, 286
Palynology. *See* Pollen analysis
Pamwak site, 36
Pandanus, 155, 159, 185, 193, 195, 198

Pandanus tectorius, 170, 175, 178, 192, 197, 268
Papeno'o Valley, 200–28
Parakeets, 71. See also *Cyanoramphus*
Parkes, Annette, 16
Parrots, 54, 60, 71. See also *Vini*
Partula, 213
Partula hyalina, 211
Partula otaheitiana, 211
Partulidae, 211
Passiflora quadrangularis, 180
Pearl shell, 126, 143–44, 145, 155
Pelea, 236
Pelicans, 280
Petrels, 54, 58, 61, 72, 120, 135, 219, 279, 280. See also Seabirds
Phaethon spp., 69
Phaethon lepturus, 77, 108
Phaethon rubricauda, 71
Phalanger spp., 15, 43
Phalanger orientalis, 36
Pigeons, 54, 65, 72, 75, 76, 108
Pigs: in Cook Is., 58, 157; elimination of, 157; exchange of, 158; feral, 243; in Hawaii, 246; husbandry of, 19; in New Guinea, 37, 39; in New Guinea Highlands, 42, 43, 48, 50; predation on birds, 51; in Samoa, 108, 117; in Tahiti, 222–23
Pinctada margaritifera. See Pearl shell
Piper methysticum, 6
Pipturus argenteus, 76
Planchonella grayana, 128
Plant introductions, 178, 267–69
Platypus, 25
Pleuropoma sp., 121
Pluvialis dominica, 60, 69, 108
Pollen analysis, 11, 12, 17, 31–32, 41, 47, 142, 152–54, 173–81, 207, 209, 228, 250, 261, 268, 269
Polynesia, 35, 52, 265–68. See also Outliers
Polyscias guilfoylei, 193

Pomarea dimidiata, 66
Pondfields, 84–85, 89–91. See also Irrigation
Population: birds, 53; growth, 284, 286; human, 20, 43, 50, 82, 151, 198–99, 277
Porphyrio sp., 159
Porphyrio mantelli, 74
Porphyrio paepae, 74
Porphyrio porphyrio, 73
Portmanteau biota, 271
Porzana rua, 159
Porzana tabuensis, 60, 76
Potoroo, 30
Premna tahitensis, 194
Pritchardia, 17, 236, 250, 254–56, 261, 264
Pritchardia vuylstekeana, 175, 176, 197
Procelsterna cerulea, 69
Progradation, of shorelines, 138–39. See also Sea-level
Prosopeia tabuensis, 72
Pseudocheiridae, 43
Psidium guajava, 180, 181, 196
Pteridium esculentum, 277
Pterodroma heraldica, 61
Pterodroma nigripennis, 58, 159
Pterodroma rostrata, 120, 135
Pterodroma ultima, 72
Pteropodidae, 67
Pteropus samoensis, 108
Ptilinopus porphyraceus, 65, 108
Ptilinopus rarotongensis, 60, 159
Puffinus sp., 61
Puffinus lherminieri, 54, 120
Pukapuka Is., 61
Putoa rockshelter, 200–28

Radiocarbon dating: of coconuts, 180–81; of cores, 167; of gardens, 98; interpretation, 85–86; of irrigation, 237; "old wood" factor, 90, 264; of pondfields, 89–91; of Putoa rockshelter, 204; of reef platforms, 134; at To'aga site, 110

Rails, 54, 60, 64, 108, 219
Rainfall, 130, 169, 242, 260
Ramu Valley, 41, 46–47
Rapanea, 176
Rapanea taitensis, 193, 194
Rapanui Is. *See* Easter Is.
Rarotonga Is., 71, 73, 126, 133, 149
Rats, 51, 117, 159, 206, 221, 246, 265, 279, 280. See also *Rattus*
Rattus exulans, 36, 65, 108, 117, 157, 159, 206, 221
Rattus mordax, 36
Rattus norvegicus, 222
Rattus praetor, 36
Rattus rattus, 65, 222
Reefs, 98, 127, 134, 151, 235. *See also* Sea-level
Religion, 19
Remote Oceania, 33, 35, 80, 81
Resource management, 273
Rhizophora mucronata, 181
Rhus, 209
Rota Is., 52
Ruderal herbs, 195. *See also* Weeds

Sahlins, Marshall, 78, 147–48, 165, 245
Samoa Is., 13, 53, 66, 77, 105–22, 125
Sarcophilus harrisii, 29
Schizostachyum glaucophyllum, 160
Scincidae, 108
Seabirds, 51, 54, 64, 69, 70, 72, 120, 158. See also *Fregata; Gygis; Phaethon; Sterna fuscata; Sula*
Sea-level: change in, 14, 25, 81, 112–17, 122, 124, 230, 242; during Holocene, 133–36; relative, 173, 197
Seals, 277, 279
Sea urchins, 109
Sediments: allogenic, 166; alluvial transport of, 241; analyses of, 171–81; authogenic, 166; budgets, 112, 116, 122, 125, 136; cores, 167; geochemistry of, 171–73, 186–92; influxes of, 83, 196; interpretation of, 84–85; lake, 166; marine, 233, 235; modifications, 223–24; processes of deposition, 138–41; rate of sedimentation, 186, 224; sources of, 136–38; terrigenous, 136, 233, 240
Selling, Otto, 258, 260
Settlement, of Polynesia, 265–68
Shearwaters, 54, 219, 279, 280. *See also* Seabirds
Shellfish, 276. *See also* Mollusks
Shifting Cultivation, 14, 19, 20, 50, 107
Sida rhombifolia, 180
Sinployea sp., 121
Society: evolution of, 147; "open," 148
Society Is., 17, 53, 71, 134, 181–96, 200–28
Soils: agricultural, 237, 245; arable, 234; degraded, 241; development of, 83; erosion of, 181; garden, 84–85, 88, 126; influx to lake, 192; lateritic, 241; mass wasting of, 245; pondfield, 87–88, 234, 237; structure, 237; swidden, 88; volcanic, 276; weathering, 241; wetland, 240. *See also* Sediments
Solomon Is., 80
Sophora sp., 276
Spondias dulcis, 5
Spriggs, Matthew J. T., 12, 16, 18, 19, 36, 265–66
St. John, Harold, 259
Stachytarpheta, 195
Steadman, David W., 12, 16, 36, 82, 117, 120, 127, 148, 261, 285
Stenochlaena palustris, 194
Stenopelmatidae, 279
Sterna fuscata, 69
Stoddart, David, 130, 136
Storms: as depositional agents, 130, 135, 188, 196; events, 141, 170, 198; high-energy, 136; storminess, 132, 243–44

Strigidae, 220
Subsidence, 127. See also Sea-level
Subulina octona, 211
Suggs, Robert C., 8
Sula sp., 69
Sula dactylatra, 75
Sula sula, 77
Swamphens, 73. See also *Porphyrio*
Sweet Potato. See *Ipomoea batatas*
Synecology, 9

Tahiti Is., 4–6, 17, 71, 73, 200–28
Tangatatau rockshelter, 58, 73, 128, 155–61
Taro, 139, 150, 160, 161, 162, 164, 169, 234, 240. See also *Colocasia*; *Cyrtosperma*
Tasmania Is., 15, 23, 25–30
Tasmanian Devil, 29
Tasmanian Thylacine, 28
Tattooing, 72, 206
Ta'u Is., 106
Tautua cave site, 161
Tavai site, 86–87
Tectonic change: sea-level and, 133–36; subsidence, 114; uplift, 97
Temples, 162, 216
Terminalia catappa, 181, 195
Terns, 69. See also *Gygis*
Terracing, 84–85, 161, 169, 237, 240, 241, 244, 245, 246. See also Irrigation
Thambetochen, 16
Thespesia populnea, 181
Thylacinus cynocephalus, 28
Thylogale spp., 30, 36
Thylogale brunii, 43
Tikopia Is., 60, 61, 64, 72, 76, 98, 116, 125, 147, 157, 213
To'aga site, 106, 110–22, 125
Tonga Is., 53, 54, 60, 64, 71, 74, 77, 114
Translocation, of animals, 15, 36, 66, 285
Trema cannabina, 175, 176
Trema orientalis, 194

Tridacna maxima, 109, 134
Trochomorpha cressida, 211
Tuamotu Is., 134, 145
Turbo, 35, 109, 143
Turbo setosus, 109, 119, 155, 159, 216
Turtles, marine, 109, 117, 207, 222, 223
Typha angustifolia, 185, 194, 196
Typha latifolia, 194
Tyto alba, 74

Ua Huka Is., 54, 56
'Uko'a Pond, 252, 258, 259, 260, 261–62, 264, 266, 270
Upolu Is., 114
Ureia site, 127–28, 139, 141

Vanuatu Is., 83, 91–98, 181
Vayda, Andrew, 9
Vegetation: anthropogenic, 153–55; burning of, 245; change in, 121, 176–78; disruption of, 198, 241; human impacts on, 30–32, 34; lowland, 249–50, 254, 270; modifications of, 207; New Guinea Highland, 41–42; of Papeno'o Valley, 207–9; pre-Polynesian, 254–58; windward communities, 259. See also Forests
Vini australis, 72
Vini kuhlii, 71
Vini peruviana, 77
Vini sinotoi, 71
Vini solitarius, 72
Vini vidivici, 54, 71

Waimanalo, 232–34
Wallaby, 25
Wallace, Alfred Russell, 1, 78
Walter, Richard, 132
War, 19, 160–61, 199
Ward, Jerome, 236, 248
Weeds, 180, 192, 195, 199, 239
Weevils, 279

Weinmannia, 153, 192, 193, 194, 195
Weinmannia parviflora, 193
Weinmannia vescoi, 194
Western Polynesia, 86, 125, 129
White, J. Peter, 23
Wiens, Harold J., 8
Wilson, Edward O., 1, 3
Wombats, 25

Xylosma, 209

Yap Is., 52, 160
Yen, Douglas E., 91, 99, 100

Zimmerman, Elwood C., 8, 17
Zooarchaeology, 12